TEACHING ARTS AND SCIENCE WITH THE NEW SOCIAL MEDIA

CUTTING-EDGE TECHNOLOGIES IN HIGHER EDUCATION

Series Editor: Charles Wankel

Recent Volumes:

CUTTING-EDGE TECHNOLOGIES IN HIGHER EDUCATION
VOLUME 3

TEACHING ARTS AND SCIENCE WITH THE NEW SOCIAL MEDIA

EDITED BY

CHARLES WANKEL
St. John's University, New York, USA

IN COLLABORATION WITH

MATTHEW MAROVICH
KYLE MILLER
JURATE STANAITYTE

United Kingdom – North America – Japan
India – Malaysia – China

Emerald Group Publishing Limited
Howard House, Wagon Lane, Bingley BD16 1WA, UK

First edition 2011

Copyright © 2011 Emerald Group Publishing Limited

Reprints and permission service
Contact: http://booksandseries@emeraldinsight.com

British Library Cataloguing in Publication Data
A catalogue record for this book is available from the British Library

ISBN: 978-0-85724-781-0
ISSN: 2044-9968 (Series)

Emerald Group Publishing
Limited, Howard House,
Environmental Management
System has been certified by
ISOQAR to ISO 14001:2004
standards

Awarded in recognition of
Emerald's production
department's adherence to
quality systems and processes
when preparing scholarly
journals for print

INVESTOR IN PEOPLE

CONTENTS

LIST OF CONTRIBUTORS

Robert Bodle	College of Mount St. Joseph, Cincinnati, OH, USA
Boris H. J. M. Brummans	University of Montreal, Montreal, QC, Canada
Ryan Busch	University of Phoenix, Phoenix, AZ, USA
Shannan H. Butler	St. Edward's University, Austin, TX, USA
Aimee deNoyelles	Ivy Technical Community College of Indiana, Indianapolis, IN, USA
Helen Farley	University of Southern Queensland, Toowoomba, Australia
Tricia M. Farwell	Middle Tennessee State University, Murfreesboro, TN, USA
Monica Flippin-Wynn	Jackson State University, Jackson, MS, USA
Scott Grant	Monash University, Melbourne, Australia
Howard M. Gregory	Kent State University, Kent, OH, USA
Lora Helvie-Mason	Southern University at New Orleans, New Orleans, LA, USA
Jennie M. Hwang	California Polytechnic State University, San Luis Obispo, CA, USA
Annie Jeffery	Boise State University, Boise, ID, USA
Andy Jones	University of California, Davis, CA, USA
Nick Pearce	The Open University, Milton Keynes, UK

Alexander Reid	University at Buffalo, Buffalo, NY, USA
Geoffrey Roth	Hofstra University, Hempstead, NY, USA
James Schirmer	University of Michigan-Flint, Flint, MI, USA
Sheila Scutter	James Cook University, Queensland, Australia
Kay Kyeongju Seo	University of Cincinnati, Cincinnati, OH, USA
Natalie T. J. Tindall	Georgia State University, Atlanta, GA, USA
Charles Wankel	St. John's University, New York, NY, USA
Richard D. Waters	North Carolina State University, Raleigh, NC, USA
Gavan P. L. Watson	University of Guelph, Guelph, ON, Canada
Corinne Weisgerber	St. Edward's University, Austin, TX, USA
Linda Wilks	The Open University, Milton Keynes, UK

PART I
SOCIAL LEARNING AND
NETWORKING APPROACHES TO
TEACHING ARTS AND SCIENCE

NEW DIMENSIONS OF COMMUNICATING WITH STUDENTS: INTRODUCTION TO TEACHING ARTS AND SCIENCE WITH THE NEW SOCIAL MEDIA

Charles Wankel

Universities are populated with a wide range of disciplines. The science disciplines and their instructors are stereotyped as tech-savvy while in the past humanities faculty have sometimes been seen as technophobic and traditional. As we advance through the second decade of the 21st century, we find instructors in all areas are embracing new technologies in their teaching. Our students have been born digital (Tapscott, 2009) and have not only experienced online games and social networking technologies such as Facebook but thrive in them. It should not be surprising that many of our colleagues are trying out the use of social media in their courses. This volume embodies a sharing of such experiences with the aim of moving you up the learning curve so that your thinking about how these new technologies might spark excitement, interaction, sharing, and enhanced work and learning by your students.

James Schirmer in *Fostering Meaning and Community In Writing Courses Via Social Media* looks at the use of quick blogging through Posterous and micro-blogging through Twitter for facilitating collaboration and

Teaching Arts and Science with the New Social Media
Cutting-edge Technologies in Higher Education, Volume 3, 3–14
ISSN: 2044-9968/doi:10.1108/S2044-9968(2011)0000003004

community in college-level courses. Students who were supervised by the author through the various writing reported having (to the instructor's surprise!) a very positive experience of using both tools and demonstrated commitment to continuous use of them after the semester was over. The continuous sharing of course-relevant information, the development of audience awareness, the ample opportunity to practice writing, and the accessibility and availability of fellow students and the instructor for commentary and questions were all benefits that students found in relation to blogging and tweeting in the first-year, advanced, and technical writing courses. Schirmer concludes that the overall ease-of-use and relative simplicity of Twitter and Posterous make for low barriers of entry for a majority of students. If proper affordances are made in terms of framing and timing, the appropriate use of social media in writing courses from first-year to graduate-level can make for a successful addition. Social media like Posterous and Twitter, both of which are rather effortless in terms of use, allow and perhaps even encourage students to think and find their way through college-level courses in ways that are meaningful and unique.

In *Learning about Media Effects by Building a Wiki Community: Students' Experiences and Satisfaction*, Jennie M. Hwang and Boris H. J. M. Brummans report the findings of an exploratory study on student experiences with creating a wiki in an upper-level undergraduate course on media effects, their reflections on functioning as a member of this wiki community, and their overall satisfaction with taking this kind of a *hybrid* course, which blends face-to-face and online interactions.

Their study shows that students enjoyed learning about media effects by collaboratively building their wiki community, but were critical about the structure of the hybrid course. Their research suggests that students are increasingly aware that it is important to learn about the effects of social media using methods that go beyond traditional course delivery (i.e., course lectures, papers, and the reading of a textbook) and give them the opportunity to use social media that form the course material itself. Hence, Hwang and Brummans's study indicates that students enjoy working within a wiki community, which they start building from day one through their online and face-to-face interactions, because it allows them to construct knowledge about various media effect theories and learn how to use a range of new media in a collaborative fashion.

However, Hwang and Brummans's research also suggests that students vary in their perceptions of a blended course when they enter this type of class and that these perceptions depend on their previous experiences with using computer technologies. For example, those who are familiar with using a computer for social interactions tend to be more comfortable with using a computer for schoolwork. In addition, students' level of confidence about doing well in a hybrid course seems to be positively related to the amount of experience they have with using a computer for schoolwork. In turn, those who lack this kind of confidence appear to view the effectiveness of teaching with technology more negatively, while those who regularly use the computer for schoolwork seem more positive about it. Those who are familiar with using different e-learning tools seem more positive about their benefits. Moreover, it seems that the more challenges with e-learning students encounter at the beginning of the term, the less likely they are to see the positive learning effects of wiki collaboration throughout the quarter or semester.

Finally, Hwang and Brummans's research indicates that in using a wiki to improve the delivery of a media effects course, an instructor also faces important challenges, some of which require significant reflection, while others have to be accepted as more or less insurmountable limitations. What is perhaps the most important challenge is to find a suitable way to structure the course so that it blends face-to-face and online interactions in a synergistic way.

Lora Helvie-Mason in *Facebook, "Friending," and Faculty–Student Communication* explores the communicative relationship between faculty members and students during a time of increasingly diverse methods of mediated communication. Social media has been embraced by higher education and Facebook, in particular, has become a marketing medium as well as a means of institutional and instructional communication with current and future students. This chapter begins by describing instructional communication and its changing nature with computer-mediated communication in higher education.

Next, the author examines student perceptions of faculty members as Facebook friends to explore the impact friending may have on the educational experiences of our students. Examining the perceptions of faculty friends is then related to current literature on immediacy, identity and disclosure, and the impact on pedagogy.

Lastly, the chapter concludes with five general areas to consider for faculty members using or contemplating using Facebook as a means of connecting to students. These areas include: before you start, know your audience, response options, collaborate and share, and alternatives to "friending" on Facebook.

In *How Twitter Saved My Literature Class: A Case Study with Discussion*, Andy Jones builds a case study by starting with the frustration that many social sciences and humanities faculty are familiar with: the lack of evidence that students have completed the assigned readings for class. Jones describes a short story class he taught recently that emphasized the instructor's high expectations for student participation. Twitter was introduced as the means by which students would collaboratively analyze assigned texts. The students soon embraced Twitter as a collaboration tool, and increasingly came to class with improved attitudes toward and readiness for class discussions. The nightly peer-review process made possible by Twitter helped students improve their spoken and written arguments and deepened their understanding of challenging texts. This chapter tells the story of the discoveries Jones made about teaching student-centered classes, and about using Twitter as a sandbox where students would share their ideas before coming to the well-attended lectures and class discussions. The chapter concludes with 10 recommended strategies for teaching with Twitter.

Robert Bodle in *Social Learning with Social Media: Expanding and Extending the Communication Studies Classroom* explores how the affordances of social media, specifically class blogs (WordPress) and microblogs (e.g., Twitter) together, help achieve social learning. Internet-based learners have various levels of proficiencies, competencies, and adoption rates. Strategies and best practices are explored to address how social media can be utilized by educators to accommodate the heterogeneity of digital learners and engage new styles of learning.

The technological and social (techno-social) affordances and communication dynamics of social media, such as visible "always on" social presence, and the ability to comment upon and rework media, can enable people to create, interact, collaborate, and learn in ways that are socially grounded and contribute to self-actualization through participation. Class blogs are useful tools for displaying and commenting on visual content and for collective and individual engagement with texts. Blogs also provide a platform for both instructors and students to publish digital media, including images, videos, and slideshows (e.g., PowerPoint slides), giving students and instructors opportunities to create, talk back, and share within a learning community. Three primary outcomes encouraged by blogs include (1) technical literacy, (2) social interaction, and (3) critical reflection. Microblogs additionally support these outcomes, as well as leverage the additional benefits of social presence and social influence, utilizing the network structure of strong and weak ties.

The very affordances that enable social learning present challenges and risks that need to be confronted and include moments of online inertia, residual issues of unequal participation, resistance to social learning as an insufficient use of time, and instructors' discomfort in sharing power or authority in the classroom. Students may overreach, abuse opportunities to talk-back, and provide excessive distractions. Utilizing social media requires flexibility, the ability to adapt, relinquish some control, and work with the unexpected. Within an augmented class there are opportunities for instructors to restore a balance of power within face-to-face meetings, and raise issues of appropriate conduct, ethical communication, and mutual respect. Additional challenges involve connecting the different modalities or learning styles together. Strategies that connect offline and online participation can maximize the benefits of social media and persuade students to view familiar online spaces as integral to the educational goals of the class. Successful augmented learning formats require maintaining focus on the linkages and mutual contingencies between online and offline work.

Blogs and microblogs together can be used to provide a social foundation that encourages social learning and allows students to express strengths that may be otherwise stifled in a traditional learning environment. The affordances of social media might enable but cannot determine student engagement and learning. Social media together with the design of diverse and meaningful group assignments and activities can bring students at various levels of talent, ability, and interest together to participate with and learn from one another. Despite the challenges and risks, instructors from many disciplines are encouraged to utilize the potential disruptions of social media to better accommodate the disparities and similarities among traditional, nontraditional, and other unique learners in higher education.

In *Teaching Social Media Skills to Journalism Students*, Geoffrey Roth looks at three areas where social media should be incorporated in teaching journalism. Social media has rapidly become deeply ingrained into the journalistic process, and in some cases is replacing traditional journalism as a means for distributing news. Therefore to effectively teach journalism at the university level, we should incorporate social media as both a learning tool and a subject for examination in our classes. Geoffrey Roth identifies media literacy and social media to be the first area that requires educators' attention. The chapter examines the tools and critical thinking skills needed to distinguish reliable from unreliable information before it is passed on to a news audience. Roth also analyzes the use of social media as a tool for gathering information. The chapter looks at how social media can be used to make and maintain contacts, dig for unique and impactful stories, and how you can your social media contacts to improve and enhance your reporting. Lastly, Roth goes over how to effectively use social media to distribute information, and the pitfalls that can occur when the personal use of social media conflicts with one's professional life as a journalist. Each section of the chapter ends with exercises that can be used to help students hone their social media skills.

Ryan Busch in *Unlocking the Voice of the University: The Convergence of Course, Content, Delivery, and Marketing through Social Media* proposes to see social media not as a variety of technological tools or services but rather as a developing ideology based around a few simple tenants: that content should be freely sharable, that technology enables sharing, and that shared content forms the basis for creating community connections. These tenants mirror many principals dear to higher education, but higher education has yet to adopt the principals in full.

Social Media ideologies represent a missed opportunity of vital importance to colleges and universities. The core tenants of these ideologies include wider and freer access to information through the use of emerging technologies. Colleges and universities should consider implementing social media ideologies to improve efficiencies in the delivery of learning and organizational operations. As an example, the chapter highlights two innovative companies founded on innovations representing a doctrine of convergence – socializing course, content, delivery, and marketing into a broader format, which not only educates the student, but also expresses the unique qualities of the organization itself. Examples include Tech University of America, eduFire, and an experimental course model developed as the result of an introduction of the leaders of these two organizations.

At their best, most colleges and universities are exploring Social Media as a marketing proposition. While the application of Social Media ideology to marketing programs is clear, this narrow focus excludes much of the unique tenor of an institution from both reaching and impacting a given audience. Colleges and universities produce a tremendous amount of new and invigorating content, but much of this content is sequestered within classroom walls (both digital and real). The true voice of an institution of higher education comes not from its marketing communications, but from within these walls. Courses, content, and delivery better represent the institution than marketing rhetoric alone. The voice of the institution can be freed by adopting social media principles. This is a process of convergence; the convergence of courses, content, delivery and marketing as a way to unlock the voice of an institution.

Several innovators have emerged with contributions based on this type of convergence. Tech University of America is developing an institution within Facebook which links coursework with a vast social community. eduFire has created a learning community through which its members gain access to all types of educational experiences. Members of the eduFire learning community can be student, teacher, or both. Members can rate and share experiences with other members and non-members and participate in the development of new learning opportunities. Piccolo International University crafted a blended-social learning model using eduFire and a traditional asynchronous online course as a way to promote the learning experience at Piccolo. Colleges and universities can gain insight from observing such innovations and considering how these small but intriguing organizations have embraced the sharing of course, content, delivery, and marketing as a result. By considering the implications of social media ideology, colleges and universities can develop customized strategies that optimize organizational efficiency, impact educational costs, improve student engagement, and promote student success.

In *Understanding Communication Processes in a 3D Online Social Virtual World*, Aimee deNoyelles and Kay Kyeongju Seo observe an undergraduate communication class where Second Life was put to use in order to accomplish learning objectives. As 3D online virtual worlds are relatively new to the educational field, it is likely that teachers and students will need guidance on how to develop learning activities in this environment. It is very important that the class feel comfortable in communicating with each other in virtual worlds. The introductory observation in the case study was that learners did not walk in with identical abilities. Those who identified as "gamers" experienced less technical difficulty and were more comfortable using avatars to communicate with others, while non-gamers struggled with technical ability and were sometimes unsure of avatar-mediated communication. Second, avatar appearance proved to be a significant factor in communication. While gamers designed fantastical avatars and often communicated about what the avatars could do, non-gamers designed typical humans that often looked like themselves in real life and talked about what their avatars looked like. This usually resulted in a slight rift, with the gamers and the non-gamers being mutually exclusive. Third, the physical proximity of the computer lab dramatically diminished the amount of online interaction in Second Life, with most students relying on the classroom context to communicate. Finally, the virtual proximity of the Second Life space also impacted communication, with social and learning spaces resembling the real world eliciting more social interaction.

These findings are important since they suggest design recommendations for teachers who are interested in incorporating Second Life into their Arts and Sciences classroom. First, it is important to assess the needs and abilities of the learners, as they will not walk in with similar abilities. It is helpful to gauge their technical abilities, but also to discuss their conceptions of the construction of online identity and online social interaction. On a related note, since avatars may look very different from their students, it is important to make sure that students can recognize other students, so they will not be concerned with communicating. It is also important to design spaces that will support them. Spaces that look like the 'real world', like the local university, dorm tower and conference hall give students a permanent space to congregate, and also subtly shape their expectations and behavior.

Annie Jeffery, Scott Grant, and Howard M. Gregory II continue the discussion in *Multi-User Virtual Environments for International Classroom Collaboration: Practical Approaches for Teaching and Learning in Second Life* by analyzing case studies from two universities, Boise State University (BSU) and Monash University in Australia, where in 2009 instructors explored the potential for a joint project between two seemingly disparate courses in Social Network Learning in Virtual Worlds and Chinese Media Studies. The Chinese language students used machinima to record in-world interviews and news reports that were posted to the Internet. This process helps both the students and the instructor to enjoy language learning in an active, creative, and purposeful way, for what is often a passive learning experience.

It was the instructors' hope that transferring control of the filming to BSU students would offer them an opportunity to collaborate on an authentic Social Network Learning project for their final assessment, to enable the Monash instructor to concentrate on teaching and learning, and to allow the Monash students to excel with their language learning. Thus the Monash students would not be forced to divide their focus by attempting to master the skills needed to successfully plan and create machinima projects. This, in turn, would strengthen their engagement and focus their energy on the language learning. At the same time it was understood that the BSU students could use the opportunity to develop networking skills and build a community of practice within their own course. In conclusion, the authors observe how communities of practice, social constructivist learning, and a range of bricoleur skills were developed among students.

Using Multi-User Virtual Environments in Tertiary Teaching: Lessons Learned through the UQ Religion Bazaar Project, by Helen Farley, examines some of the key characteristics of online multi-user virtual environments (MUVEs) and of Second Life in particular, with a view to assessing its suitability as an environment for learning based on andragogical and constructivist methodologies. Furthermore, it explores the original conception and development of the UQ Religion Bazaar project within Second Life.

Fostering an Ecology of Openness: The Role of Social Media in Public Engagement at The Open University, UK, by Linda Wilks and Nick Pearce highlights the range of ways in which one of the largest distance learning institutions in the world, the UK's Open University, uses social media within an ecology of openness. Social justice has underpinned the Open University's ethos since it was established over 40 years ago. It is this aim to remove barriers and provide learning materials to a wide audience, including those who may be excluded from other learning institutions, that has been a major strategic driver of recent changes at The Open University.

The advent of the open educational resources movement, together with the availability of new technologies, has provided the OU with new opportunities to develop innovative channels through which to offer its learning materials and research outputs to citizens across the globe. The OU has also capitalized on the appearance of new social media possibilities, using sites such as Facebook and Twitter, as well as developed in-house platforms to facilitate discussion and collaborative learning.

This chapter describes the range of initiatives taking place in the OU in 2010, many of which are in their infancy, and explores the benefits and challenges encountered in using social media to promote an ecology of openness.

Disconnect36: A Social Experiment to Teach Students to Shut Down, Turn Off, and Understand Connectivity, written by Monica Flippin-Wynn and Natalie T. J. Tindall, outlines a social media class experiment undertaken by the lead author to provide students with an opportunity to understand their reliance on social media or media in general and add to the scholastic literature on teaching and technology in the classroom. In one class, the majority of students agreed to disconnect from all communications technology and social media for 36 hours. As they started to withdraw from the media, the class assignment provided students with insights into their constant connectivity and how they manage information through various mediated channels. Students were required to complete an 800-word blog post or paper for this assignment, and from that assignment, the authors drew insight into the issues of connectivity and information overload. This chapter offers ideas on how to integrate such an experience in other class settings and provides pedagogical rationales for this type of experiment.

Sheila Scutter in *Is Podcasting an Effective Resource for Enhancing Student Learning?* gives an overview of podcasting as easy, requiring only cheap and simple technologies and little additional time on top of the provision of teaching materials. Considering very positive feedback from students, this has become one of the major drivers for providing podcasts of teaching material. This chapter discusses the way students use podcasts and the possible impacts on learning. Despite concerns about students reducing attendance at lectures, most studies have shown that lecture attendance is not diminished by the provision of podcasts. Students do not tend to use MP3 players to listen to podcasts "on the go": most students listen to podcasts direct from home computers, often while replaying PowerPoint slides. The academic staff perspective on podcasting is discussed in relation to advantages and concerns about their use.

In *Introducing Students to Micro-Blogging through Collaborative Work: Using Twitter to Promote Cross-University Relationships and Discussions*, Tricia M. Farwell and Richard D. Waters observe that while organizations are struggling to find their place in the social media world, organizational communication disciplines (e.g., advertising, marketing, and public relations) increasingly are expecting new hires to not only be aware of social media technologies but also be savvy in their strategic usage. The job market for these majors increasingly wants and expects those graduating in these areas to not only know how to create strategic plans for using social media in both one-way and two-way communication environments but also maintain proper social media etiquette and virtual culture norms for their clients. In order to assist students in learning how to navigate social media, more specifically Twitter, this chapter discusses the experiences of two faculty members at separate universities in designing and implementing a course assignment intended to promote collaboration, foster discussion and bring students to use micro-blogging through Twitter.

The courses involved in this assignment were an undergraduate senior-level advertising course that dealt specifically with the use and impact of social media in advertising and a cross-listed graduate and undergraduate-level public relations course that dealt with the applications and uses of social media in various public relations specializations. The challenge was to get the students introduced to, involved in and thinking about how to use Twitter in their various professions. Students were tasked with discussing and agreeing upon a 140 character definition of an assigned term and 10 useful strategies relating to the term. For example, if a group was assigned wikis, they would create a 140-character definition of wikis based on their conversations on Twitter with the assigned group members and then create 10 tips for using wikis in advertising and public relations.

The authors conclude that overall, educators involved in this project did feel that it was a beneficial assignment for students in both classes. While students may not appreciate the assignment as it is being conducted, many of them have attested to the value of the assignment now that it has been completed. They seem to recognize that social media is not something to be entered into lightly and needs commitment. In this regard, they are probably further along than many businesses or organizations which see social media as a quick and cheap way to send messages to their potential customers. In that sense, the assignment did reach one of the goals of having students participate as users and understand how relationships could be built.

Alexander Reid in *Social Media Assemblages in Digital Humanities: From Backchannel to Buzz* problematizes the evolution in the field of humanities. While the term "humanities" is not in itself a particularly contentious one among academics, the addition of the term "digital" creates all sorts of problems, even the contention that digital humanities are not humanities at all. The fundamental rupture between digital and print humanities lies in the turning of a materialist, object-oriented analysis on the practices of humanistic scholarship. That is, in their newness, the digital humanities are unsurprisingly self-reflective about the materiality of their scholarly practices. This self-reflection has been largely absent from traditional humanities where we had all but naturalized the material composition of dissertations, journal articles, monographs, and so on. As a result, even as we continue to pursue traditional scholarly methods, it becomes increasingly difficult to do so without a self-reflective awareness of the historical-material contingency of these practices. In short, they are no longer the same. To explore this issue, the chapter takes up assemblage theory and actor-network theory to investigate the intersection of mobile technologies and social media in the digital humanities including conference backchannels and networked research communities mediated through Twitter, Google Buzz, and similar applications. The chapter considers how, even for those who continue to publish in traditional genres on traditional subjects, the development of these digital assemblages is transforming compositional practices.

In *Social Media as a Professional Development Tool: Using Blogs, Microblogs, and Social Bookmarks to Create Personal Learning Networks*, Corinne Weisgerber and Shannan H. Butler discuss the steps involved in building, growing, and maintaining online connections made possible entirely through new technologies. The rapid growth of information and the emergence of new technologies capable of filtering information and connecting us to those we can interact with and learn from illustrate the ever-changing nature of personal networks.

The authors argue that in the context of higher education, PLNs should be viewed as an informal alternative to professional development programs that are commonplace in K-12 education. This chapter contains sagacious advice, based on real-life examples on how to plan and run personal learning networks.

Gavan P. L. Watson in *Micro-Blogging and the Higher Education Classroom: Approaches and Considerations* offers reflections on the successes and failures of integrating the micro-blogging platform Twitter into a first-year university class. Twitter, intended as a way to answer the question "What are you doing?" is now used in originally unexpected ways. Broadly speaking, Twitter's popularity can be traced to three broad factors: conversation between users, a decentralized ecosystem of third-party applications, and the distributed nature of its users. Adopted by educators in higher education, Twitter has been used as an object for study, a tool to communicate classroom announcements, a means to enable students to reflect on their learning, a chance to get instant feedback from students, and as a tool used to facilitate in-class conversations. The ongoing use of micro-blogging also appears to have the ability to change the social dynamics of a classroom, expanding the social dimensions of the classroom beyond the physical dimensions. While identifying Twitter's limitations, the chapter outlines the most significant outcome from the author's integration of Twitter: an evolution of blended learning, proposed as a plesiochronous learning model, where learning occurs outside of the classroom with learners and instructors in different places but occurring at (virtually) the same time.

As you can see, there is quite a diversity of social media to try. I suggest you not only view this volume as a collection of ideas and experience on how to proceed but also view the chapter authors as colleagues to connect with, share with, collaborate with, and learn together with. The current social media available are but slates on the path to the future. It is our hope that you will join us in proceeding into it with us in a robust way.

REFERENCE

Tapscott, D. (2009). *Grown up digital: How the net generation is changing your world*. New York: McGraw-Hill.

FOSTERING MEANING AND COMMUNITY IN WRITING COURSES VIA SOCIAL MEDIA

James Schirmer

ABSTRACT

This commentary is a reflective discussion of how to use simple social media tools in college-level writing courses, and contains research elements such as effective examples of what is attainable and possible when incorporating blogs (e.g., Posterous) and Twitter in the college classroom. In order to do this, it uses reflective writing with a focus on failures/successes in past courses, and also incorporates students' own comments on blogging and Twitter. The chapter's findings include the following: The overall ease of use and relative simplicity of certain social media tools make for low barriers of entry for a majority of students. The mobile accessibility of these online communicative technologies should also be of specific appeal. These characteristics should encourage student participation in ways that content management systems like Blackboard do not. The convenience of and allowance for quick and easy sharing of information via blogging and microblogging can also mean that each is often quicker than email for contacting someone. What makes both better than Blackboard concerns how they, when taken together, sustain class discussion, keeping it alive, present, and continuous. If proper affordances are made in terms of framing and timing, social media can make for successful additions

Teaching Arts and Science with the New Social Media
Cutting-edge Technologies in Higher Education, Volume 3, 15–38
ISSN: 2044-9968/doi:10.1108/S2044-9968(2011)0000003005

to college-level courses. Simple tools allow and encourage students to document and reflect on their own learning in ways that are meaningful and unique as they are.

INTRODUCTION

Teaching most any kind of writing is not only about aiding students in the construction of identity but also in the development of a community. The appropriate use of online communicative technologies aids the formation and illumination of both. In particular, bringing simple social media tools into college-level courses can facilitate and coordinate greater attention, encourage meaningful interaction and participation, promote better collaboration, and help students develop narratives of their own learning as well as hone the critical consumption and crafting of academic and nonacademic work. An extension of such work involves challenging students' notions of what qualifies as writing, interrogating their prior knowledge and experience while also encouraging new forms.

In essence, using online communicative technologies and social media tools can serve a dual purpose by not only scaffolding student learning in college-level courses but also supporting professors' own scholarly pursuits. Posterous, a quick-and-simple blogging service, functions as a vehicle for working through ideas in a public format and recording the directions research interests take. Twitter, a popular microblogging service, provides a way to announce as well as brainstorm new work. There is an implicit encouragement to Twitter in finding community with others, but it also functions as a launching pad to other online spaces, such as blogs, social bookmarking sites like Delicious and Reddit and even other social net-working sites like Facebook and MySpace. As such, much of this chapter draws on my own research and teaching experiences as from current documentation of the knowledge of others.

Requiring the use of blogs and Twitter in the college-level courses I guide came from a desire to streamline the communication process, to remove the tedium of email attachments and related incompatibility issues. Taking important discussions beyond the face-to-face classroom was of additional interest. Intensive course development training with Blackboard was another impetus, though, as much prior experience teaching with the content management system proved frustrating. Across the courses requiring students' use of blogs and Twitter, though, it was possible to identify progress in terms of student engagement with these technologies rivaling that of Blackboard in past courses.

Part of this might have to do with how having students engage each other online promotes collaboration and community among them, with blogs and Twitter better mirroring their more everyday actions than Blackboard. As observed by Duffy and Bruns (2006), many students already interact and comment on each other's work in Internet-based environments, indicating "a growing impetus towards personal expression and reflection, and also the sharing of personal 'spaces.'" Jessica Gross (2009) writes in the *Huffington Post* about how this is also an indication of the influence of laypeople and how they have replaced trained experts as "the people who define what's true in our world." This increasing reliance on peers for information should be acknowledged by "teaching students to learn from and with each other" (2009).

Because "students' lives are infused with each other's viewpoints" (2009), it makes sense for educators to capitalize on that, to act as facilitators and mediators of students' collaborative and community-building efforts in learning contexts. As described later, there is some disagreement over what constitutes collaboration and/or community in such contexts.

What follows, then, is some reflective discussion of the use of these online communicative technologies as manifest in a range of college-level courses, from first-year composition to a graduate class in digital rhetoric, concluding with a potential model of online engagement for not only students but also instructors.

AN INTRODUCTION TO BLOGGING

The term "blog" is the commonly accepted abbreviation of "weblog" (Paquet 2003), which is a personalized website with dated entries presented in reverse chronological order. As emphasized by Duffy and Bruns (2006), "blog" is both a noun and a verb as those who maintain a blog are called "bloggers" while the act of posting to a blog is called "blogging." Particular features of blogs include an archive of past blog entries, links to other blogs, the quick publishing of content with little technical skill required, and comments, the last of which allow others to offer feedback on the author's entries (Downes, 2004) (Fig. 1).

Because of the opportunity for reflective interaction, the ability to link to related ideas, and the freedom for readers and other bloggers to "suggest additional considerations and explorations of the idea presented and promote

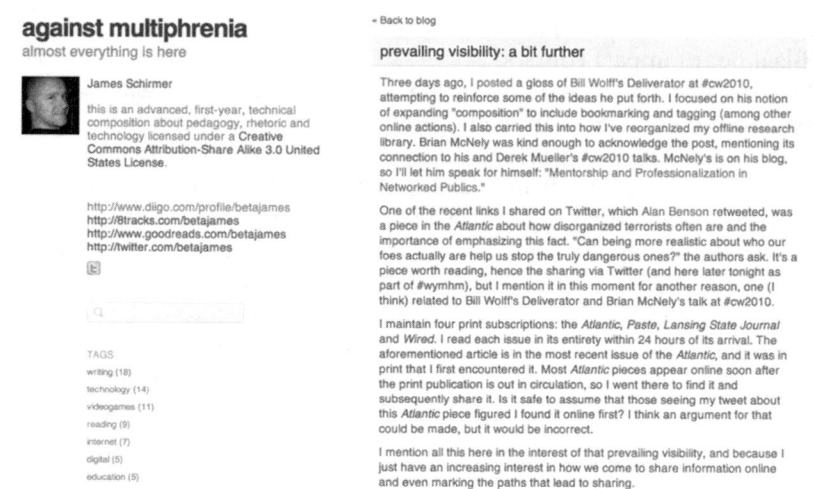

Fig. 1. Author's Blog. *Source:* Schirmer (2010a). Available at http://betajames. posterous.com/.

further reflection and thought regarding a stated viewpoint" (Duffy & Bruns, 2006), it would be unfortunate to view blogs as little more than online journals. Doing so privileges "understanding of journals as private writing spaces without considering the benefits of weblogs as public writing" (Lowe & Williams, 2004). Such a limited perspective also inhibits the use of blogs in more collaborative ways and, in "Blogs and Wikis: Environments for On-Line Collaboration," Bob Godwin-Jones (2003) points to technology-related blogs as prime examples that "form what is essentially one large, loosely interwoven net of information, as blog entries are linked, referenced, and debated." What facilitates the development of such an information net is the very act of posting online, which Godwin-Jones (2003) emphasizes as the important possibility of writing for a broader, larger audience, something "not usually possible in discussion forums" (Godwin-Jones, 2003).

Related to this is the ownership and responsibility that self-publishing online encourages on the part of students, who, Godwin-Jones (2003) suggests, "may be more thoughtful (in content and structure) if they know they are writing for a real audience." As Steve Krause (2004) puts it, "blogging opens up assignments beyond the teacher-student relationship, allowing the world to grade students and provide encouragement or feedback on their writings." I can bear witness to this as Anne Gentle, author of *Conversation and Community*, the lone required textbook for the class, once left a comment on a technical writing student's blog entry (Fig. 2). As described in a later section of this chapter, some other students also took note of this positive possibility in relation to blogging.

As Godwin-Jones (2003) notes, blogs are "very different from the rigid hierarchy to be found in [Blackboard and WebCT]," instead offering "a great deal of flexibility and the potential for creativity" on the part of both instructor and student. It is for these reasons that educational blogging can be effective for user-centered, participatory learning, contributing to "a reconceptualization of students as critical, collaborative, and creative participants in the social construction of knowledge" (Burgess, 2006, p. 105). Again, the indication here is something acknowledged by Duffy & Bruns (2006) and Gross (2009) as already taking place in non-academic contexts. The opportunities afforded by blogging should be too tempting for any educator to ignore.

QUICK BLOGGING VIA POSTEROUS TO PROMOTE COLLABORATION

Because of the relative ease in creating entries, "blogs have thus become the novice's web authoring tool" (Duffy & Bruns, 2006). With the rise of what is considered "quick blogging," online authorship is even easier to attain and

Jan 20, 2010   annegentle said...

Great post - I'm glad my book was useful to help you gel your thoughts about communicating online. You might like my former co-worker's blog post on high-context and low-context communications. It's called Group decision making with mixed high- and low-context communicators and you can read it at http://greyfiti.com/?p=69. He and I are very different communicators but we've worked collaboratively at two different companies now. It's complex, isn't it? But so important, as you recognize.

Ooo, and now I have to decide if this comment gets posted to Twitter which also means it could go to Facebook if I rewrite it so that the @ symbol doesn't come first. Considerations, considerations. Like I said, it's complicated. :)

Fig. 2. Anne Gentle's Comment on Student Kristen Bentley's Blog Entry, "Are You Mad At Me? Miscommunications in Writing." *Source:* Bentley (2010). Available at http://kbentley20.posterous.com/are-you-mad-at-me-miscommunications-in-writin. Reprinted with permission.

maintain. Identified by its creators as the "dead simple place to post everything," Posterous is a blogging platform focused on ease-of-use. Because of its relative simplicity in comparison to more hefty platforms like Blogger and Wordpress, some consider Posterous and its main competitor Tumblr qualify as microblogging. For instance, in a 2008 review of Posterous, *ReadWriteWeb* identified it as "minimalist blogging." A more accurate understanding of the capabilities of Posterous is evident in an *Ars Technica* side-by-side comparison with Tumblr, both of which Chris Foresman (2010) describes as "quick blogging tools ... characterized by two main features that set them apart from more traditional blogging tools." The first concerns a content-specific focus as text, images, videos, and links appear "in a suitable format for its content type" and the second involves the relative ease and speed of posting content. Perhaps it is because of this second feature that some might be quick to declare Posterous as a microblogging service or tool, but, as its creators boast on their FAQ, "there are no limits to what you can post."

In my experience of using Posterous and witnessing how students use the service, Posterous is quicker to begin and easier to maintain than Blogger and WordPress. In part, this is because Posterous offers multiple methods of content production, including posting options by email, mobile, and web. It also has to do with how Posterous handles the content a user sends. Text-based content, like a Microsoft Word document for example, attached to an email to Posterous appears onsite via Scribd, a free, Flash-based document-sharing service. Posterous arranges images emailed or uploaded into a web-friendly gallery; music and video files show up on Posterous in web-based Flash players, too.

This format suitability to content is important to note because not all students coming into college-level courses possess the knowledge necessary to deem the appropriate format for a particular kind of content. For instance, I required students in previous semesters to sign up for Blogger accounts. At the time, I was most familiar with this platform and I thought students would appreciate Blogger's options for customization, thereby allowing them to make their own unique, personal stamp on the blogs they created. The ability to post entries via Google Docs was an additional positive influence, and I hoped students would help me learn more about the platform through collaboration.

The potential complexity of blogger caused some trepidation, though. The great freedom of choice led to some students' poor design decisions, including use of color and font size, as well as formatting issues. Many students composed blog entries in Microsoft Word and then copied and pasted the content to blogger only to be upset by significant problems with alignment and spacing. Since the shift to Posterous, the vast majority of formatting problems have disappeared. So long as students attach their text-based efforts rather than copying and pasting direct into the body of an email, the all-important formatting of their blog entries remain intact on Posterous. As will be mentioned in a later section of this chapter, the autoposting feature of Posterous, in particular, makes for a unique success across all my courses.

While many things precipitated the move from Blogger to Posterous, the latter's overall ease-of-use was a dominant determinant. Posterous has a sharper focus on how someone might want to provide content. Personalizing a particular blog space comes as much from the content as the format and some bloggers might be more concerned with pushing content than anything else. Part of what makes Posterous successful is its inherent acknowledgement of these factors. Embedding documents, images, and videos in Posterous happens with little to no frustration. There are also a limited number of themes to choose from to eliminate the possibility of color clash and font fiasco. Posterous features a streamlined process for blogging in my writing-centered courses; it has proven to be a better choice for students.

Part of the reason for requiring blogging has to do with a certain perspective on the nature of writing and, more specifically, authorship. Contrary to popular assumptions about the writer as isolated from others, I view the act of composing as more of a collective, collaborative process than an individual endeavor. I require blogging in my courses to better illustrate this point, the necessity of interplay between writers and readers in the work of meaning-making.

An important part of the overall success of blogging, though, concerns good practice, something that Krause (2004) admits was rather absent from his early experiment with collaborative writing online. Farmer, Yue, and Brooks (2008) conclude their chapter with a list of recommendations for implementing blog technology in a college-level course, including clarity and support regarding assessment, the important invitation for students setting their own blogging goals, encouragement for and modeling of risk-taking, and early feedback on students' progress (p. 134).

Through the establishment of blogging guidelines (Fig. 3), I imparted to students the idea of the blog as a place to further explore ideas discussed in class, to write about related concepts of interest, and ask questions about both. I also encouraged students to blog in a way concerned with the regular examination of ideas presented in the course material and do so by way of incorporating images, links, and videos as means of support. Given the ease

Blogging in this course should be concerned with the regular examination of ideas and provide concise arguments via unique viewpoint and voice. Make clear to readers that there is substantive thought behind the ideas presented. Have specific references, including text, hyperlinks, video, images and audio, as means of support.

Find new ways of saying what you think you want to say. Incorporate a collage of images, audio clips and/or YouTube videos. **Push yourself to explore the ways you can get at ideas through the use of different media.** However, images, audio and video clips need explanation, too. Don't just stick them in a post and expect readers to understand why.

Your blog is a place to further explore the ideas we discuss in class, to write about related concepts of interest and ask questions about them.

When creating, designing and writing in your blog, please complete the following:
1. Choose a professional and meaningful title and subtitle.
2. Compose a detailed and relevant About page discussing who you are and the focus of your blog.
3. Choose an appropriate theme.
4. For each blog post, compose a meaningful title written for an audience beyond our class.
5. For each blog post, include 5-6 tags.

Experiment with the dashboard area, see how things work and what happens when you make changes. The more you engage with, customize and explore your blog, the more effective it will be and the more you will get out of the assignment.

There is no set requirement for the length of a blog post. One of the features of the blogging medium and the characteristics of individual posts is that length is determined by content and goals. Each post you make, though, should be thorough in discussing the subject at hand.

During the weeks regular blogging is required, be sure to post 1) an entry that extends the class discussion and 2) one that explores an area of interest particular to you. Posts that extend class discussion should take what we have discussed in class about a subject or text and continue the discussion. For example, a post might address one of the questions raised in a class discussion. Posts that explore an area of interest particular to you are just that. Ideally, these posts should serve as introductory writing toward larger, later assignments.

Fig. 3. Blogging Guidelines for Author's Courses. *Source:* http://eng112. posterous.com/pages/materials-1.

in which Posterous handles various kinds of online content, there was something of an unstated expectation and even optimism that students would go beyond text in putting forth their thoughts on the course. That many students stuck with textual production on their blogs should come as no surprise.

Still, by semester's end, many students' blogs contained an extensive record of intellectual expertise for the course. The most common entries engaged on the required readings as students latched onto particular passages that were resonant or troublesome and asked their peers for clarification. Other entries contained evaluations of in-class engagements led by fellow classmates (Fig. 4). While positive for the most part, I think these evaluations were helpful to class facilitators; knowing that peers might blog about their performance may have inspired them to put more effort into leading class discussions.

Such entries focused on in-course happenings further reveal collaborative aspects to blogging as students used their Posterous spaces to not only evaluate and reflect on the performances of their peers but to also plan toward the execution of their own. This mirrored much of what happened on a week-by-week basis regarding required readings as students responded to texts and each other's entries. These observations, though, invite the critique Steven Krause (2004) placed upon blogs in "When Blogging Goes Bad," namely that they do not have the same interactive potential of an electronic mailing list. While blogs "have the distinct advantage of allowing individuals to easily publish texts that can be responded to by others" (2004), Krause views them as no more "collaborative" than conventional print texts. There is a measure of truth to Krause's criticism, but, as Lowe & Williams (2004) note, "blogging represents the interaction of a community in the sense that all posts are subject to concerns about audience." Collaboration via blogging does not happen so much at the ground level of writing, but more in terms of furthering communication about ideas. This can be augmented to an even greater degree when blogging is coupled with microblogging, to which our attention now turns.

AN INTRODUCTION TO MICROBLOGGING

As the word implies, microblogging is a shorter, simpler variant of blogging "that people use to provide updates on their activities, observations and interesting content, directly or indirectly to others" (Ehrlich & Shami, 2010) (Fig. 5). Similar to regular blogging, microblogging can be exclusive, restricted to some, and/or available to all. There is a significant limitation to microblogging though, as updates can be no longer than 140 characters. However, this particular form of online communication has increased in popularity over the last three years. Most all of this is due to Twitter, which launched in 2006, the perhaps best-known microblogging service. Other similar microblogging services include Ident.ica, Jaiku (purchased by Google in October 2007) and Plurk, all of which share the 140-character

Group 2 Facilitation

I found the facilitation done by Group 2 to be very interesting and informative. Even after reading the text and seeing other's definitions of these various social media tools I was still not exactly clear on how they all worked and what they were used for, so I enjoyed listening to each of the presentations about various social media tools.

The information on Wikis was really helpful to me. Our group was thinking about trying to go more in depth about wikis for our facilitation and it is much easier to visualize now that I have a good idea of what exactly a wiki is. Although the video demonstration was very basic and simple it did a great job of portraying how and why one would use a wiki. I almost feel as though it would be beneficial to start one to use for our group, in order to share our ideas for the group project and class facilitation. It is nice right now being able to blog about it, since everyone can comment and leave their input in the same place. This way is better than sending a bunch of e-mails to each other but it seems that a wiki would be even more useful so that we could each edit one document rather than having pieces of it all over a blog post.

I consider my digital camera's importance to be the equivalent of one of my limbs. I love to take pictures and so I tend to have y camera in my hand ready to snap a photo most of the time. The information that Savannah gave about photo sharing was very intriguing to me. I have a lot of friends and family out of state and even more photos that I would love to be able to easily share with them and so it was nice to be given more information regarding the aspects of different photo sharing sites. I had previously heard of one of them, photobucket, but the other two were new to me and they seem to be the ones that can offer more to the user; such as editing options and a means of feedback and interaction with other users.

The part of the class that I found most interesting was the presentation about Skype. I had no idea that this technology even existed. When first hearing about it, it seems like it would be a really difficult thing to do; however, after the demonstration was given it is actually just as easy, if not easier than dialing a number on a telephone. I definitely believe that this presentation will inspire me to research skype a little more. Although at the moment I don't really have any major use for it, if one happens to come up in the future I will be familiar with using this technology.

Fig. 4. Student Kendell Davenport's Evaluation of Another Group's Class Facilitation. *Source:* Davenport (2010). Available at http://kendelld.posterous.com/ group-2-facilitation. Reprinted with permission.

limit with Twitter. While this limit forces brevity on the part of users, there is freedom within such restriction.

Status updates on Twitter are commonly known as "tweets," which often feature certain established conventions unique to microblogging, if not

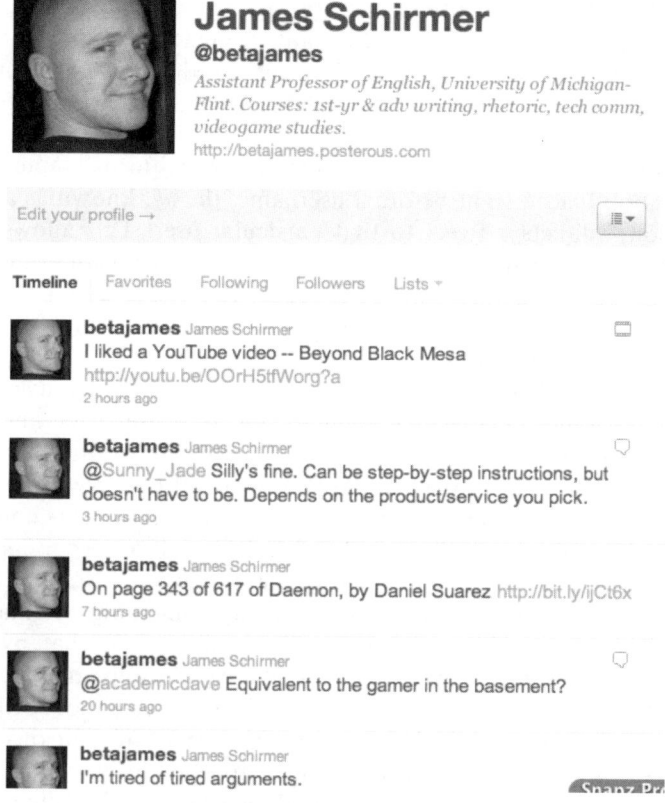

James Schirmer
@betajames

Assistant Professor of English, University of Michigan-Flint. Courses: 1st-yr & adv writing, rhetoric, tech comm, videogame studies.
http://betajames.posterous.com

Edit your profile →

Timeline   Favorites   Following   Followers   Lists ▾

betajames James Schirmer
I liked a YouTube video -- Beyond Black Mesa
http://youtu.be/OOrH5tfWorg?a
2 hours ago

betajames James Schirmer
@Sunny_Jade Silly's fine. Can be step-by-step instructions, but doesn't have to be. Depends on the product/service you pick.
3 hours ago

betajames James Schirmer
On page 343 of 617 of Daemon, by Daniel Suarez http://bit.ly/ijCt6x
7 hours ago

betajames James Schirmer
@academicdave Equivalent to the gamer in the basement?
20 hours ago

betajames James Schirmer
I'm tired of tired arguments.

Fig. 5. Author's Twitter Feed. *Source:* http://twitter.com/betajames.

Twitter itself. To group-related posts together and make them more searchable, users often incorporate hashtags (#) into their updates, following the symbol with a word or an abbreviation, for example, #edchat or #twitrhet. Some hashtags appear on a weekly basis, as with #Music-Monday in which Twitter users recommend new artists/bands and share songs. Other hashtags like #ladiespleasestop are part of the micro-meme phenomenon described by Huang, Thornton and Efthimiadis (2010) in which "clever short-lived tags catch on and then die out quickly." Still other hashtags are connected to television shows (#AmericanIdol, #glee) or news and political events (#IranElection, #obama). As observed by Reinhardt, Ebner, Beham, & Costa (2009), hashtags not only "generate a resource based on that specific thematic ... but also bridge knowledge, and knowing, across networks of interest." An important, related aspect of Twitter is the option to save searches by hashtag. Current saved searches by this writer include #cw2010, #MLA09, and #thatcamp, all discipline-specific, conference-related hashtags.

Perhaps more common than the # symbol on Twitter is employment of the @ symbol. Placed right before a username, the @, known as a "reply" or "mention," directs a tweet to that particular user. This allows Twitter users to reference each other in their own individual updates and replies/ mentions appear in the accounts of referenced users for easy tracking and responding. In other words, the addition of @ to a tweet allows for threaded conversations between users on Twitter.

"Retweeting" and sharing links are two other established conventions on Twitter. The former occurs by copying a post and username and placing "RT" before it as a status update of one's own. Retweeting happens on Twitter for any number of reasons, from a poignant post by a popular user to an announcement of important news. Many announcements come in the form of links and, given the length of website addresses, users employ a URL shortening service like bit.ly or tiny.url to stay within the 140-character limit and still have room to say something original. These two conventions have been adopted and modified somewhat by Twitter itself. There now exists the option to retweet any status update by clicking a button instead of typing "RT @username." Twitter also announced in April 2010 their intention to add a URL shortening service of their own.

A less recent, but still significant change happened in November 2009 as Twitter replaced the original status update prompt of "What are you doing?" with "What's happening?" in recognition of how the microblogging service was being used. A fuller explanation of this move was detailed on the official Twitter blog, acknowledging that it has "outgrown the concept of personal status updates" (Stone, 2009). As mentioned earlier, Twitter can be and is perhaps predominantly used for various communicative actions, from breaking news and sharing links to discussing issues and organizing events. Twitter is not just about what someone had for breakfast; Twitter is a user-driven and user-rated exchange of shared information (Reinhardt, et al., 2009).

Here is an example from my own Twitter account employing the conventions mentioned above (Fig. 6): Twitter users consume messages like this one "by viewing a core page showing a stream of the latest messages ... listed in reverse chronological order" (Naaman, Boase, & Lai, 2010). What appears in that information stream or "feed" depends on who the user

@betajames
James Schirmer

RT @DelaneyKirk "Tweeting, blogging or posting online could soon become part of your job" http://bit.ly/dmovsw #eng111 #345tw

10 Mar via Seesmic   ☆ Favorite   ↩ Reply   🗑 Delete

Fig. 6. One of the Author's Twitter Updates. *Source:* Schirmer (2010b). http:// twitter.com/betajames/statuses/10283042638.

chooses to follow. While there might be an expectation by some users, there is no requirement to follow back the person making the initial request. Unlike Facebook (whose user population still dwarfs Twitter's count by a couple hundred million), a signal of interest on Twitter need not be reciprocated.

Because of this as well as the range of third-party applications and multiple avenues of access available, Twitter users are witness to an incredible variety of messages in desktop, mobile and web-based forms. In this way, microblogging functions the way a particular user desires: "as a source of news, to listen to what people in certain groups are talking about, or to communicate with experts or leaders in certain fields" (Perez, 2009). The difference in information acquisition between a Google search and a Twitter question can be significant. Replies to the latter range from "shortened URLs containing answers ... to more intelligent responses" (Cann, Badge, Johnson, & Mosely, 2009) instead of just a list of suggested links containing the search term. Even Google is aware of this difference, having implemented tweets into their regular search results.

I mention all these to reinforce a point made by Brian McNely (2009): because Twitter is public and visible, persistent and searchable, "microblogging as a platform for backchannel communication has led to increased affordances for collaboration." Phil Beadle (2010) provides an accurate rundown of Twitter's collaborative potential, emphasizing that "having access to a ready network of peers means you have the ability to run ideas by people, get them peer-reviewed ... you can ask for and get immediate feedback as to where the best research has been done." Because of all that it is and is not, Twitter gives "a guiding hand on your shoulder within seconds of asking for it" (Beadle, 2010). This observation in particular relates to earlier observations offered by Godwin-Jones (2003) and others regarding the collaborative potential of blogging. What is possible via Posterous is also possible via Twitter; it just takes a smaller, more succinct form.

Opening such a collaborative channel in a classroom situation, though, observes Jeffrey R. Young (2009), "alters classroom power dynamics and signals to students that they're in control." Online communicative technologies can often be great equalizers, but, as described in the next section of this chapter, setting clear guidelines for their use are of near-equal importance. Further support for this observation is evident in how students approached the requirement of using Twitter in courses I guide.

MICROBLOGGING VIA TWITTER TO PROMOTE COMMUNITY

Some academics and educators on Twitter have, and often advocate for, separate accounts, maintaining personal/professional identities on Twitter in addition to accounts for each course taught in a given semester. Reasons for this separation are myriad. For instance, if there is a great diversity of subject matter and focus in a semester, a Twitter account for each course can prevent miscommunication and misunderstanding. Such an account can provide an initial focus for students as well, acting as a kind of localizing agent and guide for others in the course and those they should follow. Holding separate Twitter accounts can also be helpful when performing special course-related projects, such as Twitter user oline73's *Animal Farm* project (2010) that requires students to perform as one of the characters in George Orwell's novel (Fig. 7). Furthermore, some educators have a strong desire for some degree of anonymity online, to keep personal and professional interests apart. Maintaining a private personal account and a public professional account on Twitter is one way to exercise that anonymity.

There are also third-party applications, like Seesmic and TweetDeck, that make it easier to manage multiple Twitter accounts for multiple purposes, but I remain resistant to such differentiation. The accessibility and openness of Twitter make it difficult to keep alternate facets of one's identity hidden, even if maintaining an invitation-only account. Furthermore, I have a

Orwell_3 George Orwell3
the animals have killed Napoleon and now are in control
13 May

Orwell_2 George Orwell2
the animals just ended their revolt against napolean
13 May

Orwell_2 George Orwell2
the animals are getting ready to revolt against napolean
13 May

SSPJClover Clover the Horse
LETS GET THE PIGS!
13 May

Orwell_3 George Orwell3
the animals are revolting
13 May

SSPJNapoleon Napoleon the Pig
We are in trouble
13 May

SSPJMuriel Muriel
lets get em
13 May

SSPJtheDogs The Dogs
i am tired!
13 May

SSPJSquealer Squealer the Pig
OH MY GOODNESS!!! the animals are attacking! #animalfarm

Fig. 7. Twitter Feed of *Animal Farm* Project. *Source:* http://twitter.com/oline73/
animal-farm-project. Reprinted with permission.

sustained interest in my Twitter use fostering an identity that serves as a model for students of what is possible in terms of appropriate academic engagement within the 140-character limit. Rather than Shaquille O'Neal or Kim Kardashian being the first examples that students encounter, I want students to be witness to the ways in which I take advantage of Twitter. I want students to see the diversity in my status updates as well as in those I choose to follow. I enable students to use Twitter to, as Gross (2009) puts it, "learn from their peers, converse around issues that matter to them, and follow people they admire – [building] on students' experience rather than [encouraging] it to develop black market-style." So, in addition to @betajames, contextual hashtags and Twitter lists are localizing agents for students in the courses I guide.

The common use requirement for Twitter in the most recent semester of courses was very simple, yet specific: five course-related "tweets" a week. This was but one of the marked changes from the rather nebulous "maintain presence" requirement in the previous semester's courses. Blog entries and comments autoposted from Posterous to Twitter counted toward the "tweet" requirement. With the two required blog entries and three required blog comments every week, a student could complete blogging and tweeting at the same time. I also encouraged the posting of original thoughts, the "retweeting" of and @ replying to classmates' updates, and the sharing of links relevant to the focus of the course. I thought this range of options might allow for a low-stakes introduction to Twitter, that once students understood how it worked they would engage in microblogging to a greater degree.

However, Twitter integration met with limited success. Some students witnessed immediate benefits and others were more gradual in their appreciation, but a noticeable number held a semester-long resistance to using the microblogging service. Their dissatisfaction was apparent not only on Posterous and Twitter but also in anonymous feedback I gathered near the semester's end. Even though the minimum requirement was a mere five updates per week, students felt they did not have the time, or that Twitter constituted an annoyance, an unnecessary requirement.

Part of the reason for this lack of appreciative acquisition concerns what Twitter is. Blogging requirements make sense to many students because they are more able to view it with clarity as a form of writing, as something with which they already have a degree of familiarity. Even though Twitter is grounded in text, the 140-character limit proved to be an insurmountable barrier for some. Tweeting could not equal writing because of the strident limitations on communicating with others. Because of what Twitter does and does not allow of its users, adoption and understanding needs to be more authentic and original. Perhaps students need to be allowed to come to Twitter of their own accord and have the time necessary to do so.

Asking students to find their own way within the microblogging service could be successful for some, but not for all. Because of the structure of my courses, there was not much affordance of time for Twitter alone. Because it was supplemental to blogging and more traditional forms of writing, because of my own view of Twitter as a relatively easy course requirement for students to fulfill, I saw little reason to address its use beyond an introductory session. Even this introductory session was limited to functionality as I expected students would take the time on their own for purposes of discovery. Allowing students to find their own way through Twitter can be detrimental in another way as well, as some students took to Twitter beyond course-related purposes, participating in "tweetathons" supporting their favorite celebrities and/or charities. As a result, the five course-related tweets were often lost in the endless streams of self-indulgent content, which is perhaps what Twitter is better known for.

There was some measure of appreciation in having their instructor as a model of engagement via Twitter, but many students needed more than this. Even though they were witness to my use of it, Twitter remained difficult for some students to grasp, and there was impatience expressed in understanding the point of Twitter. So, while the continued use of Twitter in a college-level writing course might be better if coming from students instead of the instructor, the reality of this possibility is questionable. A course requirement can often do little more than reaffirm a student's initial dislike.

Not just a greater affordance of time to experiment is necessary; a better frame for using Twitter in higher education is essential. Students do need the time to find their own reasons and their own way when it comes to social media, but there should also be a clear, course-related context established for its use. For a better grasp of what I mean by this, I offer a brief overview of two Twitter-related assignments in two different college-level English courses.

Barbara Nixon (2008) implemented a 48-hour-long Twitter-related assignment early on in a course named Making Connections: Facebook & Beyond. After setting up an account and following everyone else in the class including Nixon, the assignment asks that students send at least six tweets over the next two days and respond to at least two of their classmates' tweets. As Nixon (2008) explains, students' tweets could concern pointing others in the class to "something interesting or funny you read online" or "pose a question that you'd like others to answer." The assignment concludes with a 250-word minimum blog post about the experience, including one way there might be value in continuing an account on Twitter.

In ENG 465 Reading Technology, Brian Croxall introduces a month-long Twitter-related assignment after having already spent time with students discussing media systems. With this initial foundation in place, requirements of the actual assignment include following Croxall, the class account, and other members of the class, posting at least once a day for a month and using the course hashtag #eng465. While there is a focus on experimental play with Twitter, Croxall (2010) also admits to his own interest in students using "an interconnected, mixed media system ... to see if it changes the culture or society of the class in any appreciable way." This interest forms the basis for the end of the assignment as Croxall asks students to write up an evaluation of the assignment and what Twitter teaches about community and media. Emphasized by Croxall, too, is "an honest effort to play along."

Each of these assignments is an experimental, reflective aspect as both Nixon (2008) and Croxall (2010) ask that students play with a low-entry, low-stakes form of social media. Rather than simply requiring the use of Twitter without much explanation, both instructors ask students to provide their own justification, encouraging students to take greater ownership of their performance on Twitter. In addition to considering assignments similar to those put forth by Croxall and Nixon, I am also interested in the possibility of keeping or even increasing the current Twitter use requirement while offering students the chance to opt out by composing an academic essay of equivalent length. If I might be excused for some momentary math, let us see how this would work out. Five tweets of 140 characters each leads to a total of 700 characters per week. Added up over a 14-week period, we should have around 9,800 characters total for the semester. If we can agree on an average of six characters per word, this means that a given student should produce around 1,633 words over the course of a semester. So, the requirement then becomes a question of whether a student prefers to provide a semester's worth of tweets or an additional academic essay six pages in length. I remain curious about what students will decide to perform.

An unexpected function, though, concerned how successful Twitter was as an asynchronous notification system of students' work. This happened because of a course requirement where students link their Posterous and Twitter accounts to enable autoposting, the pushing of content from one social online space to another. So, as students completed their required semiweekly blogging and commenting via Posterous, a public record of their efforts streamed through my Twitter feed (Fig. 8). I became witness to the very instances when students were not just completing coursework but also learning. This unique kind of feed was available to every student, too, required as they were to follow each other and/or the group I created for the course. This was also something for almost anyone beyond the course to see, allowing outsiders more than a peek into the course's focus and how students worked within that framework.

FrankTyra Frank Tyra
Concerns for the Final Project http://post.ly/TVDe
17 Mar

FrankTyra Frank Tyra
Well, I suppose I could limit my audience to my partner and me. I will
keep that in mind as I write up my paper. http://post.ly/RkIZ
17 Mar

FrankTyra Frank Tyra
Task Analysis http://post.ly/T2Lm
15 Mar

FrankTyra Frank Tyra
@jomacko I also agree. I think that I had mentioned it before.
Technology has grown so much that you can simply not...
http://post.ly/Ra7I
12 Mar

FrankTyra Frank Tyra
@burnsbrittany I understand that sometimes interviews can be hard
to get. Fortunately for me, I have a few people... http://post.ly/SEDt
12 Mar

FrankTyra Frank Tyra
Audience Analysis http://post.ly/RkIZ
9 Mar

FrankTyra Frank Tyra
@miblacks I could have changed them myself if I would have saw
your completed final project. Maybe next time! http://post.ly/RbWj
9 Mar

Fig. 8.   Student Frank Tyra's Twitter feed. *Source:* Tyra (2010). Available at http://
twitter.com/FrankTyra. Reprinted with permission.

My Twitter feed became a showcase of student activity, providing me with
better ideas of when students completed their work and how much time they
took to perform it. I could almost see the work happening, very nearly see
students learning. Requiring students to autopost content, both blog entries

and comments, from Posterous to Twitter, was an unforeseen success. It not only allowed me to see how and when students completed necessary work, but it also helped students in creating course-related identities. Posterous and Twitter represented additional frames for the course in which students were able to perform their own unique work and find another reason to engage with the material and each other.

SUMMARY OF STUDENTS' OBSERVATIONS ON BLOGGING AND TWEETING

Although the above sections are separated by blogging and collaboration, tweeting and community, I do not mean to indicate that community-building is exclusive to Twitter, nor collaboration to Posterous (or blogging in general). I also do not mean to leave students' own perspectives on these online communicative technologies out of this particular essay. In moving toward some semblance of a conclusion, then, I offer a sampling of students' observations about blogging and tweeting in college-level courses. I should note that these observations come from students whose academic institution is in a transition period; the first on-campus housing in the form of a 400-room dormitory became available in Fall 2008. Before that, it was a commuter campus serving a majority of nontraditional students. It is also a wireless campus and all classrooms are at least equipped with smart carts and digital projectors. Rare is the student attending this institution who has little to no access to technologies necessary for course completion. The university library and other campus buildings offer computer labs exclusively for students to use.

It was in one of these labs that I conducted critical incident questionnaire (CIQ), offered by Stephen Brookfield in *Becoming a Critically Reflective Teacher* (1995, P. 115), within the last two weeks of the semester, before official course evaluations. All students typed their anonymous responses to the CIQ in Microsoft Word and then printed them out for me to pick up at the end of the session.

For many students across all courses, from first-year composition to graduate study in digital rhetoric, there was an element of surprise to the required use of blogs and Twitter. The very introduction of blogging as a course requirement surprised one student for "most professors I've had tend to use Blackboard if any computer-related tools." Another was impressed with their own ability "to organize the use of the technology that was required" as Posterous and Twitter were never of much interest to the student before the course. A third student expressed surprise at how they ended up "liking the fact that we blogged every week" and a fourth was rather astounded about the content they were able to produce when "left to my own devices regarding material ... I surprised myself and was forced to think critically about stuff I could blog about that pertains to this class." Often partnered with this element of surprise was the expression of a desire to continue blogging and/or tweeting, either "so as not to lose writing for myself in my future" or because "I found them both to be beneficial this semester and may continue to use them even after I am not being graded on it." Initial surprise gave way to sustained value, which came from a sense of engagement over the course of the semester: "Twitter helped with sharing articles and blogs; blogging helped improve my writing and taught me what keeps readers interested and what they're looking for." Put in a more succinct manner by another student: "Blogging forced me to keep audience always in mind." This same student also echoes a point made earlier by Phil Beadle, that is, the importance of following "professionals that engage my interest, that there is a reason for [Twitter] beyond understanding what someone had for breakfast." In moving above such a superficial perspective of social media, in no longer seeing "all the blogs and posts to Twitter" as unnecessary, one additional student

> can definitely see now that it sets up a community for the students and professor to interact immediately and not having to wait until the next class. I often times forget questions I had before and never get them answered, but not in this class due to the almost 24/7 availability of either other students or the professor.

The continual sharing of course-relevant information, the development of audience awareness, the ample opportunity to practice writing, the accessibility and availability of fellow students, and the instructor for commentary and questions, all were benefits that students found in relation to blogging and tweeting in the first-year, advanced, and technical writing courses I guide.

CONCLUSION

What makes the use of social media in higher education is connection, not only among and between participants but also with regard to the social media chosen. Autoposting, the pushing of content across multiple platforms, can be a primary method of maintaining connection as well as coherence and cohesion among users, be they students or teachers. This aids in the construction and maintenance of a learning community as well as in the establishment and reaffirmation of a learning identity unique to each individual involved. This can, and often does, lead to more meaningful interaction and greater collaboration, allowing for the development of learning narratives that become a public record over the course of a semester.

Quick blogging via Posterous and microblogging via Twitter are but two possibilities for facilitating collaboration and community in college-level courses. Overall ease-of-use and relative simplicity make for low barriers of entry for a majority of students. The mobile accessibility of these online communicative technologies should also be of specific appeal. These characteristics should encourage student participation in ways that content management systems like Blackboard do not. The convenience of and allowance for quick and easy sharing of information via blogging and microblogging can also mean that each is often quicker than email for contacting someone. What make both better than Blackboard concerns how they, when taken together, sustain class discussion, keeping it alive, present, and continuous.

If proper affordances are made in terms of framing and timing, the appropriate use of social media can make for a successful addition to almost any college-level course. Social media tools like Posterous and Twitter, both of which are rather effortless in terms of use, allow and perhaps even encourage students to think and chart their own learning in ways that are as meaningful and unique as themselves.

REFERENCES

Beadle, P. (2010). As a teacher, I've realised Twitter has real potential. *The Guardian.* Available at http://www.guardian.co.uk/education/2010/mar/09/twitter-teachers-useful-resource-empathy. Retrieved on March 9, 2010.

Bentley, K. (2010). Are you mad at me? Miscommunications in Writing. Available at http://kbentley20.posterous.com/are-you-mad-at-me-miscommunications-in-writin.

Brookfield, S. (1995). *Becoming a critically reflective teacher.* Hoboken, NJ: Jossey-Bass.

Burgess, J. (2006). Blogging to learn, learning to blog. In: A. Bruns & J. Jacobs (Eds), *Uses of blogs* (pp. 105–114). New York: Peter Lang.

Cann, A., Badge, J., Johnson, S., & Mosely, A. (2009). Twittering the student experience. Available at http://newsletter.alt.ac.uk/xrctg5ovlfkimsphpsy77s.

Croxall, B. (2010). Spring 2010 ENG 465 Twitter. Available at http://briancroxall.pbworks.com/Spring-2010-Eng-465-Twitter.

Davenport, K. (2010). Group 2 facilitation. http://kendelld.posterous.com/group-2-facilitation.

Downes, S. (2004). Educational blogging. *Educause Review, 39*(5), 14–26. Available at http://www.educause.edu/EDUCAUSE + Review/EDUCAUSEReviewMagazineVolume39/EducationalBlogging/157920.

Duffy, P., & Bruns, A. (2006). The use of blogs, wikis and RSS in education: A conversation of possibilities. *Proceedings of Online Learning and Teaching Conference 2006.* (pp. 31–38). Available at http://eprints.qut.edu.au/5398/

Ehrlich, K., & Shami, S. (2010). Microblogging inside and outside the workplace. *Proceedings from the International Conference on Weblogs and Social Media.* Available at http://www.cs.cornell.edu/~sadats/icwsm2010.pdf.

Farmer, B., Yue, A., & Brooks, C. (2008). Using blogging for higher order learning in large cohort university teaching: A case study. *Australasian Journal of Educational Technology, 23*(2), 123–136.

Foresman, C. (2010). Tumblr vs Posterous: Quick blogging showdown. *Ars Technica.* Available at http://arstechnica.com/business/reviews/2010/02/tumblr-vs-posterous-quick-blogging-showdown.ars

Godwin-Jones, B. (2003). Blogs and wikis: Environments for on-line collaboration. *Language Learning & Technology, 7*(2), 12–16. Available at http://llt.msu.edu/vol7num2/emerging/default.html

Gross, J. (2009). Embracing the Twitter classroom. *Huffington Post.* Available at http://www.huffingtonpost.com/jessica-gross/embracing-the-twitter-cla_b_204463.html

Huang, J., Thornton, K., & Efthimiadis, E. (2010). Conversational tagging in Twitter. *Proceedings from Hypertext2010: 21st ACM Conference on Hypertext and Hypermedia.* Available at http://jeffhuang.com/Final_TwitterTagging_HT10.pdf

Krause, S. (2004). When blogging goes bad. *Kairos,* 9.1. Available at http://english.ttu.edu/kairos/9.1/praxis/krause/

Lowe, C., & Williams, T. (2004). Moving to the public: Weblogs in the writing classroom. In: L. J. Gurak, S. Antonijevic, L. Johnson, C. Ratliff, & J. Reyman (Eds.), *Into the blogosphere: Rhetoric, community, and culture of weblogs.* Available at http://blog.lib.umn.edu/blogosphere/moving_to_the_public.html

McNely, B. (2009). Backchannel persistence and collaborative meaning-making. *Proceedings of the 27th ACM International Conference on Design of Communication.* Bloomington, IN.

Naaman, M., Boase, J., & Lai, C.-H. (2010). Is it really about me? Message content in social awareness streams. *Proceedings from the 2010 ACM conference on Computer Supported Cooperative Work.* Savannah, GA.

Nixon, B. (2008). Assignment: 48 hours of Twitter. Available at http://makingconnectionsfye1220.wordpress.com/2008/09/09/assignment-48-hours-of-twitter/

oline73. (2010). *Animal farm project.* Available at http://twitter.com/oline73/animal-farm-project.

Paquet, S. (2003). Personal knowledge publishing and its uses in research. *Knowledge Board.* Available at http://www.knowledgeboard.com/item/253

Perez, E. (2009). Professors experiment with Twitter as teaching tool. Available at http://www.jsonline.com/news/education/43747152.html

Reinhardt, W., Ebner, M., Beham, G., & Costa, C. (2009). How people are using Twitter during conferences. In: V. Hornung-Prähauser & M. Luckmann (Eds), *Proceedings from EduMedia: Creativity and Innovation Competencies on the Web* (pp. 145–156). Available at http://lamp.tu-graz.ac.at/~i203/ebner/publication/09_edumedia.pdf

Schirmer, J. (2010a). *Prevailing visibility: A bit further.* Available at http://betajames.posterous.com/prevailing-visibility-a-bit-further

Schirmer, J. (2010b). *RT@DelaneyKirk 'Tweeting, blogging or posting online could soon become part of your job'*. Available at http://bit.ly/dmovsw #eng111 #345tw

Stone, B. (2009). What's happening? *Twitter Blog*. Available at http://blog.twitter.com/2009/11/whats-happening.html

Tyra, F. (2010). *FrankTyra*. Available at http://twitter.com/FrankTyra

Young, J. R. (2009). Teaching with Twitter: Not for the faint of heart. *Chronicle of Higher Education*. Available at http://chronicle.com/article/Teaching-With-Twitter-Not-for/49230/

LEARNING ABOUT MEDIA EFFECTS BY BUILDING A WIKI COMMUNITY: STUDENTS' EXPERIENCES AND SATISFACTION ☆

Jennie M. Hwang and Boris H. J. M. Brummans

ABSTRACT

Teachers have recently started to introduce wikis into their courses. However, comparatively few studies have looked at the actual experiences of students who are engaged in building a wiki community for a particular course. To address this limitation, this exploratory self-report study examined student experiences with using a wiki in an upper-level undergraduate course on media effects, their reflections on functioning as a member of this wiki community, and their overall satisfaction with taking this kind of a "hybrid" or "blended" course. Results show that students enjoyed learning about media effects by collaboratively building their wiki community, but were critical about the structure of the hybrid course.

☆Chapter for Charles Wankel's edited book, *Teaching Arts and Science with the New Social Media.*

Teaching Arts and Science with the New Social Media
Cutting-edge Technologies in Higher Education, Volume 3, 39–59
Copyright © 2011 by Emerald Group Publishing Limited
All rights of reproduction in any form reserved
ISSN: 2044-9968/doi:10.1108/S2044-9968(2011)0000003006

The classroom of "the Read/Write Web" signifies, according to Will Richardson (2006, p. 28), two unstoppable trends regarding the use of technologies for pedagogical purposes. First of all, as more and more content is becoming available on the Internet, everyone, including teachers and students, needs to learn how to access and assess knowledge more effectively and efficiently. Second, digital content is increasingly created through collaboration. A wiki or "piece of server software that allows users to freely create and edit Web page content [and hyperlinks] using any Web browser" (http://wiki.org/wiki.cgi?WhatIsWiki) is a powerful data sharing and construction tool that is being used progressively more for collaborative purposes in various settings. For example, several scholars (see Li & Bernoff, 2008; Tapscott & Williams, 2006) have pointed out that wikis are becoming popular tools for enhancing collaboration in business contexts. In addition, wikis are increasingly being used in courses, because they are presumed to foster interactive collaboration between peers and improve student learning (see Bruns & Humphreys, 2005; Hsu, 2007; Richardson, 2006).

Since teachers have only recently begun to introduce wikis into their courses, comparatively few studies have looked at the actual experiences of students engaged in building a wiki community for a particular course. Hence, this chapter explores student experiences with creating a wiki in an upper-level undergraduate course on media effects, their reflections on functioning as a member of this wiki community, and their overall satisfaction with taking this kind of a hybrid course, which blends face-to-face and online interactions (see Jackson & Grimes, 2010), based on a small-scale self-report study. In the next section, this study will be situated in extant literature on the use of hybrid courses to improve student learning. After this, the process of building a wiki community in the media effects course that was studied will described, followed by an explanation of the research methods that were used to conduct this investigation and its results. To conclude, the lessons that teachers, interested in using wikis in their courses, can learn from this study will be discussed. Furthermore, several directions for future research will be provided, based on a discussion on the limitations of this research.

WHAT DOES EXTANT LITERATURE TELL US ABOUT USING HYBRID COURSES TO IMPROVE STUDENT LEARNING?

Much research has looked at the advances of technology and their impacts on the ways knowledge is created and diffused. As Leadbeater (2000) noted,

Knowledge sharing and creation is at the heart of innovation in all fields [e.g., science, art, and business] and innovation is the driving force for wealth creation ... Information can be transferred in great torrents, without any understanding or knowledge being generated. Knowledge cannot be transferred; it can only be enacted, through a process of understanding, through which people interpret information and make judgments on the basis of it ... Great tides of information wash over us every day. We do not need more information, we need more understanding. (p. 12)

In classroom settings, the majority of the time is devoted to learning various subjects through different modes of communication. Traditionally, teachers were believed to convey information to students, yet a more modernized learning model focuses on constructive learning through interactive student participation or project-based learning, which fosters knowledge creation and team collaboration (see Bruffee, 1999). As Leadbeater (2000) indicated, focusing on knowledge creation rather than information transfer has important implications for the current dynamics of "the network [or network*ed*] society" (Castells, 2000; see also Haythornthwaite & Wellman, 2002).

A number of studies (e.g., Bryant, 2006; Hsu, 2007) have suggested that using so-called social software or social media (i.e., blogs, social networking, etc.) in higher education enables the knowledge creation to which Leadbeater refers, and provides hands-on, practical experience by allowing students to form and take part in their own learning community. Considering the changing landscape of higher education, it is not surprising that hybrid or blended courses are becoming more and more popular, since they are presumed to enhance teaching effectiveness (see Jackson & Grimes, 2010). Some studies have provided support for this claim. For example, the Department of Education recently conducted a meta-analysis that examined over 51 effects reported by studies published between 1996 and 2008. The study concluded:

In recent experimental and quasi-experimental studies contrasting blends of online and face-to-face instruction with conventional face-to-face classes, blended instruction has been more effective, providing a rationale for the effort required to design and implement blended approaches. (Dept. of Education, 2009, p. xvii)

Another study, conducted by Delialioglu and Yildirim (2007), found that students developed intrinsic motivation to learn and interact with their teacher and peers in a blended learning environment. Students especially liked the well-designed course website that enabled them to process fairly large amounts of information and engage more actively in classroom activities. Hence, this study suggests that the use of technology may benefit teaching and learning, because it improves the ways in which content is delivered and enhances collaborative learning. In spite of these benefits, Delialioglu and Yildirim's investigation also raises questions about the appropriateness of specific types of technologies for particular courses and the perceptions students have of their learning needs in a blended course (see also Atan, Rahman, & Idrus, 2004; Hsu, 2007; Lim & Morris, 2009). The next section looks at the potential benefits of using of wikis to improve student learning.

What Are the Potential Benefits of Using Wikis in Courses?

A wiki is a (relatively) new Web 2.0 tool that is easy to use because it allows people to edit, delete, or modify website content without having to do HTML coding or set up a server (see Richardson, 2006, p. 59; Vossen & Hagemann, 2007). Thus, a wiki enables both individual contribution and group collaboration. In business settings, wikis have been used to enhance collaboration through virtual communication, a great asset for industries in which innovation, creativity, and problem-solving are key (see Tapscott & Williams, 2006). The use of wikis in classroom settings has also shown promising results. For example, Bruns and Humphreys (2005) demonstrated that a wiki can been used as part of a social constructivist pedagogy in order to advance information and communication technology (ICT) literacy and stimulate creative ways of collaborating on projects among university students. Bruns and Humphreys' research showed that a wiki is especially useful for giving students first-hand experience with "advanced new media technologies in authentic exercises" (p. 29). In addition, this research indicated that a wiki facilitates tutorial work and student interactions outside of the classroom. For example, in the project that was studied, tutors set up

> discussion groups in the wiki for each of their classes, and used these groups to coordinate their students' emerging group projects … [T]hrough these discussion groups students were able to nominate topics of interest to them and thus find like-minded collaborators, and student teams could begin to flesh out their project ideas (while at the same time providing the tutor with a permanent record of their project ideas and overall participation in class). (p. 29)

Minocha and Roberts (2008) investigated the use of wikis in a university setting as well, focusing particularly on engineering students. This study showed that by participating in a course wiki, students developed a better, more comprehensive understanding of concepts and built their team-working skills. Through their wiki-based collaboration, students learned to appreciate the challenges of the real-world team-working process and to adapt themselves to different roles and apply different rules to accomplish effective virtual participation.

In another study (Xiao & Lucking, 2008), the use of a wiki in an undergraduate writing program was found to facilitate peer assessment. This research showed that students were very satisfied with their wiki collaboration environment, because it enabled them to provide effective peer feedback and improve their critical thinking skills.

Extant research thus suggests that blended courses can help improve student learning, since they stimulate students' intrinsic motivation to learn and interact with their teacher and peers, and they help them process large amounts of information and engage more actively in classroom activities and collaboration with their peers. Moreover, the research literature suggests that wikis are well-suited for these kinds of courses, because they encourage individual student contribution and collaboration; give them hands-on experience with a new communication technology; allow for student interactions inside and outside the classroom; help students develop a better understanding of concepts and theories, and work in teams; and provide feedback and develop critical thinking skills.

What Questions Remain to Be Explored?

As this brief review of literature shows, previous studies offer many insights into the use of wikis to improve student learning. What remains to be explored in further detail, though, is how a course wiki is actually constructed "from scratch" and how students perceive and experience this process (for a rare example in medical education, see Otter, Whittaker, & Spriggs, 2009), since most research on web instruction has centered on content or student outcomes without paying attention to the methods used to deliver this kind of instruction and the way specific delivery methods affect students' motivation to learn (see Small, 1999). In light of these observations, the current chapter explores how students perceived their participation in the process of constructing a course wiki from start to end, the effectiveness of their wiki collaboration, and their overall satisfaction with a blended course on media effects, based on a small-scale self-report study. The next section provides insight into the design of the course that formed the context for this study by focusing in particular on the use and construction of the course wiki. After this, the research methods that were used to collect and analyze the data for this research will be described, followed by its results.

BUILDING A WIKI COMMUNITY IN AN UNDERGRADUATE MEDIA EFFECTS COURSE

In the course that formed the context for the current study, undergraduate students of a southwest university learn about the effects of media on individuals and society from an empirical point of view. By exploring several mass media topics, including the role of traditional media (e.g., radio, television, and newspaper) in creating and shaping stereotypes, the ubiquity of violent and sexual media content and its impact on society, and the influence of the news on public opinion, the course introduces students to the complexities of traditional media effects. Furthermore, the course gives students a thorough introduction to various new media (e.g., the Internet, social-networking, the use of cellular phones) and their far-reaching effects on users.

A wiki was introduced into this course to improve student learning by giving students first-hand experience with being part of a wiki community. The course wiki was first used in the class in the spring quarter of 2008 and has been used ever since. Because this is a course on the communicative effects of traditional and new media, the course wiki allows students to use and integrate electronic media content, such as audio, video, and text files, into their group projects, and thereby pushes the class beyond the limitations of its existing paper-based assignments. Thus, the new instructional model encourages students to be active learners through participatory knowledge construction, ongoing assessment, and collaboration (see also Wallace, 2003).

During the course term, different media effect theories are introduced on a weekly basis via regular class lectures and learning modules on the course Blackboard, an online learning/course management system used by many colleges and universities for the purpose of e-learning. Each week, the learning module incorporates learning objectives and activities to help students keep up with class readings and develop good organization and time-management skills.

To build their wiki community, students form small "wiki groups" of three to four people at the start of the term. After they have formed their groups, students are asked to assign a specific role (e.g., team leader, researcher, editor, presenter) to each group member and specify their individual responsibilities. Each wiki group's task is to selects one class-related media effects theory (e.g., uses and gratification, agenda setting, or social cognitive theory), describe its main tenets, provide a brief literature review of studies that draw on the theory, and offer an insightful critique. To accomplish this, group members use the course wiki to conduct collaborative research and create an informative online portrait of the theory they have selected. More specifically, each group is required to use the course wiki to: (1) organize multiple academic sources (e.g., journal articles, book chapters, newspaper articles) as well as real-world examples of media effects (e.g., images, audio/video clips); (2) synthesize data online; (3) support an in-class group presentation on their theory; and (4) engage in online discussions and provide group members and other classmates with feedback. Through these activities, each wiki group develops a specific section of the overall course wiki on a particular media effects theory – that is, the overall course wiki is a collection of group wikis on different media effects theories, accessible through the course Blackboard. Each group wiki is comprised of wiki pages that contain the following information: (1) an introduction page/table of contents; (2) a background description of the theory; (3) an in-depth discussion of specific aspects of the theory; (4) a discussion of specific studies that use the theory; (5) a set of critical questions for classmates; and (6) a bibliography.

A grading rubric is used to evaluate the individual and collaborative contributions of each student. This allows the teacher to assess the quality of the online content that students produce, including the clarity and logic of its organization, the selection and citation of credible sources, and creativity. Classmates are also asked to use these grading criteria to engage in peer reviewing of each other's group wikis.

To gain insight into the ways students perceived their participation in the process of constructing their group wiki and their overall course wiki, the effectiveness of their wiki collaboration, and their overall satisfaction with this blended course, a self-report study was conducted during one course term (i.e., the winter quarter of 2009). The next section describes the data collection and analysis methods that were used to conduct this study.

DATA COLLECTION AND ANALYSIS

Data Collection

Since this research was exploratory in nature, closed and open-ended survey questions were used. Data were collected using a pretest and a posttest survey on the first and last day of class.

The pretest questionnaire investigated students' (1) technology use experiences; (2) perceptions of e-learning; and (3) basic demographic data. As far as their technology use experiences were concerned, students were asked to indicate whether they had used an online learning/course management system (e.g., Blackboard or WebCT) and various e-learning tools (e.g., discussion board, blog, online test/survey, grade book) before this class, as well as the time they spent online to engage in social interactions or schoolwork during the school year (e.g., "When enrolled in school, on average, how many hours per week do you spend using a computer for schoolwork?"). To investigate students' perceptions of e-learning, 12 statements on teaching and learning with communication technologies were adapted from studies that investigated students' and teachers' perceptions toward distance education or e-learning (see Tao, 2008; Tao & Yeh, 2008). A 7-point Likert-type scale (1 = strongly disagree; 7 = strongly agree) was used to measure these perceptions. Example statements included: "E-learning rapidly delivers knowledge and information to learners;" and "E-learning produces better learning results than traditional teaching." The pretest questionnaire concluded with a set of demographic questions, eliciting information about a respondent's gender, age, race, major, and class standing.

The posttest survey focused on evaluating students' perceptions of learning through wiki collaboration, as well as their overall learning experience and satisfaction with the course. Statements were rated based on a 7-point Likert-type scale (1 = strongly disagree; 7 = strongly agree). Example statements included: "I was able to learn from participating in this hybrid course;" "Collaborative learning in a wiki environment works better than in a face-to-face learning environment;" "Collaborative learning in my wiki group was effective;" "As a result of my experience with this course, I would like to take another hybrid course in the future;" and "Overall, this course met my learning expectations." Students also reflected on their wiki experiences through a set of qualitative questions, such as: "What did you like about this course?" "What was not working in this course?" "What things could be improved?"

As is custom, the conduct of the study was approved by the Human Subjects Review Committee of the first author's university. Accordingly, students were assured that the collected data would remain confidential.

Data Analysis

To analyze the quantitative portion of the data, descriptive statistics (i.e., means and standard deviations) were used. Subsequently, although the number of students who participated in this study was low ($N = 26$), exploratory factor analysis was employed to determine the main factors of students' perceptions, and Pearson correlations were computed to analyze the relationships between the studied variables. All quantitative data were analyzed using SPSS with a significance level for statistical tests of .05.

Thematic analysis was used to analyze the qualitative portion of the data. This involved repeatedly reading the responses to the open-ended questions, which enabled the identification and labeling of recurring points of reference in the data. In turn, looking at the regularity with which these points of reference resurfaced allowed for the definition of specific themes in students' experiences and satisfaction with building their wiki community.

STUDENTS' EXPERIENCES AND SATISFACTION WITH BUILDING A WIKI COMMUNITY: SUCCESSES AND CHALLENGES

Who Participated in Building the Wiki Community?

The students who participated in this study ($N = 26$) were predominantly female (77%). Their average age was 20 years ($SD = 1.06$). Two-thirds of the participants were upper-level communication major students and nearly 90% of the students were Caucasian. On the basis of the pretest survey conducted on the first day of class, all the students had used an online learning/course management system (i.e., Blackboard) and about half (46%) of them had taken a blended course before entering the class. In terms of their use of e-learning tools in the classroom, using a discussion board (92%), submitting assignments online (92%), and downloading files (81%) were reported most frequently. Less than one in four (23%) students had used a wiki before. On average, students reported to spend 6–10 hours per week using the computer for social interactions, and about 3–5 hours per week for schoolwork. Moreover, the majority of students (81%) reported that they were comfortable using technology for coursework, and most of them were fairly confident that they would perform well in a hybrid course ($M = 5.30$, $SD = 0.96$, based on a 7-point Likert-type scale).

How Did Students Perceive E-learning?

As mentioned, 12 statements, employing a 7-point Likert-type scale, were used to measure students' perceptions about e-learning. Table 1 details the descriptive statistics for these statements.

Exploratory factor analysis was used to determine the main factors. Only factors with eigenvalues bigger than one were selected. Four factors were found (74% of the total variance). The first factor represented *e-learning benefits* and included four items (21.6% of variance explained, $\alpha = .75$): "easy to provide a personalized learning environment," "better learning results than

Table 1. Descriptive Statistics for Students' Perceptions of E-learning.

	Mean	SD
E-learning is important for overcoming geographical constraints	6.08	1.04
E-learning effectively integrates teaching resources	5.88	0.97
E-learning makes it easier for schools/teachers to provide students with a personalized learning environment	5.23	1.19
E-learning produces better learning results than traditional teaching	4.47	0.91
E-learning improves a teacher's teaching strategies	5.96	0.98
The value of e-learning lies in its capability to integrate technology into teachings	6.00	0.89
E-learning boosts students' learning interests	5.00	0.82
E-learning rapidly delivers knowledge and information to students	5.87	0.97
E-learning helps in understanding students' individual preferences	4.90	1.38
E-learning boosts the teacher's teaching performance	4.32	1.20
E-learning environment construction is not easy	4.55	1.28
E-learning is a long-term investment strategy	4.87	1.49

Note: Mean scores reflect perceptions based on a 7-point Likert-type scale (1 = strongly disagree; 7 = strongly agree).

traditional teaching," "helps in understanding individual learning preferences," and "boosts teaching performance." The second factor represented *e-learning challenges* (19.6% of variance explained, $\alpha = .79$) and included items such as "environment construction is not easy" and "requires a long-term investment strategy." The third factor represented *effectiveness of teaching with technology* (17.8% of variance explained, $\alpha = .70$), including items like "capability to integrate technology into teaching," "increases learning interests," and "rapid delivery of content." The fourth factor represented *convenience of teaching with technology* (15% of variance explained). This factor included the following items: "overcomes geographic constraints," "effectively integrates teaching resources," and "increases flexibility in teaching strategies." However, since the Cronbach's α for this factor was low ($\alpha = .47$), it was not included in the correlation analysis.

How Did Students Perceive Learning through Wiki Collaboration?

Students reported on their wiki collaboration by evaluating ten statements through the use of 7-point Likert-type scales. Table 2 provides the descriptive statistics for these statements.

Table 2. Descriptive Statistics for Students' Perceptions of Wiki
Collaboration in a Hybrid Course.

	Mean	SD
I was able to learn from participating in this hybrid course	6.04	0.81
I was stimulated to do additional readings or research on topics discussed in this hybrid course	4.96	1.92
Online discussions helped me understand other points of view	4.96	1.42
Collaborative learning in a wiki environment works better than in a face-to-face learning environment	3.67	1.52
I felt part of a learning community in my group	5.54	1.02
I actively exchanged my ideas with group members	5.92	1.18
I was able to develop new skills and knowledge by interacting with group members	5.87	1.08
I was able to develop problem-solving skills through peer collaboration	5.54	1.10
Collaborative learning in my wiki group was effective	5.92	1.05
Collaborative learning in my wiki group was time consuming	5.87	1.04

Note: Mean scores reflect perceptions based on a 7-point Likert-type scale (1 = strongly disagree; 7 = strongly agree).

Three factors were found through exploratory factor analysis (72% of the total variance). The first factor represented *positive learning effects of wiki collaboration* and included six items (40.3% of variance explained, $\alpha = .90$): "being able to learn from participating in wiki collaboration," "feeling part of a learning community," "active exchange of ideas," "being able to develop new skills by interacting with others," "being able to solve problems through collaboration," and "wiki allows for effective collaborative learning." The second factor represented *cognitive stimulation through engagement in wiki collaboration* (19.1% of variance explained, $\alpha = .69$) and included items such as "being stimulated to do additional research beyond class requirements," "understanding others' points of view through online discussions," and "requiring more time." The third factor (12.5% of variance explained) consisted of only one item, namely "wiki collaboration works better than face-to-face learning."

How Satisfied Were Students with the Hybrid Course?

Eight statements were used to measure students' satisfaction with the hybrid course. Table 3 shows the descriptive statistics for these statements.

Exploratory factor analysis revealed only one factor in this case, that is, satisfaction (70.6% of the total variance, $\alpha = .93$).

Table 3. Descriptive Statistics for Students' Satisfaction with the
Hybrid Course.

	Mean	SD
Overall, I am satisfied with my collaborative learning experience in this course	5.70	1.11
As a result of my experience with this course, I would like to take another hybrid course in the future	5.00	1.38
This course was a useful learning experience	5.74	1.05
I put in a great deal of effort to learn the hybrid learning/computer-mediated communication system to participate in this course	5.52	1.41
My level of learning in this course was of the highest quality	5.17	1.50
Overall, the learning activities and assignments of this course met my learning expectations	5.61	1.16
Overall, the instructor for this course met my learning expectations	5.96	1.19
Overall, this course met my learning expectations	5.65	1.27

Note: Mean scores reflect satisfaction based on a 7-point Likert-type scale (1 = strongly disagree; 7 = strongly agree).

How Were the Studied Variables Related to Each Other?

To understand the relationships between students' technology use experiences, perceptions of e-learning, their experiences with learning through wiki collaboration, and their overall satisfaction with the blended course, Pearson correlation analysis was used. Table 4 shows the outcomes of this analysis.

The time students spent using a computer for social interactions (e.g., chat rooms, MySpace/Facebook, downloading music, blogs) was positively related to their level of comfort with using technology for schoolwork ($r = .54$, $p < .01$), while the time spent using a computer for schoolwork was positively related to their confidence of performing well in a hybrid course ($r = .40$, $p < .05$) and their perceptions of the effectiveness of teaching with technology ($r = .59$, $p < .01$). Moreover, students' belief that they would do well in a hybrid course was positively related to their perceptions of the effectiveness of teaching with technology ($r = .47$, $p < .05$).

Looking closer at their perceptions of e-learning, students' perceptions of the benefits of e-learning were positively correlated with their experience of using different e-learning tools (i.e., discussion board, submitting assignments online, and downloading online files) ($r = .51$, $p < .05$). In addition, their perceptions of e-learning challenges were negatively correlated with their perceptions of the positive learning effects of wiki collaboration ($r = -.49$, $p < .05$).

Table 4. Pearson Correlations for Studied Variables.

	1	2	3	4	5	6	7	8	9	10	11	12
1 Social use of a computer (hours per week)	—	.29	.05	.54**	.23	-.01	-.16	.25	.15	.08	.36	.25
2 Schoolwork use of a computer (hours per week)		—	-.07	.13	.40*	.09	-.08	.59**	.21	-.09	-.13	.13
3 Range of e-learning tools usage			—	.20	-.33	.51*	.14	-.38	.01	.04	.33	-.10
4 Comfortable using technology for course work				—	.36	.16	.08	.39	-.02	.07	.35	-.08
5 Confident to perform well in a hybrid course					—	-.25	-.36	.47*	.18	.07	-.21	.19
6 E-learning benefits						—	-.11	.21	.19	-.34	.35	-.19
7 E-learning challenges							—	-.04	-.49*	-.16	-.10	-.46
8 Effectiveness of teaching with technology								—	.04	-.03	.13	-.09
9 Positive learning effectives of wiki collaboration									—	.24	.23	.76**
10 Cognitive stimulation through engagement in wiki collaboration										—	.12	.47*
11 Wiki collaboration works better than f2f learning											—	.06
12 Satisfaction												—

*$p < .05$, **$p < .01$.

In terms of satisfaction, the more students perceived positive learning effects of wiki collaboration, the more they were satisfied with the hybrid course ($r = .76$, $p < .01$). Students' course satisfaction was also positively related to their perceptions of cognitive stimulation through engagement in wiki collaboration ($r = .47$, $p < .05$).

What Were the Successes and Challenges Involved in Learning about Media Effects through Building a Wiki Community According to Students?

On the basis of the thematic analysis of their written statements, students seemed very positive about their experiences with the hybrid course. Some students used simple, fairly nondescript statements such as the following to express their views on the course: "I like the group wiki, I think it works well to teach the assignment." "I enjoy learning to use the wiki." "I like the wiki collaboration task." "It's been a lot of fun!" More insightful comments centered on the fact that the blended course enabled "hands-on learning" and created a "comfortable, open learning environment." Some students also enjoyed the "newness" of the hybrid class, which made the experience very stimulating. One student remarked, for example, that she had "never taken an online course or made a wiki." Hence, it was "a great new educational experience for [her]."

Besides these kinds of remarks, several students commented on the conveniences of the blended course. As one student wrote, "The reason I like this class is because this course is a hybrid course. It really makes it easy to learn when you can go back online and look up notes after class or do assignments." Many students also liked the fact that using the course wiki allowed them to digest class materials in different ways. "[Wikis] grab my interest," one student stated, "and I have been learning a great deal of information from each wiki!" Another student mentioned:

> I find the wikis to be very helpful. It is nice having someone who understands the topic put the definitions and ideas in their own words, especially when the textbook uses abstract language which make concepts difficult to understand. The use of videos and other forms of media also help one to understand the topic better and remember it for the test.

Along the same lines, yet another student indicated that "the wiki presentations reinforce the ideas learned in the readings and lecture, but allow for you to have visual associations to further reinforce the topic and ideas being presented."

Finally, some students were happy to learn by participating in building the wiki community, because it allowed them to learn skills that would be useful to them in their future jobs. According to one student, "The wiki collaboration really forces you to adjust to group collaboration and really makes you an expert at particular media effects and new theories that you would not ordinarily know." Someone else stated that "the wiki is a skill-enhancing project, advancing our knowledge of the Internet and html design." Similarly, one student wrote,

> I like the wiki project because it introduced me to a new aspect of technology and widened my skill set. Nowadays technological skills are so important and this was the first class I've had that actually taught me a skill that can be used in the real world.

Comparatively few critical comments were offered. However, these remarks provided important insights into the challenges that students experienced with the blended course. For example, some students commented on the fact that getting used to the hybrid format required extra time and effort. In a similar vein, a few students mentioned that they had wanted "more initial instruction" on what to expect in terms of the "wiki content." Surprisingly, only one student reported that it was difficult to check things online due to the hardware limitations of his computer: "Initially, I was not used to the learning module on the Blackboard. I was used to have a syllabus with all the assignments listed because my computer is slower and sometimes a pain to check things on."

The most important critique focused on the way the blended course was structured in terms of face-to-face and online interactions. In the course, Tuesdays were days when the teacher explained a particular media effects theory through a traditional lecture, while Thursdays were days when a wiki group presented their group wiki on this theory. A number of students questioned this set-up, indicating that it, as one student stated, "stole their thunder":

> Having a lesson on Tuesday and a presentation on Thursday of each week is an effective structure. However, I have some criticism about covering the topic that the group will be presenting the following class session because it 'steals their thunder.' In other words, the presentation becomes repetitive or they can't include certain things that are main concepts in their topic because the professor has just gone over them.

Another student reinforced this idea by writing:

> I like how every week is thoroughly planned out and you know what to expect. Also the wiki presentations reinforce the ideas learned in the readings and lecture, but allow for you to have visual associations to further reinforce the topic and ideas being presented. Sometimes the presentations become repetitive with what is presented in lecture and

what the group wiki covers. Maybe have the group presentation go first because it is frustrating when you see a large portion of your presentation already covered, and then have the class lecture be used as a follow up to reinforce particular ideas that are important and to cover information that was left out.

Challenges like these, which are related to the structuring of the hybrid course, require further thought, as will be discussed in the next section.

WHAT CAN TEACHERS LEARN FROM THIS STUDY?

As the study that formed the basis of this chapter shows, what teachers need to keep in mind before starting their hybrid course is that students vary in their perceptions of a blended course when they enter this type of class, and that these perceptions depend on their previous experiences with using computer technologies. For example, those who are familiar with using a computer for social interactions tend to be more comfortable with using a computer for schoolwork. Moreover, students' level of confidence about doing well in a hybrid course seems to be positively related to the amount of experience they have with using a computer for schoolwork. In turn, those who lack this kind of confidence appear to view the effectiveness of teaching with technology more negatively, those who regularly use the computer for schoolwork seem more positive about the effectiveness of teaching with technology, and those who are familiar with using different e-learning tools seem more positive about its benefits. Finally, it seems that the more challenges with regard to e-learning students perceive at the beginning of the term, the less they see the positive learning effects of wiki collaboration throughout the quarter or semester.

In line with Bruns and Humphreys' (2005) research, these results suggest that a teacher needs to focus on explaining the benefits and challenges of a blended course, especially at the start of term. While students are more and more familiar with e-learning tools, many of them still need guidance about how to use these tools to reap their benefits, both in courses as well as in future professions. Thus, the role of the teacher remains central – if not becomes *more* central – in a hybrid course, and it is his or her task to make sure that students do not "fall off the boat," so to speak, especially if they do not have access to a powerful computer outside of school and/or are not very computer savvy. Students also need to be given clear information about what is expected from them in terms of grading in a blended course, since traditional grading is still held as the standard. In the course that was investigated, the teacher developed elaborate grading criteria for evaluating students' wiki presentations as well as their final product – the final version of their group wiki. These criteria allowed the teacher (and students) to evaluate the quality of the online content that the students produced, including the clarity and logic of its organization, the selection and citation of credible sources, and creativity. Despite these efforts, some students were still not sure how they were going to be evaluated, most likely due to the fact that they had comparatively little experience with using new technologies.

Furthermore, similar to Otter, Whittaker, and Spriggs' (2009) research, this study suggests that since students vary in their degree of Internet literacy, it is important to monitor students in their use of the Internet to find appropriate, high-quality sources and examples, like online research articles, credible websites, and so forth, to build their course wiki. A wiki allows a teacher to perform this kind of monitoring. An added benefit is that the wiki simultaneously allows him or her to track social loafing by monitoring the modifications that specific students have made to their group wiki and tracing the history of each group wiki (i.e., each new version of the group wiki is saved in a chronological order and each student's individual contributions/changes are highlighted).

To conclude, this study shows that in using a wiki to improve the delivery of a media effects course, a teacher also faces important challenges, some of which require significant reflection, while others have to be accepted as more or less insurmountable limitations. As will have become clear, in comparison with a traditional course, teaching a blended course that involves the use of a wiki requires a great deal of additional time and energy. What is perhaps the most important challenge, though, is to find a suitable way to structure the course so that it creates synergy between face-to-face and online interactions. This study suggests that students are very motivated (even excited) to present the results of their group wiki collaboration to their classmates. Hence, a teacher should find a way to structure the class so that course lectures do not lower this excitement, because they "steal the thunder" of students' presentations.

Factors beyond the immediate control of the teacher of a blended course pertain to technical limitations inherent in technologies like Blackboard and wiki. What is inconvenient, for instance, is that multiple students cannot log into, and thus work on, their own wiki at the same time. Although many different kinds of wiki software are available nowadays, a teacher has to experiment with these platforms to find the one that best suits the course.

What is even more difficult to manage or influence, of course, is the technological infrastructure that is available at a given college or university. Thus, the availability of a sound technological infrastructure and the necessary technical support, as well as the general attitude toward teaching with technology on a particular campus, may greatly affect the successful adoption and outcomes of a blended course (see Slevin, 2008; see also Jackson & Hwang, 2010).

LIMITATIONS AND DIRECTIONS FOR FUTURE RESEARCH

While this study only investigated a small group of communication majors' perceptions of their participation in the construction of a course wiki on media effects, the effectiveness of their wiki collaboration, and their overall satisfaction with the hybrid course, during a single term in the context of a southwest university where hybrid courses are not the norm, it provides insights into the successes and challenges involved in this kind of course from the point of view of students. It will be important to use more rigorous, experimental (or quasi-experimental) designs in future studies to examine how differences between traditional and blended course delivery affect student learning and their overall satisfaction with hybrid learning. This is especially important because, as Bruns and Humphreys' (2005) research has shown, more and more courses, if not entire educational programs, require hybrid formats to prepare students for the changing needs of different professional fields. As this study illustrates, students *themselves* are the first to acknowledge these needs, and many of them are keen to immerse themselves in blended course environments.

Because this research focused on one hybrid course, future studies also need to examine how variations in structuring a blended course or combinations of online (e.g., group wikis) and offline (e.g., traditional exams) assignments affect student learning. In addition, future research will need to be conducted to understand whether certain kinds of courses lend themselves more to the hybrid course format than others. For example, to what extent would this kind of course format be suitable for teaching courses on social science research methods, English literature, social psychology, physics, or engineering? In addition, since only communication majors participated in this research, it will be worthwhile to compare and contrast the experiences of communication students with those of non-communication majors.

CONCLUSION

This chapter has provided a description of how a wiki can be integrated into a course to improve student learning and explored how students perceive their participation in the construction of a course wiki on media effects, the effectiveness of their wiki collaboration, and their overall satisfaction with a blended course. It demonstrates that an undergraduate course on media effects lends itself well to the integration of new technologies, such as Blackboard and wiki, because it gives students first-hand experience with the impact of new media. As this chapter indicates, students are increasingly aware that it is important to learn about the effects of new media through ways that go beyond traditional course delivery (i.e., course lectures, papers, and the reading of a textbook) and give them the opportunity to use the new media that form the very subject of the course. Hence, students enjoy working within a wiki community, which they start building from day one through their online and face-to-face interactions, since it allows them to construct knowledge about various media effect theories and learn how to use a range of new media in a collaborative fashion. A also wiki allows them to move beyond the limitations of existing presentation software, such as PowerPoint or Keynote, and create more dynamic, more engaging presentations that combine written content, YouTube videos, images, hyperlinks, and so on, in creative, integrated ways. The course that was studied in this chapter illustrated this by showing that the wiki enabled students to select relevant media events and news clips to demonstrate their ability to apply a specific theory to real-life events. "The beauty" of using wikis in a blended course on media effects therefore lies in the fact that their structure is "shaped from within – not imposed from above" (Lamb, 2004, p. 40), which gives students the opportunity to shape the course content according to their own interests, insights, and so on (see also Engstrom & Jewett, 2005).

This chapter may have only provided a glimpse into the possibilities of using wikis in blended courses. Nonetheless, it shows us how important it is to conduct this kind of research, because it helps us learn to use technology "for *delivering* instruction" rather than simply "for preparation and communication" (Russell, Bebell, O'Dwyer, & O'Connor, 2003, p. 297; emphasis added), and because it helps us develop innovative teaching methods that meet the demands of today's networked society.

REFERENCES

Atan, H., Rahman, Z. A., & Idrus, R. M. (2004). Characteristics of the web-based learning environment in distance education: Students' perceptions of their learning needs. *Educational Media International, 41*(2), 103–110.

Bruffee, K. (1999). *Collaborative learning: Higher education, interdependence, and the authority of knowledge* (2nd ed.). Baltimore: Johns Hopkins University Press.

Bruns, A., & Humphreys, S. (2005). *Wikis in teaching and assessment: The M/Cyclopedia project.* Proceedings of the International Wiki Symposium, San Diego. ACM. Available at http://eprints.qut.edu.au/archive/00002289/

Bryant, T. (2006). Social software in academia. *Educause Quarterly, 2,* 61–64.

Castells, M. (2000). *The rise of the network society* (2nd ed.). Malden, MA: Blackwell.

Delialioglu, O., & Yildirim, Z. (2007). Student's perceptions on effective dimensions of interactive learning in a blended learning environment. *Educational Technology & Society, 10*(2), 133–146.

Engstrom, M. E., & Jewett, D. (2005). Collaborative learning the wiki way. *TechTrends, 49*(6), 12–15.

Haythornthwaite, C., & Wellman, B. (2002). The Internet in everyday life: An introduction. In: B. Wellman & C. Haythornthwaite (Eds), *The Internet in everyday life* (pp. 3–41). Malden, MA: Blackwell.

Hsu, J. (2007). Innovative technologies for education and learning: Education and knowledge-oriented applications of blogs, wikis, podcasts, and more. *International Journal of Information and Communication Technology Education, 3*(3), 70–89.

Jackson, L. D., & Grimes, J. (2010). The hybrid course: Facilitating learning through social software. In: T. Dumova & R. Fiordo (Eds), *Handbook of research on social interaction technologies and collaboration software: Concepts and trends* (pp. 220–232). Hershey, PA: IGI Global.

Jackson, L. D., & Hwang, J. M. (2010). Teaching with technology: Faculty experiences. *Academic Exchange Quarterly, 14*(2), 77–81.

Lamb, B. (2004). Wide open spaces: Wikis, ready or not. *Educause Review, 39*(5), 36–48.

Leadbeater, C. (2000). *Living on thin air.* London: Penguin.

Li, C., & Bernoff, J. (2008). *Groundswell: Winning in a world transformed by social technologies.* Boston, MA: Harvard Business School Press.

Lim, D. H., & Morris, M. L. (2009). Learner and instructional factors influencing learning outcomes within a blended learning environment. *Educational Technology & Society, 12*(4), 282–293.

Minocha, S., & Roberts, D. (2008). Social, usability, and pedagogical factors influencing students' learning experiences with wikis and blogs. *Pragmatics & Cognition, 16*(2), 272–306.

Otter, M. E., Whittaker, S., & Spriggs, S. (2009). Using wikis and peer evaluation to teach medical students how to find and assess evidence based resources: A pilot study. *New Review of Academic Librarianship, 15*(2), 187–205.

Richardson, W. (2006). *Blogs, wikis, podcasts, and other powerful Web tools for classrooms.* Thousand Oaks, CA: Corwin Press.

Russell, M., Bebell, D., O'Dwyer, L., & O'Connor, K. (2003). Examining teacher technology use: Implications for preservice and inservice teacher preparation. *Journal of Teacher Education, 54*(4), 297–310.

Slevin, J. (2008). E-learning and the transformation of social interaction in higher education. *Learning, Media, and Technology, 33*(2), 115–126.

Small, R. V. (1999). An exploration of motivational strategies used by library media specialist during library and information skills instruction. *School Library Media Research*. Available at http://www.ala.org/ala/mgrps/divs/aasl/aaslpubsandjournals/ slmrb/slmrcontents/volume21999/vol2small.cfm. Retrieved on February 10, 2010.

Tao, Y.-H. (2008). Typology of college student perception on institutional e-learning issues – An extension study of a teacher's typology in Taiwan. *Computer & Education, 50*, 1495–1508.

Tao, Y.-H., & Yeh, C.-C. R. (2008). Typology of teacher perception toward distance education issues: A study of college information department teachers in Taiwan. *Computers & Education, 50*, 23–36.

Tapscott, D., & Williams, A. D. (2006). *Wikinomics: How mass collaboration changes everything.* New York: Portfolio.

U.S. Department of Education, Office of Planning, Evaluation, and Policy Development. (2009). *Evaluation of evidence-based practices in online learning: A meta-analysis and review of online learning studies,* Washington, DC. Available at http://www2.ed.gov/ about/offices/list/opepd/ppss/reports.html. Retrieved on February 10, 2010.

Vossen, G., & Hagemann, S. (2007). *Unleashing Web 2.0: From concepts to creativity.* Burlington, MA: Elsevier.

Wallace, R. M. (2003). Online learning in higher education: A review of research on interactions among teachers and students. *Education, Communication & Information, 3*(2), 241–280.

Xiao, Y., & Lucking, R. (2008). The impact of two types of peer assessment on students' performance and satisfaction within a wiki environment. *The Internet and Higher Education, 11*, 186–193.

FACEBOOK, "FRIENDING," AND FACULTY–STUDENT COMMUNICATION

Lora Helvie-Mason

ABSTRACT

This chapter explores the communicative relationship between students and faculty members through Facebook. Since its inception in 2004, Facebook has become an avenue not only for student–student connections, but increasingly for faculty–student communication. This chapter explores the impact on pedagogy and instruction when faculty members "friend" their students and/or create class groups on Facebook. Emphasis focused on student perceptions of faculty, identity, and disclosure, communication patterns, educational impact, and guidelines for faculty and students communicating through Facebook.

THREE-PARAGRAPH OVERVIEW

This chapter explores the communicative relationship between faculty members and students during a time of increasingly diverse methods of mediated communication. Social media has been embraced by higher education and Facebook, in particular, has become a marketing tactic as

Teaching Arts and Science with the New Social Media
Cutting-edge Technologies in Higher Education, Volume 3, 61–87
Copyright © 2011 by Emerald Group Publishing Limited
All rights of reproduction in any form reserved
ISSN: 2044-9968/doi:10.1108/S2044-9968(2011)0000003007

well as a means of institutional and instructional communication with current and future students. This chapter begins by describing instructional communication and its changing nature with computer-mediated communication in higher education.

Next, the author examines student perceptions of faculty members as Facebook "friends" to explore the impact "friending" may have on the educational experiences of our students. Examining the perceptions of faculty friends is then related to current literature on immediacy, identity and disclosure, and the impact on pedagogy.

Lastly, the chapter concludes with five general areas for faculty members to consider if they have or contemplate friendships with students. These guidelines also work if Facebook is used for class assignments or part of class discussions. These areas include: before you start, know your audience, response options, collaborate and share, and alternatives to "friending" on Facebook.

FACEBOOK, "FRIENDING," AND FACULTY–STUDENT COMMUNICATION

The red balloon appears over your "friend requests" icon as you login to your Facebook account. Interested, you click the balloon to see one of the freshmen in your introductory course requested you as a friend. Students are continually reaching out on Facebook and you feel conflicted. You find yourself thinking, "Do I want students to have access to my personal life including pictures of my family and friends? Do I want to hear about their lives? How would this impact our instructional relationship?" Faculty members on Facebook have added a new communication option for students. Is it appropriate?

Social network sites (SNS) are Internet spaces that often allow members to personalize their pages and post activities, pictures, videos, personal information, or thoughts visible to others in the network. Facebook is a rapidly growing SNS. Students flock to social networking sites. According to the Pew Research Center, 75% of Millennials have a profile on a social networking site (Pew, 2010). The Millennial Generation includes people born after 1980 and the role of technology is apparent in shaping the lives of that generation. They are described as confident and connected using technology as a forum for self-expression and communication. As Millennials enter college, their traits shape the learning environment.

This chapter explores the communicative relationship between faculty members and students during a time of increasingly diverse methods of mediated communication. Results from a closed- and open-ended survey about Facebook are reported. On the basis of the survey and the author's own Facebook use, guidelines for using Facebook in the classroom are provided. Social media is embraced by higher education, and Facebook, in particular, is used as a marketing tactic as well as a means of institutional and instructional communication for current and future students. This chapter explores the impact on pedagogy and instruction when faculty members are Facebook "friends" with their students. Owing to its popularity and continued use, reviewing the perceptions of students on faculty use and the perceptions of faculty on students is important to the understanding of teaching and learning in the online environment. Understanding identity, disclosure, communication patterns, and educational impact shapes best practices and pedagogy. The chapter concludes with guidelines for faculty and student communicating through Facebook.

INSTRUCTIONAL COMMUNICATION

Instructional communication is the process by which educators and students stimulate meanings in the minds of each other using verbal and non-verbal messages (McCroskey, 1968; Mottet, Richmond, & McCroskey, 2006). Examining the communicative relationship between students and faculty is important as it directs the development of our learning environment. Teaching involves more than just subject familiarity; it involves effectively communicating that subject to others (Lane, 2009; Hurt, Scott, & McCroskey, 1978). The communication we engage in with our students within education shapes our instructional practices. Today, instructional communication is responding to the changing face of higher education. As higher education is more technologically enmeshed, so is our personal communication with students. We increasingly rely upon computer-mediated communication. New technology and new uses for existing technology changes the way communication takes place between educators and students. The study of instructional communication describes and informs the educational environment and the relationship between students and educators.

Students and faculty today embrace multiple avenues of communication. We no longer rely solely on face-to-face communication and office hours. Many professors offer blended course designs linked to Blackboard or similar online learning platforms. Our syllabi are a reflection of this change, offering not only traditional office hours and locations, but including email addresses, blogs, our faculty web pages, and instant message options. Our manner of communicating and approaching communication with our students can impact if and how they seek assistance during our course.

As a graduate student in 2006, I began using Facebook when on a fellowship over 2200 miles from home. I marveled at the connections and the way "friends" were suggested and networks were slowly built. I posted pictures of my new campus, announced what I was learning in status updates, and did not think too much about it. After all, I began keeping a personal blog in 2005, stored my pictures online, used instant messenger, and felt comfortable with technology and my ability to control privacy settings. The first time I really thought about a negative element to Facebook was when a former student requested me as a friend. As I transitioned to faculty life, I often faced the decision: to "friend" or not to "friend" current and former students. What might this friendship look like? Why did they request a friendship? Could I still post my family pictures? What if my brother says something on my Wall that embarrasses me? In the end, I developed a personal policy about Facebook use. I accept any friend request from a student, but I avoid initiating student friendships. I also avoid clicking on student profiles, although I can still see status updates and profile pictures in the "news feed." I reply if they post on my wall or send me a message. I respond to them without seeking them out. This policy worked well for me and soon I realized the potential for Facebook to serve as a powerful tool to communicate with my class and explore communication constructs. I was hesitant, however, as I saw status updates declaring, "2 drunk 2 function!" and "Sitting in a boring lecture – will this teacher ever shut up?!" I began to wonder what it meant, communicatively and educationally, for us to be "friends" on Facebook. Would my students think differently of me once they viewed my profile and would I find myself thinking differently of students after seeing their status updates?

Over the past three years as a faculty member, I saw numerous benefits. Facebook provides new inroads into the student–faculty relationship. I used Facebook as a way to reach my predominantly commuter-based students. When scattered for a hurricane evacuation the first week of our 2008 semester, it was amazing to see students connecting to check on our university and when we would return. The information spread faster than our administration could update our webpage. I explored using Facebook as a way to reach students on course issues. I will occasionally post reminders about class assignments or notes on public speaking events students can attend for extra credit. Although my activity has been steady, so too has my curiosity about ethics, identity, privacy and disclosure. Today many faculty members use Facebook to communicate: We create group pages as advisors of student clubs; we develop study groups or even class groups through Facebook. We use Facebook to foster positive relationships with students (which can help student learning outcomes) and stay connected to those who once sat in our classrooms. We might even feel that using Facebook increases our credibility with our students as we show them we understand their computer-mediated culture.

FROM BRICKS AND MORTAR TO BLOGS AND WIKIS

Many educators are encouraged to explore new technology. We update information on university-hosted web sites, provide online points of access through learning platforms, and answer emails. But is this really where our students are? Students today are more connected through technology than ever before (Pew, 2010), but they are not passive in their technology (Tapscott, 2009). Indeed, they co-create their technology through blogs, microblogs, social networking sites, and wikis. Students want instant responses bemoaning email as "too slow" (Tapscott, 2009), expecting rapid replies and immediate attention. Some professors add Blackboard to their on-ground courses, link their instant message accounts and Twitter widgets to their course pages, and even offer virtual office hours, meeting students in real-time from distant locations (Edwards & Helvie-Mason, 2010). Email is less utilized. Students do not often utilize their institutional email accounts, prompting Boston College to stop issuing addresses to students in 2008. New students come to campus already linked in to a pre-existing email and are on Facebook in record numbers. In fact, Millennial students in general do not use email as their primary means of communication but favor texting, then email, and then social networks. Smart phones provide new levels of connectivity. Now that our mobile devices have leapt from simple phones to complex computers, Facebook can be accessed from anywhere and communication and education are not limited to office hours or bounded by geographic location. However, this can also lessen the professional feel and response in communication.

Considering these trends in higher education, it is important to ask, should we use social networking as a means to communicate with our students? Instructional communication, the creation and fostering of learning environments and personal issues of privacy collide with this one question. Therefore, we should examine social networks and their use in higher education. We should be fully informed as we consider how we connect with our students virtually.

FACEBOOK

Facebook began in 2004 and targeted an academic audience (students, faculty, and staff) with the requirement that members have a .edu email. Its use has ballooned in recent years to include all sectors of society. With more than 400 million active users (Facebook, April 2010), Facebook is a communication phenomenon. Most people join Facebook to keep in touch with friends (Valenzuela, Park, & Kee, 2009) and are linked to others who share similar interests.

People, groups, companies, religious organizations, universities, and political parties have adopted its use as a means of marketing and communicating. Facebook is a social networking site that allows users to create a profile, upload pictures or videos, and update their status. A status update consists of a brief comment that is created by a user and then shown on a "news feed" appearing on friends' pages as well as the user's profile page. Users can "tag" individuals to link them to pictures, videos, comments, and posts. One can also be "tagged" by others, losing the individual ability to determine what is linked to your profile, though it is easy to remove a tag. Facebook members are also quite active. More than 35 million update their status each day and the average user has 130 friends, sends 8 friend requests a month, and spends more than 55 minutes per day on Facebook (Facebook, April 2010). Our students are using Facebook and so are many of our professors, despite age ranges. Not all students are part of the "NetGeneration" (Tapscott, 2009); however, many of our students have "grown up digital" according to Tapscott. He describes the NetGeneration (or Millennials) as a generation that finds communicative speed normal and innovation appealing (p. 7). Facebook offers instant gratification. One can gather large quantities of the latest information while communicating via wall posts, comments, messages, and instant messaging. It provides an electronic network of contacts.

With Facebook's large and constantly growing presence in the world of higher education (Bugeja, 2006; Ellison, Steinfield, & Lampe, 2007), it is important to understand if and how these worlds interact and how the communication relationship between students and faculty members connecting through Facebook may shape the learning environment. However, the use of Facebook for higher education through group pages, class pages, student and faculty "friends," and club associations may have implications for faculty–student relationships and learning. Many may add a "friend" who is a student without considering the full impact such a move could have. Indeed there are benefits and drawbacks to establishing such a relationship with students.

FACEBOOK IN THE UNIVERSITY

Our students are increasingly comfortable and reliant on technology; we see educators meeting the students where they are: within the social networking world. As Ajjan and Hartshorne (2008) describe, the decision to adopt new technology into the classroom is dependent on many factors such as ease of use, comfort, and colleague/student influence. They note, "The use of Web 2.0 technologies has significant potential to support and enhance in-class teaching and learning in higher education" (p. 79). SNS use in higher education has been studied to explore the impact of social interactions online on education (Boyd, 2006; Madge, Meek, Wellens, & Hooley, 2009; Selwyn, 2009). These interactions are often begun offline. Lampe, Ellison, and Steinfield (2006) found that students use Facebook to learn more about people they meet offline, such as classmates, and not to foster new relationships. Educational Facebook use by students, according to Selwyn (2009) happens in five areas: recounting and reflecting on the university experience, exchanging practical information, exchanging academic information, displaying supplication or disengagement, and lastly for banter or humorous exchanges. This information led Selwyn to note, the best role for the educator may be one that is "backstage" (p. 173) to avoid coercive or surveillance-based use of the SNS. The presence of faculty or mandated Facebook use may feel coercive, unappealing, or an invasion of student privacy. However, Facebook was further explored by Ophus and Abbitt (2009), who found that students responded favorably to using Facebook in university courses despite their concern about privacy. The students also worried about distractions on Facebook when completing school work. Students were much more comfortable than faculty members with Facebook as support for classroom concepts (Roblyer, McDanial, Webb, Herman, & Witty, 2010). Educators struggle with this decision to utilize SNSs for educational purposes.

The literature shows a dearth of research showing privacy remains a concern for students (Boyd & Ellison, 2007; Gross & Acquisti, 2005; Hodge, 2006; Mendez et al., 2009), although the concept of privacy differs for many Millennial students, many of whom struggle to see a difference between personal and public spheres (Tapscott, 2009). Faculty members' concerns for personal privacy is not as richly studied, although Mazer, Murphy and Simonds' (2007) research concluded students' perceptions of faculty members can be influenced by the amount of information the faculty member has disclosed online – leading some students to determine whether or not to enroll in the professors' courses simply based on profiles.

Social networks are increasingly present in the educational environment and offer benefits for the faculty–student relationship. Instructional communication relationships appear strengthened as students are more disposed to communicate with faculty whom they know through Facebook (Sturgeon & Walker, 2009), and faculty Facebook use can have a positive effect on the face-to-face relationship with students (Haspels, 2008).

Instructor Immediacy

Facebook interactions can help foster a new sense of connection between students and faculty members. Students viewing the personal life (or what elements of a personal life the faculty decides to share on his or her profile) may find a greater sense of immediacy with the instructor. Described as communicative behavior which fosters psychological and physical closeness (Mehrabian, 1971), immediacy is a common topic in instructional communication literature (Christophel, 1990; Frymier, 1993; Frymier & Houser, 2000; Gorham, 1988; Rocca & McCroskey, 1999). Educators foster immediacy by sharing experiences, examples, non-verbal appearance, knowing and using student names, student feedback, and even through small talk in the traditional face-to-face classroom environment. Immediacy is related to many positive student outcomes as Rocca (2007) noted, including affect toward the professor (Gorham, 1988), learning (Chesebro & McCroskey, 2001; Titsworth, 2001), higher student evaluations (Moore, Masterson, Christophel, & Shea, 1996), attendance and participation (Rocca, 2004), and out-of-class communication (Jaasma & Koper, 1999). Christophel (1990) noted that teacher immediacy is positively associated with student motivation and increased learning. Recently, Mazer, Murphy, and Simonds (2007) noted educators can "increase mediated immediacy by including forms of self-disclosure on personal web pages" (p. 2). To this end, the faculty and student communication through Facebook can serve as a way to increase immediacy and promote positive immediacy outcomes.

Therefore, faculty revealing elements of their personal lives fosters immediacy and can make them more approachable or more "human" in the eyes of the student. As students noted in this survey,

> I think it makes me feel more comfortable with the professor because it takes away a communication barrier for me. (It makes me feel as though we are all human and approachable.)

Determining what personal elements to share and what to withhold is one of the fascinations of social-networking. Educator self-disclosure can ripen student-faculty relationships by encouraging communication (Fusani, 1994). When faculty members offer examples, share narratives the students find the personalized teaching can motivate increased communication both inside and outside the classroom.

The issue of faculty self-disclosure has been explored in computer-mediated communication (O'Sullivan, Hunt, & Lippert, 2004) who noted the importance of communication cues within mediated channels to influence the psychological closeness of students and educators. As Mazer et al. (2007) claimed, "The findings suggest that high teacher self-disclosure as operationalized in the present study may lead students to higher levels of anticipated motivation and affective learning and lead to a more comfortable classroom climate" (p. 12). They continued, "Those students who access their teacher's Facebook page may feel more comfortable communicating in the classroom and will approach the teacher with course questions and concerns, which might have a positive influence on important learning outcomes" (p. 13).

The intriguing thing about social networks and their use in education is that you can formally craft your identity and manage the faculty image. As Toma (2010) noted, "Social networking sites enable users to connect with important people in their lives by creating virtual representations of the self" (p. 1749). Considering which "self" we want to put out to our varied audience may be part of the controversy over using Facebook in higher education. We are able to meaningfully and purposefully construct the identity we want to show the world through the strategic selection of photos, posts, personal information, and group pages or affiliations. But when faced with multiple audiences this constructed identity may be problematic. Some faculty members have a separate "student-friendly" profile on Facebook they use for students to see while the "real" profile connects them with friends and family members. Others adopt a policy not to "friend" students. Those who friend students (and students who friend faculty members) need to know the impact of this relationship and how it carries into the learning environment. Research in user identities and social networks shows that our identities are carefully constructed to portray how we want others to view us (Peluchette & Karl, 2010).

Like any relationship, the computer-mediated relationship faces the issues of identity, disclosure, and purpose. Peluchette and Karl (2010) noted that students and faculty members are able to intentionally manage their desired identity through strategically selecting pictures, posts, and which personal information is visible to others. Students' impression of teachers were shaped by the use of font, language, punctuation (O'Sullivan et al., 2004) as well as more immediate communication such as students' first names and even the use of emoticons (emotional icons) such as smiley faces in emails and other computer-mediated communication (Waldeck, Kearney, & Plax, 2001).

"The use of computer-mediated communication (CMC) in the instructional context could ultimately have a positive effect on the student-teacher relationship, which can lead to more positive student outcomes." (Mazer et al., 2007, p. 2). Facebook allows faculty to engage with students in new ways. As I recently updated a status post proclaiming, "Preparing to speak at the conference and excited to share my work," students saw I not only teach public speaking but I take part in speaking events outside of my institution. They responded, "Good luck, Dr. Lora" and "Just do what you tell us every day." When I returned to class, students asked how the talk went and we discussed techniques for handling unexpected changes in speaking venues. Facebook allowed me to share my experiences, and include my students in real-time activities outside the classroom that shaped our discussion of course content.

Additionally, Mazer, Murphy, and Simonds (2009) noted that students viewing a professor high in self-disclosure led to higher levels of teacher credibility than those with low self-disclosure profiles. The crux of SNSs can be this issue of disclosure.

Disclosure

Sharing personal stories to facilitate learning is not new in higher education; Facebook simply offers a new medium for the transmission of information. Nosko, Wood, and Molema (2010) examined disclosure of personal information on Facebook and found "as age increased, the amount of personal information in profiles decreased" (p. 406). When considering Facebook and higher education, disclosure concerns go beyond how much privacy we have, though, to what the impact of the disclosure could be on our learning outcomes. Mazer et al. (2007) examined student–teacher relationships through Facebook to determine whether teacher self-disclosure impacted the students' learning to find that teacher self-disclosure has a positive influence on the learning environment. By joining a social network, we determine how much we self-disclosed to our audience, but the issue can become confused with multiple audiences and multiple identities we want to put forward. Educators disclose in personal examples and stories within the classroom. We share our experiences with our students to help them understand the course content. Disclosure, as Cayanus and Martin (2002) note, can be studied in terms of valence (positive and negative nature of the disclosure) and relevance (the way the disclosure relates to the content). Peluchette and Karl (2010) wrote,

> Many students make a conscious attempt to portray a particular image, and, as predicted, their intended image was related to whether they posted inappropriate information. Those who believed they portrayed a hardworking image were unlikely to post inappropriate information, whereas students who felt they portrayed an image that was sexually appealing, wild, or offensive were most likely to post such information. (p. 30)

While faculty may debate about their work and personal worlds interacting or about issues of privacy, today's Millennial generation (1977–1997) and Generation Next (1998 to present) see less concern about personal disclosure. Tapscott (2009) described our upcoming generation as eager to share and connect, but lamented they give away personal information too freely. He stated,

> Of all my concerns, one big one stands out. NetGeners are making a serious mistake and most of them don't realize it. They're giving away their personal information on social networks and elsewhere and in doing so are undermining their future privacy. They tell me they don't care; it's all about sharing. But here I must speak with the voice of experience. Some day that party picture is going to bite them when they seek a senior corporate job or public office. I think they should wake up, now, and become aware of the extent to which they're sharing parts of themselves that one day they may wish they had kept private. (p. 7)

West, Lewis, and Currie (2009) noted that Millennial students found it difficult to see two spheres of public and private life, "Social networking sites such as Facebook are associated with new ways of construing some of the notions surrounding the traditional public/private dichotomy" (p. 615). They further state that "Computer-mediated communication appears to make this fuzziness more apparent than has hitherto been the case" (p. 615). However, Subrahmanyam and Greenfield (2008) remind us that privacy settings continue to offer control over who views profiles and how much of a profile is visible to various audiences.

Despite such controls, merging our social sphere with our academic sphere continues to take place as students request faculty members as friends; it can carry the benefit of reaching students in a new, more accessible way.

This semester, I taught a very shy student in my Fundamentals of Public Speaking class. She sat by the wall, never made eye contact and avoided our scheduled speaking days. After missing several speeches, she was failing the course at midterm. I could not reach her during class – she seemed distant, uninterested, and uncaring. However, she "friended" me on Facebook after midterm grades were posted and immediately sent me a message asking, "help, what can i [sic] do 2 [sic] boost my grade?" I was able to quickly respond through my mobile phone, and although she never emailed, never phoned, never stopped by my office or stayed after class, she asked many questions, sent rough drafts, and sought reassurance about her progress through Facebook, telling me how shy she was. Despite the many concerns my colleagues express about Facebook and students, this student reminds me there are ways we can communicate that reaches some students and not others. She never commented on my posts, pictures, or profile but by the end of the semester I knew how to help her succeed and could count on several Facebook messages before each speech. I knew she would find the answers she needed.

Expectations and Violations

Demonstrating an understanding of the SNS and being in that culture does not mean that computer-mediated relationships provide the best opportunity for student engagement or for successful instructional communication. Faculty presence on Facebook may create tension with students. "When communicating with others, people have preexisting expectations regarding the others' behavior" (Levine et al., 2000, p. 124). When our behavior does not match expectations, an expectancy violation takes place. As Burgoon, LePoire, and Rosenthal (1995) noted our expectations come from social norms, stereotypes, and our known behaviors based on previous interactions with the individual or those like the individual. Violating an expectation, however, is not inherently negative. If we observe behavior that is unexpected, but which we find appealing, it is a positive violation. If we observe unexpected behavior which is unappealing to us is a negative violation (Levine et al., 2000).

Our students expect certain behavior from faculty members and as educators we should know that when expectations are not met the violator falls under increased scrutiny (Levine et al., 2000). The sheer presence of us as faculty members in the Facebook realm, within *their* social world, may be a violation of expectations. As Rotman (2010) found, students handled professional and academic matters through email or face-to-face interaction and left social networks for their friend-groups, noting, "Professional communication was limited to conservative, structured channels such as email or face-to-face meetings, not just because of the breadth of expression they allow, but mostly because they are considered 'ok', 'appropriate', 'suitable' for maintaining respectful distanced relationships with faculty" (p. 4336). Entering the realm of Facebook and "friending" students may be seen as a violation of the expected professional relationship.

This violation of student expectation has led to the term "creepy tree house." This term describes the unsavory, coercive or inauthentic use of SNSs by adults or outsiders. However, Melanie McBride (2008) noted that the use of SNS is not inherently coercive or inauthentic in higher education. She writes that the labeling of "creepy tree house" is based largely on bad pedagogy when implementing Web 2.0 strategies, not the inherent presence of an instructor.

Faculty "friending" could lead to increased inspection and critique of the faculty member, a common result of socially violated expectations. Although some may not mind this closer inspection, it is worth noting before embarking upon a SNS relationship with our students. We might break the social expectations the student holds (such as the stereotypical "professor" persona) through our Facebook profile. They might see pictures of the professor on a motorcycle, at concerts, jumping out of a plane, or with a child on her hip (all images one could see on my profile) and such violations of expectations (both positive and negative) will impact how we are viewed. The way we verbally and non-verbally communicate impacts how we are perceived. It can alter our perceived credibility and authority within the classroom and even throughout our career.

Additionally, as faculty expectancy violations can alter the way we view our students. Our expectations may also be violated once we interact with the student through Facebook. When we see a status update, picture, or post which violates our expectation, we may find ourselves scrutinizing the students more than before our interaction through the computer-mediated communication.

An overview of the literature showed the importance of instructional communication, the changing realm of higher education, the Facebook presence in the university, and the positive and negative impacts on instructor immediacy, disclosure, and expectations. As faculty members considering using Facebook, it is imperative we understand the student perspective on the presence of faculty members. To this end, the following research question was devised:

How do students perceive faculty members on Facebook?

METHOD

To explore the research question, an open- and closed-survey was developed. The survey was created using an online survey-hosting site where the survey could be sent to the participants as a link. The survey consisted of a brief demographics section, a section exploring Facebook use (sample questions include: why did you join Facebook, how many "friends" do you have, how much time do you spend on Facebook a day, have you "friended" a faculty member, who initiated the friend request?), and reflective questions regarding faculty members and Facebook (sample questions include: why did you "friend" a faculty member, has your view of the faculty member changed, do you limit or change the pictures, video, or content of your account to consider faculty members, why/why not?). The survey was announced in four introductory communication courses as an optional avenue for extra credit. The survey was posted on the course Blackboard web sites and emailed to the students' campus email accounts. Data was collected over a two-week period. Analysis involved computing statistical responses to closed-ended questions and the thematic analysis of open-ended responses. Seventy-four participant Facebook users from the introductory public speaking course at a Historically Black University expressed their thoughts on "friending" faculty members, yielding a 77% response rate. Participants included 22 males and 51 females with one respondent leaving the question of "sex" blank. The majority, 93%, were African-American, 4.1% Caucasian, 1.4% Latino/a, and 1.4% Haitian. Ages ranged from 18–51 with a median age of 26. Of the participants, 30.4%, or 21 participants, had "friended" a faculty member in the past. Of those, 60.9% requested the friendship. The results inform faculty and student relationships on Facebook. Of the open-ended questions, two general categories emerged (students who friend faculty members and students who do not) resulting in several themes within each category.

Student Perceptions of Faculty on Facebook

The survey resulted in two categories: students who do not friend faculty members and students who friend faculty members. Within each category themes emerged describing student perceptions (Fig. 1).

The majority of students surveyed had not nor do not friend faculty members (69.6%), although not everyone felt strongly about it. One student who did not friend faculty members noted, "I think that friending a Faculty Member is a good idea. It could be useful for both the student and the

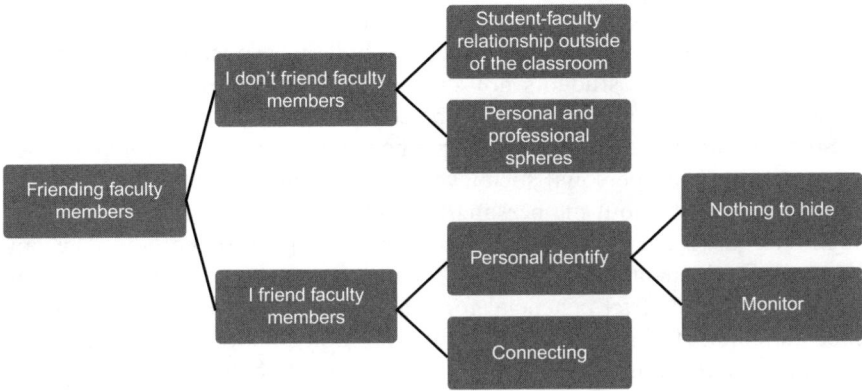

Fig. 1. Student Perceptions of Faculty Members as Friends

Faculty Member." But many responses showed a desire to keep their academic and social lives separate.

There were two themes emergent from those who are not friending faculty members. The first theme is student–faculty relationships and the second is the idea of personal and professional spheres.

First, students noted that Facebook was a place where they did not expect to encounter faculty members and that "friending" would not be appropriate in the student–faculty member relationship. Students noted,

> I don't befriend faculty members because it would be a conflict of interest.

> I think there should be a line drawn between faculty members and students being friends, especially when the student is in that faculty members [*sic*] class. Many students are or at least should be mature enough to handle a friendship with a faculty member on & off campus. When in school it's [*sic*] business outside of school they can do whatever they choose to.

> I think its [*sic*] inappropriate to know a faculty member on a personal level.

This result is not uncommon in the literature, as Hewitt and Forte (2006) found over a third of their participants felt faculty should not be present on Facebook. The expectation that faculty are not on Facebook or that friending violates the professional relationship for some students may relate back to the student sense of privacy and "creepy tree house" phenomenon, which was seen more clearly in the desire for academic and personal spaces to be separate. However, in a study done by Sturgeon and Walker (2009) two thirds of respondents had a faculty/student friendship on Facebook. With nearly 70% of students not sharing the Facebook friendship in this study, clearly more research should be conducted.

Students noted a desire to keep their academic/professional worlds separate from their personal/social worlds. This was especially clear with the comment, "FB is about me personally so i [sic] don't have faculty members as my friends. It's my own little space where i [sic] get to vent about certain things." Students added, "I don't want my teachers in my [p]ersonal business unless its [sic] beneficial for school. I use Facebook as a personal tool and lots of things I don't want everybody to know or see." And, "Certain things you may not want to share with Faculty members. I think people should stay in the age appropriate bracket when it comes to fb."

One student responded that he has not friended faculty members before, but that it could be beneficial if the realms of personal and professional were respected, "I don't think it would be a problem "friending" a faculty member as long as you keep it strictly professional."

Students who did not or have not "friended" faculty members desired the ability to define and control where and how instructional communication transpired. They sought to keep an "appropriate" connection with the faculty member where their personal activities were separate from the classroom.

A third (30.1%) of the respondents "friended" faculty members, resulting in themes of personal identity and desire to connect with the faculty member.

First, students described issues of their identity in two ways: "nothing to hide" and "monitored identity." Many students felt that their Facebook profiles represented who they were and that if you did not like it, you "could delete me" or simply "don't look." They claimed, "I am who I am!" and responded, "I feel this is my facebook and have the right to do anything to my page." While others noted, "I believe in staying true to who I am, and not changing for anyone." And, "everything I post on my facebook page is a part of me showing people "My Friends" who I am and I will not change that for anyone."

These expressions describe the idea of a personal identity that does not change for a faculty member or other audience within the social network. Because they did not change for other audiences or consider the views of others, they did not see a negative to "friending" faculty members, noting "It's just Facebook, nothing personal."

Other students responded their content on Facebook was self-monitored, claiming the content was fine for faculty members' eyes, "Everything is internet friendly" and "I only put what I want everyone to see." Others concurred noting, they self-censor Facebook materials, "I limit what I post on facebook, because I will have to see that faculty member the next day or later that day." Although not everyone is diligent or consistent, as one student claimed, "sometimes [I change my profile] because i [*sic*] don't want the faculty getting the wrong idea about me." These students saw Facebook as one way of developing an identity and freely modified content to consider social and professional ramifications to their image.

Second, the respondents noted faculty as "friends" on Facebook fostered connections. It allowed the student to work with the faculty member through student clubs, through mentoring, or as a way to personally connect to their instructor. As one student noted, "I think it makes me feel more comfortable with the professor because it takes away a communication barrier for me. (It makes me feel as though we are all human and approachable.)" Other respondents claimed they could use Facebook for "extra help for college assignments," "mentoring," and "to keep in touch" both during class and to "keep in contact" after class is over. One student remarked, "I suppose I view it as the cyber version of the 'open office door' policy. It is like taking a minute to ask a prof. a question in the hall, except at their convenience in terms of responding."

While another student felt Facebook helped to foster their relationship, "i [*sic*] would request them all because i [*sic*] have great relationships with them and would love to keep in contact." These students embraced Facebook as a way to communicate and connect with the faculty members.

Viewing a faculty member's profile, which 26.5% of respondents admitted to doing, made that instructor seem, "more relatable and comfortable" and humanized the faculty member. "i [*sic*] see that they are not just professors but they have stuff going on in their lives just like students do." This personalization of the professor can benefit the students as it creates that important sense of immediacy.

In general, the students were divided on their views of faculty members on Facebook. The majority did not friend faculty members, but not all were against the option. Most expressed a desire to keep the social and academic spheres separate. Of those who "friended" faculty members, they had favorable responses to the relationship and its carry-over to the educational environment. On the basis of these responses, Facebook use between faculty and students should be considered carefully by both parties. I believe the students should decide if they want to embark on such a connection using Facebook. Owing to the concerns noted above, mandatory connections through Facebook should never transpire. But there are benefits many of us, especially at commuter institutions, can embrace. Once we realize these caveats, Facebook use can serve as a way to interact with our students, to encourage communication, to stay involved in campus events, and to work as an additional point of contact. Valenzuela, Park, & Kee (2009) noted Facebook fills a communicative need, "Facebook can fulfill the informational needs of users, a key ingredient for strengthening weak ties and promoting collective action" (Kenski & Stroud, 2006; Shah, Kwak, & Holbert, 2001). It is with this view that I encourage faculty to consider their position on Facebook in the classroom.

"FRIENDING" FACEBOOK

Facebook can be a convenient forum for student dialogue. When considering if or how to join Facebook, faculty and students alike will do well to remember that Facebook content can and has been used as evidence in legal cases (Woo, 2005). Additionally, institutions may have their own policies for faculty or student use of Facebook. Required "friending" and class use of Facebook remains controversial (Parry, 2010) so I avoid any mandatory requirement in my courses. The optional nature of Facebook communication helps those who prefer to keep social and academic roles separate and respects the decision of the student.

Just like transitioning to online instruction, using Facebook as an educational tool requires pedagogical intent. As Lehmann and Chamberlin (2009) noted, "Educators use of Web 2.0 social technologies, like blogging, webcasting, photo sharing, streaming video, and tagging, has created a change in the industry requiring software developers to incorporate these social networking tools" (p. 33). A Facebook "badge" serves as an example. A badge is a set of html code linking any site (such as Blackboard) to our Facebook page. It shows up as our current profile, updated status, and when clicked links to our profile. We should not simply add the badge to our course page without first reflecting on its use, how we want to interact through Facebook, and consider if we want to create student guidelines. We should also avoid limiting Facebook to a purely communicative tool and embrace its collaborative abilities. Our students want to collaborate. Tapscott (2009, p. 11) noted, "In education, they [Millennials] are forcing a change in the model of pedagogy, from a teacher-focused approach based on instruction to a student focused model based on collaboration." Facebook can serve as a platform to link our students to new trends, data, and also learning opportunities beyond the classroom. Skerrett (2010) described the use of Facebook in literacy education where students and teachers collaborated in book clubs on Facebook. In class, students created Facebook profiles of characters they read about to show detailed understanding of the characters' personalities. In communication courses we can examine different types of communication (interpersonal, group, gender, intercultural, organizational) or even bridge to business and marketing course to examine messages to consumers and public identities of corporations. Hard sciences can use the course to follow real-time events such as the oil spill in the Gulf and foster dialogue for classroom consideration. Political science courses can examine the content and trends to understand student engagement in political issues. Using Facebook for communication or as a pedagogically crafted tool for course content, faculty members should carefully consider drawbacks and benefits to the overall student–educator relationship.

Faculty members considering Facebook as a means to communicate with past and current students should craft a personal policy on the instructor–student relationship, considering the following guidelines.

INTEGRATING FACEBOOK INTO INSTRUCTIONAL COMMUNICATION

To ease the transition to social media as a part of the educational journey and the instructor–student relationship, five areas of guidelines are offered: before you start, know your audience, response options, collaborate, and alternatives to "friending."

Before You Start

As much of the immediacy and disclosure literature reflects, it is important to be genuine with your students. When you first set up your Facebook page or when you first determine you want develop a strategy for "friending" try to communicate your expectations about computer-mediated interactions with your current students. A colleague noted she shares with her students that she is on Facebook. She tells her class if Facebook is their preferred mode of communication, they should follow university guidelines for emailing (which include providing a name and course section). I am not that formal but agree heartily with sharing expectations for students. Let them know what you expect so both parties are successful. I do not over-emphasize Facebook although I have included it (along with my Twitter and Instant Messenger handles) as a "contact me" option on the course syllabus. In the past, announcements about Facebook serve nicely for those who want to give students the option of Facebook without pushing it as a requirement for the class. Consider including the students' input – ask them how they would like to proceed and bring that dialogue into the classroom. Consider adopting your own personal policy on "friending" and students. It might differ from mine greatly, but whatever it is should work with your comfort level and privacy concerns. I hesitate to craft a "professor" page that differs from my personal page, although that is an option. It does not feel genuine to me. There are a few options at the end of this section that might work for those who want the benefits of Facebook connections without fully "friending" their students. No one policy on student friending works for everyone, just be consistent in your treatment of students. I avoid being inconsistent by friending some students and not others, for example. If they ask, I confirm. I give equal treatment to both past and current students. When declining a friendship request, you should do so by responding, "Thank you for the friend request, however I do not friend (current) students," or otherwise explain the decision to your student. Again, consistency is key so some students do not feel favoritism is taking place. Denying a student request for friendship is a gap in the literature which should be addressed in future research, although some K-12 institutions require faculty and staff deny friend request. Stern and Taylor (2007) followed how often student–student denials take place, but higher education has not weighed in on the ramifications of a denial on the instructor–student relationship.

One final note before you begin "friending" students: Beware of the time. Adding one more method of communication may seem like a small choice, but it does add to your overall work time. You might find yourself responding to students on your mobile Facebook application at random points in the day or night. All those minutes add up and we can easily find that our time is taken up in this communicative endeavor. As current students are my Facebook friends, I respond to their questions on Facebook just as I would an email or any contact initiated by a student. I do not ignore a student inquiry regardless of medium choice (message, instant message, or wall post). Guide students on what you prefer through the syllabus or class discussion. For this reason, time spent on Facebook for work purposes may be kept manageable by only addressing student concerns when you would answer student emails or within set office hours.

Know Your Audience

If you are friending co-workers, alumni, students, or staff, remember that your audience is multifaceted. It is rewarding to have friends, family, employers, and students converge in your life, but it can also lead to miscommunication. Simple messages like "I'm so frustrated" might be construed as dissatisfaction with your job, disappointment in your students, or even anger by your colleagues. Multiple audiences require careful communication. Be a model for responsible computer-mediated communication for your students. Show them the importance of a positive, professional image and discuss the permanence and implications of our online images/communication with students. Finally, remember that the institutions, companies, and group memberships you are linked to can be viewed by those who "friend" you, so click on your links with care.

Response Options

Communicating through Facebook can be rewarding, but I was initially confused on where to respond to students. Should I post on their Wall? Send a message? Instant message when I see he or she is also on Facebook? I follow the student's lead and respond as I was contacted. If the student sends a private message or instant message that is the way I answer. If they post publicly on my wall or as a comment, I reply in kind. However, this is one area where your own policy could come into play. Tell your students if you will only respond through messages or if you prefer instant messaging. Consider using the "news feed" or your status updates for larger-scale communication: "COMM 210 students: Persuasive speech outlines are due this week" may show up briefly on student pages. Generally, my students have many friends and posting announcements in this manner is not as effective since the news feed is a rolling update based on most recent posts and as their 600 friends' status update my announcement is not likely viewed. Using Facebook groups can help this as a "message" can be sent to all group members easily landing in their in-box.

Collaborate and Share

Facebook offers so many ways for students and faculty to network. Why not serve as a liaison encouraging your alumni to mingle with your current students? If a student interested in an internship in broadcasting contacts me, I immediately contact a graduate who is working at the local radio station asking if the student can chat with her for advice. Or I message students (and even post as a status) asking, "Does anyone know of openings in area radio or television stations-I have an interested and capable student." Friending does allow a unique opportunity to link students to one another (informally or formally), open doors to organizational partnerships and even internships. After a student posted "I'm graduating! Thanks, Doc" on my page, other undergraduates congratulated him and received advice and encouragement from the graduating senior on how to successfully matriculate. Little opportunities to link students might have big impact. As you examine your role in Facebook, you might embark upon collaborate, learner-centered teaching (Weimer, 2002) where students direct their education through choosing projects and even content. This could be done in various ways in different disciplines.

Alternatives to "Friending"

If "friending" seems too personal or if you find yourself with reservations, it does not mean you cannot link with your students and use the many collaborative tools on Facebook. Two options work nicely for those of us who shy away from "friending." First, consider using a "Fan" page. This feature allows you to reach out to many people, update information, and participate on Facebook without "friending." Many organizations take advantage of the Fan page option; however, it is a public page and you cannot restrict who is a fan. The best option might be the Group function, which allows you to have a public or private group. Groups that are private can be created easily and messages and posts are available only to the members. Park, Kee, and Valenzuela (2009) noted the Groups option is "a particularly popular and useful module that allows discussion forums and threads based on common interests and activities" (p. 729). Within group pages, individuals can find exclusive information while interacting with group members without "friending" them. Park et al. (2009) continued, "Students join Facebook Groups because of the need to obtain information about on- and off-campus activities, to socialize with friends, to seek self-status, and to find entertainment" (p. 732). I personally use the group page option to advise my campus organization and have found it a great tool to reach students almost instantly. During a campus event we needed students to host our table after one was sick and another had to work. A quick message through the group's page was sent, and students were able to show up and help within minutes.

As you consider your communicative relationship with your students, examine new technology and popular media for its relevance within your pedagogy and teaching philosophies. Be prepared to explore new directions, but do not forget to examine what you are doing and how it is going. Reflect! Ask yourself if your approach is working. Explore what could be changed by seeking student input. Ask yourself, "Would I say this in class or at a faculty meeting?" If the answer is no, do not post it if you have "friended" students.

CONCLUSION

This research described student perceptions of faculty members on Facebook and offers many in-roads to future studies. The issue of student age and concerns about privacy and the division of academic spheres and personal spheres would inform the research. Additionally, the second phase of this research study is underway examining faculty members' perceptions on "friending" students and use of Facebook. Further research may consider the impact of academic disciplines and student majors on Facebook use as well as the impact of a "denial" of a student request for friend-status on Facebook.

This chapter examined faculty friending and instructional communication. Computer-mediated communication has become a companion of our higher education environment. As our students' needs change, so to must our approaches. Facebook offers a cyber-culture where we can communicate with our students and where our instructional communication can be enhanced or undermined. Understanding how our instructional communication is shaped by expectancy, immediacy, and disclosure allows us to be better informed as faculty, as "friends," and as educators.

REFERENCES

Ajjan, H., & Hartshorne, R. (2008). Investigating faculty decisions to adopt Web 2.0 technologies: Theory and empirical tests. *Internet and Higher Education, 11*(2), 71–80.

Boyd, D. M. (2006). Friends, friendsters, and top 8: Writing community into being on social network sites. *First Monday*, 11(12). Available at http://www.firstmonday.org/issues/issue11_12/boyd/

Boyd, D. M., & Ellison, N. B. (2007). Social network sites: Definition, history, and scholarship. *Journal of Computer-Mediated Communication, 13*(1), 210–230.

Bugeja, M. J. (2006). Facing the Facebook. *Chronicle of Higher Education, 52*(21), C-1.

Burgoon, J. K., LePoire, B. A., & Rosenthal, R. (1995). Effects of preinteraction expectancies and target communication on perceiver reciprocity and compensation in dyadic interaction. *Journal of Experimental Social Psychology, 31*, 287–321.

Cayanus, J. L., & Martin, M. M. (2002, November). Development of a teacher self-disclosure scale. Paper presented at the meeting of the National Communication Association, Atlanta, GA.

Chesebro, J. L., & McCroskey, J. C. (2001). The relationship of teacher clarity and immediacy with student state receiver apprehension, affect, and cognitive learning. *Communication Education, 50*, 59–68.

Christophel, D. M. (1990). The relationships among teacher immediacy behaviors, student motivation, and learning. *Communication Education, 39*, 323–340.

Edwards, J. T., & Helvie-Mason, L. (2010). Technology and instructional communication: Student usage and perceptions of virtual office hours. *Journal of Online Learning and Teaching, 6*(1), 174–186.

Ellison, N. B., Steinfield, C., & Lampe, C. (2007). The benefits of Facebook "friends:" Social capital and college students' use of online social network sites. *Journal of Computer-Mediated Communication, 12*(4). Available at http://jcmc.indiana.edu/vol12/issue4/ellison.html. Retrieved on April 3, 2010.

Facebook. (2010). *Statistics.* Available at http://www.facebook.com/press/info.php?statistics. Retrieved on April 4, 2010.

Frymier, A. B. (1993). The impact of teacher immediacy on students' motivation: Is it the same for all students? *Communication Quarterly, 42*, 454–464.

Frymier, A. B., & Houser, M. L. (2000). The teacher-student relationship as an interpersonal relationship. *Communication Education, 49*, 207–219.

Fusani, D. S. (1994). "Extra-class" communication: Frequency, immediacy, self-disclosure, and satisfaction in student-faculty interaction outside the classroom. *Journal of Applied Communication Research, 22*, 232–255.

Gorham, J. (1988). The relationship between verbal teacher immediacy behaviors and student learning. *Communication Education, 37*, 40–53.

Gross, R., & Acquisti, A. (2005). Information revelation and privacy in online social networks (the Facebook case). In: *Proceedings of ACM workshop on privacy in the electronic society* (pp. 71–80). Alexandria, VA: Association for Computing Machinery.

Haspels, M. (2008). *Will you be my Facebook friend? Presented at the 4th annual GRASP symposium.* Wichita, KS: Wichita State University.

Hewitt, A., & Forte, A. (2006). Crossing boundaries: Identity management and student/faculty relationships on the Facebook. Presented at the Computer Supported Cooperative Work Conference, Banff, Alberta, Canada.

Hodge, M. J. (2006). The Fourth Amendment and privacy issues on the new Internet: Facebook.com and Myspace.com. *Southern Illinois University Law Journal, 31*, 95–123.

Hurt, H. T., Scott, M. D., & McCroskey, J. C. (1978). *Communication in the classroom.* Reading, MA: Addison-Wesley.

Jaasma, M. A., & Koper, R. J. (1999). The relationship between student-faculty out-of-class communication to instructor immediacy and trust, and to student motivation. *Communication Education, 48*, 41–47.

Kenski, K., & Stroud, N. J. (2006). Connections between Internet use and political efficacy, knowledge, and participation. *Journal of Broadcasting & Electronic Media, 50*(2), 173–192.

Lampe, C., Ellison, N., & Steinfield, C. (2006). A Face(book) in the crowd: Social searching vs. social browsing. In: *Proceedings of ACM Special Interest Group on Computer-Supported Cooperative Work*, ACM Press (pp. 167–170).

Lane, D. R. (2009). Communication with students to enhance learning. In: E. M. Anderman & L. H. Anderman (Eds), *Psychology of classroom learning: An encyclopedia* (pp. 222–227). Detroit, MI: Macmillan Reference.

Lehmann, K., & Chamberlin, L. (2009). *Making the move to e-learning: Putting your course online*. New York: Rowman & Littlefield.

Levine, T. R., Anders, L. N., Banas, J., Baum, K. L., Endo, K., Hu, A. D. S., & Wong, N. C. H. (2000). Norms, expectations, and deception: A norm violation model of veracity judgments. *Communication Monographs*, 67(2), 123–137.

Madge, C., Meek, J., Wellens, J., & Hooley, T. (2009). Facebook, social integration and informal learning at university: It is more for socialising and talking to friends about work than for actually doing work. *Learning, Media and Technology*, 34(2), 141–155.

Mazer, J. P., Murphy, R. E., & Simonds, C. J. (2007). I'll see you on Facebook: The effects of computer-mediated teacher self-disclosure on student motivation, affective learning, and classroom climate. *Communication Education*, 56(1), 1–17.

Mazer, J. P., Murphy, R. E., & Simonds, C. J. (2009). The effects of teacher self-disclosure via Facebook on teacher credibility. *Learning, Media and Technology*, 34(2), 175–183.

McBride, M. (2008, April 26). Classroom 2.0: Avoiding the "creepy treehouse." Available at http://melaniemcbride.net/2008/04/26/creepy-treehouse-v-digital-literacies/ Retrieved on June 24, 2010.

McCroskey, J. C. (1968). *An introduction to rhetorical communication*. Englewood Cliffs, NJ: Prentice-Hall.

Mehrabian, A. (1971). *Silent messages*. Belmont, CA: Wadsworth.

Mendez, J. P., Curry, J., Mwavita, M., Kennedy, K., Weinland, K., & Bainbridge, K. (2009). To friend or not to friend: Academic interaction on Facebook. *International Journal of Instructional Technology & Distance Learning*, 6(9), 33–47.

Moore, A., Masterson, J. T., Christophel, D. M., & Shea, K. A. (1996). College teacher immediacy and student ratings of instruction. *Communication Education*, 45, 29–39.

Mottet, T. P., Richmond, V. P., & McCroskey, J. C. (2006). *Handbook of instructional communication: Rhetorical and relational perspectives*. Boston: Pearson.

Nosko, A., Wood, E., & Molema, S. (2010). All about me: Disclosure in online social networking profiles: The case of Facebook. *Computers in Human Behavior*, 26(3), 406–418.

Ophus, J., & Abbitt, J. T. (2009). Exploring the potential perceptions of social networking systems in university courses. *Journal of Online Learning and Teaching*, 5(4), 639–648.

O'Sullivan, P. B., Hunt, S. K., & Lippert, L. R. (2004). Mediated immediacy: A language of affiliation in a technological age. *Journal of Language and Social Psychology*, 23, 464–490.

Park, N., Kee, K. F., & Valenzuela, S. (2009). Being immersed in social networking environment: Facebook groups, uses and gratifications, and social outcomes. *CyberPsychology & Behavior*, 12(6), 729–733.

Parry, M. (2010). Course requirement: Friend your professor on Facebook. *Chronicle of Higher Education, Wired Campus Blog*. Available at http://chronicle.com/blogPost/Course-Requirement-Friend/8827/. Retrieved on April 1, 2010.

Peluchette, J., & Karl, K. (2010). Examining students' intended image on Facebook: "What were they thinking?!". *Journal of Education for Business*, 85(1), 30–37.

Pew. (2010, February). The millennials: Confident. Connected. Open to change. *Pew Research Center*. Available at http://pewresearch.org/pubs/1501/millennials-new-survey-generational-personality-upbeat-open-new-ideas-technology-bound. Retrieved on June 27, 2010.

Roblyer, M. D., McDanial, M., Webb, M., Herman, J., & Witty, J. V. (2010). Findings on Facebook in higher education: A comparison of college faculty and student uses and perceptions of social networking sites. *Internet and Higher Education, 13*(3), 134–140.

Rocca, K. A. (2004). College student attendance: Impact of instructor immediacy and verbal aggression. *Communication Education, 53*, 185–195.

Rocca, K. A. (2007). Student motivations and attitudes: The role of the affective domain in geoscience learning. Presentation made at Carleton College. Available at http://serc.carleton.edu/NAGTWorkshops/affective/immediacy.html. Retrieved on April 2, 2010.

Rocca, K. A., & McCroskey, J. C. (1999). The interrelationship of student ratings of instructors' immediacy, verbal aggressiveness, homophily, and interpersonal attraction. *Communication Education, 48*, 308–316.

Rotman, D. (2010). Constant connectivity, selective participation: Mobile-social interaction of students and faculty. *Conference on Human Factors in Computer Systems*. Atlanta GA, April 10–15.

Selwyn, N. (2009). Faceworking: Exploring students' education-related use of Facebook. *Learning, Media and Technology, 34*(2), 157–174.

Shah, D. V., Kwak, N., & Holbert, R. L. (2001). "Connecting" and "disconnecting" with civic life: Patterns of Internet use and the production of social capital. *Political Communication, 18*, 141–162.

Skerrett, A. (2010). Lolita, Facebook, and the third space of literacy teacher education. *Educational Studies, 46*, 67–84.

Stern, L. A., & Taylor, K. (2007). Social networking on Facebook. *Journal of Communication, Speech, & Theatre Association of North Dakota, 20*, 9–20. Available at http://www.cstand.org/UserFiles/File/Journal/2007.pdf#page=9. Retrieved on June 26, 2010.

Sturgeon, C. M., & Walker, C. (2009). Faculty on Facebook: Confirm or deny? Presented at the 14th Annual Instructional Technology Conference, Murfreesboro, Tennessee.

Subrahmanyam, K., & Greenfield, P. (2008). Online communication and adolescent relationships. *The Future of Children, 18*(1), 119–146.

Tapscott, D. (2009). *Grown up digital*. New York: McGraw-Hill.

Titsworth, B. S. (2001). The effects of teacher immediacy, use of organizational lecture cues, and students' note-taking on cognitive learning. *Communication Education, 50*, 283–297.

Toma, C. (2010). Affirming the self through online profiles: Beneficial effects of social network sites. *Conference on Human Factors in Computer Systems*. Atlanta GA, April 10–15.

Valenzuela, S., Park, N., & Kee, K. F. (2009). Is there social capital in a social network site?: Facebook use and college students' life satisfaction, trust, and participation. *Journal of Computer-Mediated Communication, 14*(4), 875–901.

Waldeck, J. H., Kearney, P., & Plax, T. G. (2001). Teacher email message strategies and students' willingness to communicate online. *Journal of Applied Communication Research, 29*, 54–70.

Weimer, M. (2002). *Learner-center teaching: Five key changes to practice*. San Francisco: Jossey-Bass.

West, A., Lewis, J., & Currie, P. (2009). Students' Facebook 'friends': Public and private spheres. *Journal of Youth Studies, 12*(6), 615–627.

Woo, S. (2005, November 3). The Facebook: Not just for students. *The Brown Daily Herald (Providence, RI).* Available at http://www.browndailyherald.com/2.12231/the-facebook-not-just-for-students-1.1679665#. Retrieved on April 2, 2010.

PART II
SOCIAL MEDIA PEDAGOGIES
FOR THE FUTURE OF ARTS
AND SCIENCE LEARNING

HOW TWITTER SAVED MY LITERATURE CLASS: A CASE STUDY WITH DISCUSSION

Andy Jones

ABSTRACT

Like many faculty teaching in the social sciences or humanities, I've often been frustrated when students show no evidence of having completed assigned readings for my discussion-centric literature classes. I recently taught a short story class that emphasized my high expectations for student participation, and the means by which students would collaboratively and nightly analyze assigned texts: Twitter. My students soon embraced Twitter as a collaboration tool, and increasingly came to class with improved attitudes toward, and readiness for, class discussions. The nightly peer-review process made possible by Twitter helped students improve their spoken and written arguments, and deepen their understanding of challenging texts. This chapter tells the story of the discoveries I made about teaching student-centered classes, and about using Twitter as a sandbox where students would share their ideas before coming to the well-attended lectures and class discussions. The chapter concludes with ten recommended strategies for teaching with Twitter.

Teaching Arts and Science with the New Social Media
Cutting-edge Technologies in Higher Education, Volume 3, 91–105
ISSN: 2044-9968/doi:10.1108/S2044-9968(2011)0000003008

I.

More than anything else, what distinguishes a great class from an adequate class is the attitude of the participants. As Shakespeare, Mowat, Werstine, and Folger Shakespeare Library (2002) reminds us in *Hamlet*, "There is nothing either good or bad, but thinking makes it so." When an instructor's course objectives are clear, when a professor brings enthusiasm to lectures and to class discussions, and when he or she sets high expectations for all of the class's participants, then learning can take place. But primarily it is the participants' reaction to these necessary elements of a class that can make that class truly successful. As teachers, we treasure students who are curious, enthusiastic, optimistic about the next class meeting, and reflective about course content after the proverbial bell has wrung. We can inspire such students to pursue a lifelong intellectual relationship with the content of our courses, and perhaps also inspire enthusiasm about the importance of learning itself. When we identify and adopt the methods and tools to sustain that motivation and enthusiasm, we make the learning experience richer and more satisfying for our students and for ourselves.

In order for such a course to exist, in order for this transformational learning to take place, the instructor and students together must face down an array of obstacles and distractions. Many of our students are overworked, committed to taking too many units, and tempted by a great number of electronic, networked and video-based distractions. Some students resent having to take our class if it is a requirement, while others may question its relevance if it is an elective. For these and many other reasons, some students bring to class an attitude of borderline complacency, a hope that "just enough" work in the class will be sufficient to allow them to pass, or even do well. The lecture format of the larger classes we teach may contribute to this attitude; in many large enrollment classes students slouch in their seats comfortably, almost languidly, as they might do on the couch at home, and expect to be entertained or at least informed. We oblige them with our preformulated lectures and our slide shows, hoping that they will stay awake, and that some of the students in the front rows may ask or answer questions at the end of the lecture. Many of us leave such a class confident that teaching has taken place, if not always learning. Often we settle for the mode of teaching that was used with us, rather than considering to the modes of learning that appeal to our students. I will suggest that a professor can fruitfully adopt the ubiquitous microblogging tool Twitter as a means to inspire students to achieve a course's learning objectives, and to eagerly anticipate and enjoy the classroom experience.

II.

Perhaps recent research in industrial psychology could help us understand what motivates our students. In 2009, in his book *Drive*, Daniel Pink posits three primary motivations of all employees: the need for autonomy, the search for mastery, and the quest for a shared purpose. If we apply Pink's theories to the classroom for a moment, we can see why a typical lecture-dependent course may quickly sap the attention, motivation, and drive of our students. Before encountering Daniel Pink, I learned similar lessons from conversations with UC Davis professor emeritus Dick Walters (personal communication, 1997). Walters argues that the most successful university classes are student centered, activity driven, and informed by frequent and casual instances of instructor feedback. And indeed recent research into best practices of teaching the 21st-century learner confirms Walters' thinking in the field of college instruction, emphasizing that the classroom leader should focus on the facilitation of learning, rather than on lecturing. Consider especially the work of Marc Prensky, coiner of the term "digital native," and his 2010 book, *Teaching Digital Natives: Partnering for Real Learning*. Reflecting on these principles, we might admit that a lecture-format course would satisfy neither Pink nor Walters nor Prensky, for such a course is typically instructor centered, it depends on silent note-taking by students, and it often neither requires nor rewards input from the participants. In the context of Pink's thoughts about motivation, we might say that the lecture-style class places all the autonomy with the single lecturer, gives students no opportunities to practice intellectual tasks in the classroom, and forsakes casual feedback in the name of coverage. How, then, might we shift the format and function of a large-enrollment class, and its activities, so that it allows students to participate more actively in their own learning process?

I had this question in mind as I was preparing for a short fiction class for 30 undergraduates that I taught during the summer of 2009 at the University of California, Davis. This intense six-week review of mostly 20th-century American, British, and continental short stories required significant reading, fast-paced analysis of the stories, and multiple writing assignments. I surmised from my students' bleary eyes and crossed arms on the first day of class that they were unenthused about the 8 a.m. start time, skeptical that I could hold their attention for the full two hours of our class meetings, and concerned that the syllabus was too ambitious. When on the second day of class some students said that they hoped I would "tell them what that last story means," I explained to the class that my goal was to provide students the tools and approaches to make independent and meaningful discoveries about the stories we would read; even if a particular story had "a meaning," it wouldn't be my job to present that meaning to them. As we were talking about approaches to literary analysis, I became increasingly convinced that the lecture approach students were expecting would not provide them adequate opportunities to participate in class. I also knew that, in order for the class to succeed, in order for our class meetings and class discussions to be meaningful and engaging for all the participants, I would need to motivate my students to complete and digest the readings before we met three times a week at 8 a.m.

In effect, I decided to run my literature class like the student-centered writing classes that I teach in computer classrooms. To this end, I learned my students' names on the first day of class, gave mini reflection-essay pop quizzes in the first five minutes of our meetings, and called on students to share their interpretations and ideas about assigned texts. We also completed impromptu in-class writing assignments, divided into groups when wrestling with different elements – characterization, themes, the function of plot choices, etc. – and responded to interactive PowerPoint presentations that functioned more like quiz shows than backdrops of lectures my students might be accustomed to in other disciplines. But my greatest ally in this summer class was an unexpected one: Twitter. By inviting students to discuss class texts outside the classroom, by requiring students to show evidence of their having taken responsibility for their own learning as college students, and by incentivizing absolute readiness for participation in class discussions, I found that Twitter could save my English class.

My thinking about how best to use Twitter was informed by a teaching technique that I borrowed from a mentor of mine, Professor emeritus Jon Wagner of the School of Education at UC Davis. Wagner requires that students enrolled in some of his large enrollment classes submit daily responses to assigned readings. Rather than creating study questions for each assigned text, Wagner requires only that students submit a single one-page document that includes salient quotations from the text, one or more questions about difficult concepts or obscure passages, and one or more comments about a crucial or central topic or concern raised by the previous night's reading (personal communication, 2007). These documents are called QQCs, standing in this instance for *Quotation, Question, and Comment*. The best of these comments function as the equivalent of thesis statements: interesting and debatable assertions that require supporting evidence from the text. By collecting these QQCs, and offering commentary on students' thinking, the instructor can provide necessary feedback on a student's understanding of concepts encountered in the assigned texts as well as answer questions posed directly to the instructor.

The QQC approach helps to ensure that students complete the reading for class and gives the instructor a sense of students' discoveries and misconceptions; however, a number of concerns remain: first, at least in a summer class, the course moves so quickly that timing alone makes the instructor's comments decreasingly relevant and perhaps decreasingly helpful. General concepts can be illuminated by an instructor's comments, but generally an instructor will find it challenging to offer responses that are both substantive and still current. The second concern is that faculty cannot provide feedback in time to improve classroom discussion, the class activity that determines the success of a class. Students who dutifully complete their QQCs would come to class better prepared to speak, but they would not arrive with the benefits of having those ideas subjected to peer or instructor review. The time-delayed insights are revealed only to the instructor, and his insights sharpen student thinking only on stories that had already been discussed. I decided for my summer literature class that it was time for a new tactic, a new tool to make the daily assignments of the submission of QQCs more interactive, social, and immediate. For our fast-moving summer class, I told students, they would be using Twitter as a means of submitting and sharing their QQCs.

I first encountered Twitter as a journalist. Having hosted a humanities computing public affairs radio show since 2000 (called "Dr. Andy's Poetry and Technology Hour"), I knew about the importance of Twitter from a social networking and business standpoint. In 2007 *Time Magazine* had written that Twitter was "on its way to becoming the next killer app" (Hamilton, 2007), and the *New York Times* said that Twitter was "one of the fastest growing phenomena on the internet" (Pontin, 2007). Twitter's growth has been sustained over the last few years, as evidenced by the 1,100% year-over-year growth from the beginning to the end of 2008. Seventy-five million people visited Twitter.com in January of 2010, and all those Twitterers sent about 1.2 billion tweets, each 140 characters or less, and most of them read by fans and followers (Schonfeld, 2010). During the summer that I taught the short fiction class, huge numbers of Iranian dissidents had used Twitter to circumvent state-run media and their government's attempts at censorship to communicate with each other and with the world. The evening news ran "illegal" YouTube footage of millions of protesters marching through the streets of Iran, and newscasters and other journalists began describing the unrest as a "Twitter revolution"

(Morozov, 2009). I was covering the Twitter phenomenon and the use of Twitter by Iranian dissidents on my radio show, so I was sold on the idea that this new social networking interface was becoming increasingly important, as well as adaptable. If millions of dissidents could hatch complex protest strategies using Twitter, I thought, then my students and I should be able to use the same tool to discuss Edgar Allan Poe and Anton Chekhov. I hoped that I had discovered a way to make QQCs more relevant and helpful, and that we could use this trendy tool to share our thoughts, paperlessly, with everyone in the class.

While I was optimistic that with Twitter I could address my timeliness and relevance concerns with the QQC assignment, I hadn't anticipated my students' initial objections. Student pushback was immediate, and actually rooted in impressions they had formed from media representations. Few of them had Twitter accounts, and many of them originally thought of Twitter as picayune, specious, and faddish – a needless means to update friends on meal choices, or a way for Hollywood celebrities to blather about their whims to vapid fans. How could students represent or quote a meaningful or substantive portion of a literary text using only 140 characters? How could that same character limit allows students to respond to each other's questions? What thesis statement could be that short and still make a meaningful assertion? These were all valid concerns, and many of them were debated in class. And the debate was fruitful, for as we talked I realized that my students were learning important lessons about communication, clear thinking, and clear writing. Having taught writing classes for 20 years, I was comfortable leading debates about the tools, processes, and strategies for writing clear and substantive prose, but in the literature classes I teach I usually leave far less time for this focus on process. I enjoyed hearing my students' comments about rhetorical strategies, the paths to successful writing, for I knew such discussions would help them deepen and clarify their thinking, and submit stronger essays about assigned stories.

As this was the first time that I taught with Twitter in a college class, I had a number of misconceptions about possible uses of the tool that my students helped me address and resolve. At first I thought that Twitter would function merely as a place to share among the students the QQCs that otherwise they would share only with me. I imagined that all the students could benefit from the opportunity to review each other's insights, each other's favorite quotations, and each other's questions about the text. In class some students would reflect on another student's submitted QQCs, but they also began registering complaints with Twitter. When Twitter was used *only* to list my students' QQCs, interest in reading the other students' tweets sank. But when certain enterprising students responded to each other's questions, and many conversations broke out, students became deeply engaged, and ended up writing much more than was required about the stories we read. This practice by the students corroborates many lessons that we know about learning, such as the need to inspire intrinsic motivation for students to take on assigned tasks and projects (Malone & Lepper, 1987). In their discussion of an organismic (or active and volitional) motivational theory, Deci and Ryan (1985) assert that "The active-organism view treats stimuli not as causes of behavior, but as affordances and opportunities that the organism can utilize in satisfying its needs." From my perspective as a teacher, I should have recognized that Twitter should not have been used merely as a means of enforcing preparation through the recording of QQCs. Twitter gave me another chance to focus on learning rather than teaching, and I almost missed the opportunity. Discussions with students helped me realize that this social impetus to read, to ask questions, to hear them answered, and in short to socialize while engaging with sometimes difficult texts, would appeal to my students and drive them to make discoveries that I might not have anticipated. I listened and learned, and the class policy sheet on the desired function for Twitter evolved alongside our understanding of the texts.

On the second day after we discussed this more social and casual approach to Twitter as a learning and collaboration tool, I noticed some visible improvements in the classroom. One was that attendance increased. We would sometimes go three or four classes without a single student missing class. On occasion, students would write me e-mails apologizing in advance, and profusely, for having to miss one of the lectures because of a family trip, a medical appointment, or a funeral. As teachers, we are used to hearing such excuses, if students bother to make excuses at all for missing a lecture. There's an old joke among writing teachers that we sometimes feel guilty for assigning research papers because of how such assignments seem to threaten the health and well-being of our students' grandparents. As all my students managed to attend most of the lectures, I remember remarking to a colleague that assigning Twitter work, by contrast, actually improved the health of my students' grandparents, and that its use might be recommended if only for this reason.

Another of Twitter's profound effects was not only the consistent number of the students in the class but also the improved quality of our discussions. As teachers, we have often faced long and awkward silences after asking open-ended questions about a topic that might have come up in the day's reading. I learn the names of all my students on the first day of class, not only to make my students feel valued and comfortable but also to make them feel uncomfortable: that is, to be able to call on them by name in class, and thereby require their participation and their contributions to the class discussion. In this class, I rarely had to put students on the spot. A few classes after encouraging a more casual and conversational style using Twitter, I found myself feeling more like orchestra leader Leonard Bernstein than like Ferris Bueller's high school history teacher Ben Stein, for I would ask a question about a story that we had read the night before and see at least a dozen hands go up (Jacobson & Hughes, 1986). Impressed and almost overwhelmed, I used my own hand gestures to recognize, welcome, and celebrate the many raised hands. Students felt not only like they had an opportunity to speak but also a *compulsion* to speak. At times I would even have to list the names of students on the board so that we could remember in which order their comments, questions, objections, and clarifications would be heard. All of us could tell that something magical was going on, and I could hardly keep from smiling.

As my students continued to read assigned texts more and more closely, I discovered why the class discussions were so productive and elicited such thoughtful responses from the participants. Tweeting together about assigned texts the night before a class discussion, students would ask each other clarifying questions about the basics: questions about plot points and questions that sought explanations of the differences between key characters. My discussion of the Hemingway (1927) story "Hills like White Elephants," for example, did not have to begin with the standard question, "What are these two characters talking about?" Everyone who had read the tweets from the night before knew that the two characters were discussing an abortion; in fact, they could see all the tweeted evidence that supported that conclusion. We could stipulate this fact and move on to deeper analyses of the story. The class conversations benefited from the many voices heard, with students responding to each other, rather than engaging in a series of dialogues with me. With the support and investment of my students, we could take on more interesting challenges, and focus on higher-order thinking tasks.

In his taxonomy of thinking tasks, Benjamin Bloom (1956) argued, among other things, that while it was often appropriate for middle school and even high school students to focus on recall and comprehension, by the time students came to college, they should be focusing on thinking tasks such as application, analysis, synthesis, and evaluation. Indeed, as a teacher, I have often considered what strategies would help move my students away from memorization and recall, and move them toward a more holistic practice of advanced critical and creative thinking. Twitter helped me dramatically with this endeavor, for students would, among themselves, review the basics of assigned stories, with some students leading some of their peers to a basic understanding of assigned text hours before the class met with me. As a result, in my class students eagerly presented insights that reflected a deepened and mature understanding of the short stories they read, and I rarely had to take time to clarify the plot of the story that some students might have read hurriedly, or not at all. Because of the extra research and discussion that my students were completing via Twitter, my undergraduates came to my class prepared with the sort of reflective wisdom that one would be pleased to see from advanced English Majors sitting around the table of a senior seminar.

Whereas I appreciated that students were using Twitter in class conversations to create these somewhat interesting assertions that might make up a thesis statement for a submitted essay, students understandably were mostly excited about the social opportunities and rewards of talking about literature outside of class. Starting about midway through the summer session, I would sometimes find almost half the class in their seats already talking with each other as I arrived, each of them with our class anthology already open, rather than working through crossword puzzles or thumbing through email on their cell phones. Because of Twitter, and because of the group collaborative presentation assignments, those minutes before class resembled smiling reunions, with students alluding to each other's tweets, as well as the jokes, the asides, and the plans for study groups. Once class discussion began, the peer pressure moved toward the direction of contribution rather than silence. Students sometimes complain that professors in a large lecture classes will welcome questions from their smartest or most talkative half-dozen students, while the rest of the class sits watching the conversation, sometimes hoping that it ends soon so that the professor can return to the business of covering content. I have heard from students in office hours that sometimes the silent majority even resents the talkative minority, and wishes that these garrulous "A" students would succumb to peer pressure and pipe down. In my class, that dynamic was reversed. Students who chose not to speak felt left out. Students who missed the Twitter conversations of the night before might feel as if they had missed a great pizza party, or at least a TV show that all their friends had watched without them. I enjoyed calling on many students that summer, but I found myself pressuring fewer silent Bartlebys (Melville, 1856) than ever before.

Sometimes in class we would take a break from all of the deep textual analysis to have a meta-conversation about the class, and these discussions helped students better understand how it is they learn, what motivates them to learn. I would ask a number of questions, and delight in their answers:

Why does Twitter work better than Facebook for the sort of social media-assisted, outside class work that we do in this course?
Facebook offers too many distractions. Using Twitter and the specific hash tag for this class (in our case, #ucdf), we can form our own interpretive and mutually supportive community without having to worry about looking like dorks in front of our Facebook friends.

To what extent does the use of Twitter improve your writing, even though you have only 140 characters to work with?
Well, we have to be so careful and precise with the words we choose.

Why must you make every word count?
So we can do a better job sharing assertions with our friends. Plus we appreciate all the practice writing thesis statements, and hearing the arguments of our peers.

When do you have your most fruitful discussions on Twitter?
Between 11 p.m. and 2 a.m.

Why during that time?
Because that's when we do our work, and we like being able to share our discoveries as soon as we make them.

Why do we have so many confident speakers in this class?
Because we have rehearsed our thoughts at home with our friends.

What makes a tweet valuable?
A valuable tweet is clear, insightful, assertive, and obviously succinct.

What quotations should I ask you about on the midterm?
Those we have discussed in class, and those we have most discussed on Twitter.

Could you name all (30 of) the students in our class if I asked you to?
Yes.

Have you made any friends in this class?
I know more people in this class than I do from the rest of the classes I've taken at UC Davis, put together.

I loved this approach to teaching a literature course, and I feel using Twitter judiciously could enliven any class that depends on, or would benefit from, class discussions. Recent research into academic uses of Twitter suggests that this new communication medium is helping many faculty engage with their students (Posetti, 2009; Wesch, 2008; Young, 2009), and reach classroom goals. From my perspective, I noted at the end of my summer experiment that all my students became more autonomous learners; all of them improved their ability to analyze texts, and share insights verbally and in written form; and all of them saw how the extra work they chose to do with their peers improved their grades and heightened their commitment to our course objectives. I also felt that the class validated the multiply-mediated way in which Millennials live and learn, but without sacrificing the sort of deep and sustained thinking that was necessary for students to excel in the class. Our students are ubiquitously connected, and wise use of Twitter can help a faculty member harness the opportunities provided by this reality, rather than lament or try to ask students to suspend those connections while doing academic work.

III.

When teaching with Twitter, faculty can adopt a number of strategies to augment the class discussions and assignments without hijacking them. I will list here six approaches that I tried in 2009, and then four more that I will try the next time I am teaching a large enrollment course that would benefit from more student interaction, reflection, and engagement.

1. **Establish a Twitter account just for your teaching.** One can establish as many Twitter accounts as one has email addresses (free Gmail accounts can help with this). I use an account just for my teaching (andyatucdavis) that is separate from the personal Twitter handle that I have used for a couple of years (andyojones) to stay connected with the social and professional communities that are important to me. Students are also encouraged to create dedicated class personae to tweet from (though they can include mention of their personal Twitter account on their class Twitter account bio, in case people want to find out more about them). This way, class participants can follow their peers without having to worry about reading through irrelevant tweets.

2. **Choose a hash tag that works for your class or subject area**. For my class, I chose #ucdf (even though that was later discovered to be the acronym of the United Cheer and Dance Foundation). Using a hash tag made it easy for anyone who wished to follow the class conversations about great fiction to do so. Conduct a bit of hash tag research to ensure that a current academic or professional conference isn't using your same hash tag; otherwise both your students and the conference goers could become confused (see Parr, 2009).

3. **Set clear expectations of how you expect your students to use Twitter**. Discuss the number and quality of tweets that you would like to see, as well as some options of rhetorical approaches of their tweets. At first I asked students to tweet QQCs. After we realized that the QQC approach was too formulaic for readable tweets, I encouraged students to offer at least three valuable tweets before each class meeting, one of which must be a response to a statement or a question presented by a classmate. This approach encouraged the sort of positive peer pressure that made students so eager to share in class (see Comm, 2009 for an excellent review of the varieties of tweet categories and functions).

4. **Reread the class Twitter stream as part of your final preparations for class.** Note areas of student enthusiasm and confusion. Demystify what deserves demystification, but also lead the class in a discussion of what students feel to be the crux of a text. If you can, allude to Twitterers by name to involve them in the conversation and to validate their out-of-class work.

5. **Validate and reward substantive and helpful Twitter participation.** Suggest ways that students can present the quality and quantity of Twitter participation at the end of the term in the form of a portfolio. Recognize the importance of class participation and preparation by rewarding those who sustain the spirit of enthusiasm and critical inquiry in your class. Make your policies clear beforehand so students know at the beginning how you expect them to invest in your class.

6. **Tweet multiple media.** My class appreciated my tweeted links to pictures of Rome at the time of *Daisy Miller* (James, 1878), and Oxford, Mississippi at the time of *A Rose for Emily* (Faulkner, 1930). Challenge students to direct their peers to relevant images, audio, and video that might help them understand the settings and context of assigned texts. The Web site http://www.tvider.com, for example, makes it easy to share video and other media with Twitter followers.

Examples of more ambitious uses of Twitter as a teaching and learning tool.

1. **Set up a RSS feed of class tweets on the course Web page.** I saw productive use of this strategy at the Computers and Writing Conference in 2009. Conference participants were encouraged to respond to and ask questions about the presentations they were observing, and all the participants could watch the instant reporting in real time. Having such a Twitter stream on the course Web page allows all class participants to check in on the most recent posts that include the shared hash tags.

2. **Share your work as a faculty member with your students.** Students look up to faculty not only for what they communicate in the classroom but also for how they conduct research, solve problems, and consider the intellectual challenges of preparing for class. That said, these processes are largely hidden from our students, and thus they represent a missed opportunity for students to learn about the multiple layers of focus and commitment required to do the work we do. Generally speaking, students who see evidence of a professor's commitment to the class reward that faculty member with stronger course evaluations, and with their own increased focus on course objectives. If you want to point students toward what books you are reading, or what news story that you are following, use a URL shortener such as http://www.bit.ly or http://www.ow.ly. They will allow you to fit long links within your 140-character tweets.

3. **Use a time-delay application to front-load your tweeted wisdom.** If you are teaching a class where students would benefit from a series of lessons or nuggets of wisdom that you would like to upload once a week and have them be tweeted on a schedule that you determine, consider one of the many simple tools that allow one to give the illusion of your Twitter industry. For instance, as a writing teacher, I would love to augment my literature class with a half-hour critical thinking or writing lesson once a week, but we wouldn't usually have time. Instead I could upload pre-tweeted links to relevant handouts from online writing labs (the OWLs at Dartmouth and UNC Chapel Hill are two of my favorites), and watch as they are tweeted at predetermined intervals over the course of the week (or the semester). This approach would help students recognize their writing deficits, and address them in a systematic fashion with the online help of expert resources (see California State University East Bay, 2008 and Guajardo, 2009).

4. **Consider using Twitter as a personal response system (aka "clickers").** Most of your students currently carry cell phones that allow them to tweet. In large classes, faculty could invite students to tweet responses to a polling question during lecture, and then ask a student or TA to monitor the responses. In the coming years, more of our students will be bringing to

class tablet computers to which assigned texts have been uploaded. These same networked computers will provide student multiple opportunities to interact with faculty in real time using a variety of media, including instant messages and Twitter (see Wetzel, 2009).

I hope this narrative of discovery and these resources are helpful to you. The British biologist and mathematician Jacob Bronowski (1973) once said "We are all afraid for our confidence, for the future, for the world. That is the nature of the human imagination. Yet every man, every civilization, has gone forward because of its engagement with what it has set itself to do." My hope for my classes and for yours is that the right mix of strategies and teaching tools will help our students find the practice and then the confidence to engage meaningfully with our lectures, our class activities, and our class discussions, and thus develop the effective communication skills and higher-order thinking skills to commit to a lifetime of curiosity and independent learning.

REFERENCES

Bloom, B. S. (Ed.) (1956). *Taxonomy of educational objectives: The classification of educational goals; handbook I, cognitive domain.* New York: Longman.

Bronowski, J. (1973). *The ascent of man.* New York: Little Brown & Company.

California State University East Bay. (2008). Writing guides. Available at http://www20.csueastbay.edu/library/scaa/writing-guides.html#General

Comm, J. (2009). Fourteen types of tweets, June 25. Available at http://joelcomm.com/fourteen-types-of-tweets.html

Deci, E. L., & Ryan, R. M. (1985). *Intrinsic motivation and self-determination in human behavior.* New York: Plenum Press.

Faulkner, W. (1930). *A rose for Emily.* New York: Random House.

Guajardo, Y. (2009). 5 time saving twitter tools for managing your friends and tweets. Successfool, January 29. Available at http://successfool.com/5-time-saving-twitter-tools-for-managing-your-friends-and-tweets/

Hamilton, A. (2007). Why everyone's talking about Twitter. *Time*, March 27. Available at http://www.time.com/time/business/article/0,8599,1603637,00.html

Hemingway, E. (1927). *Hills like white elephants. Men without women.* New York: Charles Scribner's Sons.

Jacobson, T., (Producer) & Hughes, J., (Director). (1986). *Ferris Bueller's Day Off*, June 11. California: Paramount Pictures.

James, H. (1878). *Daisy Miller: A study in two parts.* New York: Harper and Brothers.

Malone, T. W., & Lepper, M. R. (1987). Making learning fun: A taxonomy of intrinsic motivations for learning. In: R. E. Snow & M. J. Farr (Eds), *Aptitude, learning and instruction III: Conative and affective process analyses.* Hillsdale, NJ: Erlbaum.

Melville, H. (1856). *Bartleby, the scrivener: A story of Wall Street. The Piazza Tales.* New York: Dix & Edwards.

Morozov, E. (2009). Iran elections: A Twitter revolution? *The Washington Post*, June 17. Available at http://www.washingtonpost.com/wp-dyn/content/discussion/2009/06/17/DI2009061702232.html

Parr, B. (2009). How to: Get the most out of Twitter #hashtags. *Mashable*, May 17. Available at http://mashable.com/2009/05/17/twitter-hashtags/

Pink, D. (2009). *Drive. The surprising truth about what motives us.* New York: Riverhead Books, Penguin Group.

Pontin, J. (2007). From many tweets, one loud voice on the Internet. *The New York Times*, April 22. Available at http://www.nytimes.com/2007/04/22/business/yourmoney/22stream.html

Posetti, J. (2009). Twitter as a journalistic tool: Drilling beneath the rhetoric. *J-Scribe*, November 8. Available at http://www.j-scribe.com/2009/11/twitter-as-journalistic-tool-drilling.html

Prensky, M. (2010). *Teaching digital natives: Partnering for real learning.* Thousand Oaks, CA: Corwin Press.

Schonfeld, E. (2010). Nearly 75 million people visited Twitter's site in January. Techcrunch. Available at http://techcrunch.com/2010/02/16/twitter-75-million-people-january/

Shakespeare, W., Mowat, B. A., Werstine, P., & Folger Shakespeare Library. (2002). *The tragedy of Hamlet, Prince of Denmark.* New York: Washington Square Press.

Wesch, M. (2008). Teaching with Twitter. Digital ethnography. Available at http://mediatedcultures.net/ksudigg/?p=170

Wetzel, D. (2009). 10 personal response systems teaching strategies. Teaching & Technology-Suite101, October 14. Available at http://bit.ly/9Ygbwn

Young, J. (2009). Professor encourages students to pass notes during class – via Twitter. *The Chronicle of Higher Education*, April 8. Available at http://chronicle.com/blogPost/Professor-Encourages-Students/4619

SOCIAL LEARNING WITH SOCIAL MEDIA: EXPANDING AND EXTENDING THE COMMUNICATION STUDIES CLASSROOM

Robert Bodle

ABSTRACT

Recent studies suggest that many of today's students are highly proficient in their use of digital media and are developing new learning styles heavily dependent on social media and the Web. Theories of social learning seem to address these new learning styles, which are interest and friend driven, and occur in contexts that are outside of class and within the flow of students' everyday lives. Social learning emphasizes participation, group interaction, and utilizing collaborative environments. This chapter explores how using social media, specifically class blogs (WordPress) and microblogs (e.g., Twitter) together, help achieve social learning. Internet-based learners have various levels of proficiencies, competencies, and adoption rates. Strategies and best practices are explored to address how social media can be utilized by educators to accommodate the heterogeneity of digital learners and engage new styles of learning.

Teaching Arts and Science with the New Social Media
Cutting-edge Technologies in Higher Education, Volume 3, 107–126
ISSN: 2044-9968/doi:10.1108/S2044-9968(2011)0000003009

INTRODUCTION

Recent studies suggest that many of today's students are highly proficient in their use of digital media and are developing new learning styles heavily dependent on social media and the Web. Theories of social learning seem to address these new learning styles, which are interest and friend driven, and occur in contexts that are outside of class and within the flow of students' everyday lives. Social learning emphasizes participation, group interaction, and utilizing collaborative environments. This chapter explores how using social media, specifically class blogs (WordPress) and microblogs (e.g., Twitter) together, help achieve social learning. Internet-based learners have various levels of proficiencies, competencies, and adoption rates. Strategies and best practices are explored to address how social media can be utilized in higher education to accommodate the heterogeneity of digital learners and engage new styles of learning.

Social media sites and services provide innovative tools for dynamic forms of communication and participation. These tools have shifted the ways information is circulated and shared, challenging traditional gate-keepers, disrupting the relationship between experts and amateurs, and "reconfigure[ing] the ways that people can exert influence in the world" (Rainie, 2010, p. 17). The technological and social (technosocial) affordances and communication dynamics of social media offer new ways to create, interact, collaborate, and learn. Social media also provides important tools for asserting one's voice and connecting to the world. The social network layer underlying many forms of online interaction has applications for student engagement as well as civic engagement (and the two might be related).

Educators are discovering that social media can engage learners by facilitating the social processes required for learning (Dunlap & Lowenthal, 2009). A "social turn in learning theory" (Ito et al., 2010, p. 13) suggests that learning most often occurs outside of formal educational settings, and in informal, friend- or interest-driven contexts; with learning happening in the everyday flow of learners' online and offline interactions (Ito et al., 2010). Educators buttressed by social theories of learning are searching for ways to harness social media and go where students already are and live. As a result, more online spaces are identified as potential locations for learning and opportunities for instruction. Yet, challenges arise when utilizing colla-borative online spaces to encourage student engagement that includes privacy and accounting for learner disparities.

When exploring opportunities for using social media in the classroom, it is also important to acknowledge the need to respect the privacy of learners' online lives. A potential risk of educators is colonizing or invading learners' social spaces and compromising online privacy. Instructional approaches that emphasize online sharing and open exchange should also emphasize privacy as part of new media education.

Current studies identify young people's adoption of and proficiency in new media along generational lines. For example, a Pew Internet and American Life Project report found that teens and Generation Y (18–32 year olds born 1977–1990) are more likely than other generations to seek entertainment online (video, games, virtual worlds), to read and write blogs, to use social networking sites, and to instant message on their handheld devices (Jones & Fox, 2009). Perhaps new media disrupt more than define generational identity, disrupting traditional notions of how learners of different generations interact and relate (Ito et al., 2010, p. 5). Regardless, if adult learners are the fastest growing population in higher education, as some research suggests (Tierney & Hentschke, 2007), then we need to discover new models that can enhance intergenerational interaction and participation to address generational differences. Technological usage might not fully account for the diversity among learners, but they can point to the need to better understand other important disparities.

An overlooked aspect of Internet use is the disparity in information literacy skills and user savvy or "know how" (Hargittai, 2002, 2008). An important aspect of social learning is participation, which not only "involves 'learning about' the subject matter but also 'learning to be' a full participant in the field" (Brown & Adler, 2008, p. 4). Students may be participating and learning online, but their skill levels are not even (Lenhart & Madden, 2005). Hargittai's research shows a "statistically significant relationship" between Internet user proficiency and socioeconomic status, finding that parental education and Internet skill are linked (2008, p. 101). This research suggests that although most students are constantly wired and have been for several years, their usage and connectivity does not suggest user savvy (p. 95). In the classic study, *Diffusion of Innovations* (2003), Rogers also relates varying rates of adoption of new educational technologies to a number of social and other factors. Socioeconomic insights into this participation gap can help account for differences in students' use of new media, as well as guide the implementation of social media in a diverse classroom.

In this chapter, I share some of my experiences and observations using a class blog and microblog together, exploring how social media can

(1) temporally and spatially augment the classroom, (2) build a community of learners beyond quarters and semesters, and (3) address special challenges posed by the diversity of traditional and nontraditional learners' needs. Technology is utilized as a means "to extend a physical classroom" (augmented), not to replace in-classroom learning. However, other categories of teaching with technology – either blended or online courses – share some of the same challenges and opportunities (Siemens & Tittenberger, 2009, p. 16). To better assess how social media might accommodate new styles of learning, I will first discuss methodology, complimentary theories of social learning, outline new learning styles, and identify the affordances of new media. Insights into the affordances and limitations of social media can help identify their potential for effective teaching and learning, as well as indicate the risks and how to manage them. This chapter ultimately demonstrates how social media expands and extends an augmented classroom and supports social learning by offering new social contexts for sharing, engagement, dialogue, and interaction.

METHODOLOGY AND APPROACH

I teach a variety of Communication Studies courses including Visual Communication, New Media and Society, New Media Ethics, and Film Studies at a small liberal arts college (12:1 student-faculty ratio) located in the Midwest. The majority of students are traditionally college aged, with a growing number of adult learners each year. Over half of incoming freshmen are first-generation college students who live at home. Students have access to laptops (loaned or purchased), three computer labs, and WiFi on campus. Personal experiences of teaching the course Visual Communication were used to illustrate the central argument of this chapter – how social media can support social learning in a diverse augmented classroom. Visual Communication is an upper-level undergraduate course required for Communication Studies majors. Recent studies examining digital learners and new media literacies help guide and inform an empirical assessment and analysis of utilizing social media in the classroom.

As the latest research suggests, social media is inextricably linked with everyday practices of participation, which encourages an ecological approach to the application of social media in teaching and learning. A media ecology approach recognizes the "structure and context" of new media use (Ito et al., 2010, p. 31) that occurs in "contexts of group interaction" (p. 18), within networks that have a mix of weak and strong ties

(Christakis & Fowler, 2009, p. 163), and throughout the rhythm and flow of everyday life. Social learning theory suggests that social learning styles benefit from the ecological nature of social media, which is interactive, ongoing, and where the construction of meaning and value is cocreated through these interactions.

NEW LEARNING STYLES AND THEORIES: AN OVERVIEW

Mutually reinforcing insights into social learning styles and theories seem to support the turn to the social in education (Ito et al., 2010). Learning happens, according to Brown and Adler's situational theory, when students are actively involved in doing, either in conversation with one another or the instructor (Brown & Adler, 2008, p. 4). This theory reflects a shift in focus on social context rather than content; how learning content is "socially constructed through conversations about that content" (p. 2). Lev Vygotsky's "social development theory" similarly focuses on social interaction suggesting that doing and interacting lead to higher thinking skills (Vygotsky, 1978). Emphasis on how and where students learn, as opposed to what they learn, challenges the Cartesian perspective that separates content from delivery, and seeks to focus on bringing the two together in new ways that are interactive, social, and participatory.

Traditional students seem to learn from their peers in social contexts that are informal and support "multiple modes of learning" (Ito et al., 2008; Brown & Adler, 2008). In the "most extensive US study of youth new media use," researchers stress that young people gain media and technical literacies through self-directed learning involving play and experimentation (p. 2). These playful modes of participation, according to Ito et al., represent "different levels of investments" and include (1) "hanging out" (being together and sharing experiences and space), (2) "messing around" (unstructured experimentation and play), and (3) "geeking out" (intense level of investment and frequency of use, with status or reputation tied to validation and feedback from demonstrating specialized knowledge) (2010, p. 35). Palfrey and Gasser (2008) similarly identify distinct but related learning practices among "digital natives" (18–23 year old students born between 1977 and 1990): (1) perfecting the art of grazing (gathering snippets of information from many sources), (2) the deep dive (learning more about a topic of interest), and (3) the feedback loop (engaging with information by talking back on blogs, social network sites (SNS), and word of mouth). The learning styles mentioned above are greatly enabled by the affordances of new online sites and services collectively identified in this study as social media. I will now address their specific affordances.

TECHNOSOCIAL AFFORDANCES OF SOCIAL MEDIA: USES, FUNCTIONS, AND FEATURES

Technology can influence, but not determine, everyday life by creating possibilities or "social affordances" (Wellman et al., 2003, p. 7). Some affordances of social media include social presence (Dunlap & Lowenthal, 2009), "enhanced social connectedness" (Rainie, 2010, p. 8), "access," "expression," "creation," "interaction," and "aggregation" (Siemens & Tittenberger, 2009, p. 41). Social media provides users with dynamic ways to interact, create, and share, which encourages interaction and participation. Social media or "mass self-communication platforms" (Castells, 2009) include macroblogs (WordPress, Blogger, Blogspot), microblogs (Twitter, Yammer, Yelp), content-sharing sites (YouTube, Flickr, Delicious), and SNS (Facebook, MySpace, Orkut).

Individual and group uses of social media include self-expression, creativity, intimacy building, impression management, identity construction, recreation and entertainment, therapy, advertising/branding (building audiences and consumers), political activism, advocacy, democratic participation, and governance (using social power embedded in social networks to influence decision-making).

Social media can provide the potential for democratic forms of participation, empowering people to exercise their voice. The implications of social media for preparing an active citizenry have not been lost on online community pioneer Howard Rheingold who asserts that today's college students "need to use various Web 2.0 tools to be good citizens because those modes of communication are increasingly the way political discourse and activism take shape" (Young, 2008). The most prominent account of social media enhancing democratic participation is the often cited postelection protests in Iran in June of 2009, with street-level developments disseminated on Facebook, YouTube, and Twitter, prompting some to call it a "Twitter Revolution" (Ambinder, 2009). Social media has also been hailed as the long-awaited set of applications that will enable "Athenian style direct democracy," where "every citizen is connected to the state" and can participate directly in policy-making (Giridharadas, 2009; Hickins, 2009).

The democratic aspects of social media can also benefit social learning, evident when comparing social media tools to traditional learning management systems (LMS). Social media sites and networks are more informal, connected to other social spaces and platforms (mobile, Web, desktop), integrated into the learner's daily information flows, and are relatively open to the rest of the World Wide Web. An LMS, on the other hand, requires deliberate navigation to a more formal and closed context, removed from the immediacy of everyday spaces and cut off from the broader public sphere. Additionally, social media is not bound by term limits of semesters or quarters, and can enable "faculty and students to maintain on-going relationships after a course ends" (Dunlap & Lowenthal, 2009, p. 4). This allows open access to valuable resources beyond physical contact, important for lifelong learning, and for maintaining one's network.

A prominent feature of SNS, a subset of social media, is their ability to "enable users to articulate and make visible their social networks" (boyd & Ellison, 2007, p. 2). Articulated networks enable users to view each other's extended links, providing a form of transparency that reveals larger connective patterns for future networking across communities of interest. Transparency can encourage networking, enhance social presence, and also elicit participation. Students' social status can be related to level of participation (e.g., blogs posts and comments, uploading videos and photos, linking, and tagging; Skog, 2005). People gain a sense of empowerment and "enhanced social connectedness when they contribute material online" (Rainie, 2010, p. 8). Social media enables participation that can be used to build reputations, grow networks, and gain social influence and power.

Blogs and microblogs (WordPress and Twitter) together provide a wide range of affordances that can be utilized for varying modes of participation. Twitter and blogs provide flexible and informal communication dynamics, interoperability with one other, and the ability to embed text, images, sound, and video. The communication dynamics of social media include a combination of direct messaging (one to one), self-mass communication (one to many), and narrowcasting (one to few or interactive). Blogs (Web logs) are a form of self-publishing on the Web; individual or collaborative Web pages that usually consist of short and frequently updated entries (video, image, text), that can vary from "personal diary to journalistic community news," and are arranged in reverse chronological order (Gillman, 2006). Blogs provide a visibly engaging platform for a wide variety of expressive forms of communication, including longer essays, multimedia content, and comments and may support multisensory learning.

Studies in perceptual learning suggest students learn optimally in a multisensory environment as opposed to a single stimuli or "single sensory modality" (Seitz & Shams, 2008). Using multimedia elements such as sound and images might prove more efficient in learning environments (Neo & Neo, 2004, p. 10), as well as assist students with learning disabilities (Miller, 2001).

Text-oriented microblogs like Twitter enable short text messages of 140 characters or less, or "tweets," that can be sent privately or broadcast to a public timeline and to a network of followers. Short messages that are read in a consecutive list of "tweets" or "social awareness streams" can be characterized as (1) public, (2) short, and (3) highly connected (Naaman, Boase, & Lai, 2010, p. 1). Twitter allows users the option of locking their "tweets," setting them to private so that only approved followers can view them. Most Twitter users, however, do not lock their tweets in order to contribute to the pubic forum or "Twitterverse" and to gain more followers.

Social media promotes participation by enabling sharing, commenting, and responding. Commenting features enable interaction and encourage participation in the form of dialogue-talking back. Blogs typically enable comments with the ability to share blog posts, images, and embedded videos and links, collectively forming a "blogosphere." Twitter users comment through direct messages (DM) and by reposting or "retweeting" (RT) other's messages, circulating short messages, media, and links. Messages are often retweeted, with accrediting indicated by placing the @ symbol before a prior sender's account name (e.g., RT @msjviscom). In this way information circulates from network to network, across communities of followers. The act of RT or copying and resending "is not simply to get messages out to new audiences but also to validate and engage with others" (boyd, Golder, & Lotan, 2010, p. 1).

Social media also encourages sharing from one service to another (or across platforms and applications), helping to integrate with learners' everyday information flows. A technological feature that contributes to social media's adoption is achieved through open APIs (application programming interfaces). APIs are programming tools that Web companies share with one another in order to integrate with other social media services. Twitter's open API encourages its use in different contexts and enables a number of applications (TweetDeck, Seesmic, Twitteriffic, twhirl, Twitpic) to send and receive tweets, photographs, and other media in a variety of contexts (SNS, search engines, mobile devices, television shows, Second Life, and software platforms). Blogs are integrated by offering syndication using Real Simple Syndication (RSS), although their integration into a broader range of social media sites and services is not as ubiquitous. For example, Twitter can feed tweets through a widget on the blog site, whereas the blog does not feed to the twitter account.

Blogs and microblogs enable both individual and group interaction. A collective blog, for example, enables members to join, post, comment, and associate with other members within a community built around the blog. Group blogs help provide a collective space for community building. Twitter does not consist of members or (authors) but enables community interaction by forming lateral ties based on reciprocal following. To enhance the shared ecology of group interaction on Twitter, everyone should follow everyone, creating a large-scale dense network (Shirky, 2008), which can easily be adapted to sparser, small-scale networks of few to few as the culture within social networks can change and users can un-follow at will (Christakis & Fowler, 2009). Social interaction is initiated by following and being followed back. Reciprocated links provide an equal basis for relationship building and help form a sense of community. The "always on" aspect of online life and the frequent exchanges afforded by Twitter also enhance the perception of social presence.

Microblogs can enhance social interaction through the affordance of social presence, which can help motivate student engagement (Dunlap & Lowenthal, 2009). Twitter puts students in touch with each other and with instructors through quick, informal exchanges, providing opportunities. Twitter, in particular can enhance an instructor's social presence, encouraging student–faculty interaction, critical for student engagement, satisfaction, and commitment to learning (Chickering & Gamson, 1987; Dunlap & Lowenthal, 2009). Short exchanges enable students and instructors to express themselves uniquely, projecting "their characteristics into the community" (Garrison, Anderson, & Archer, 2000, p. 89 as cited in Dunlap & Lowenthal, 2009, p. 2).

A dynamic feature of Twitter is the use of hashtags to provide synchronous, interactive, and subject-specific information streams. Live tweeting, or following tweets in real time, is utilized by the search function on Twitter – placing hashtags (# symbol) before topics, which enables users to follow topic-specific messages in real time and augment participation at events. A recent study of Twitter use at conferences reveals how live tweeting enhances the learning experience as users share information with each other, post links to related Web resources, and interact with other Twitter users who are unable to attend (Perez, 2009). Live tweeting can provide a backchannel that can embolden student participation as well as "hold the interest of today's students" (Young, 2010, p. 10).

In the next section, I will match the technosocial affordances of blogs and microblogs with the desired outcomes and learning activities of the course Visual Communication, engaging diverse learning styles using social learning methods.

SOCIAL MEDIA IN ACTION

The undergraduate course, Visual Communication, develops students' visual literacy and understanding of visual communication theories and their application. Students learn how to engage in critical analysis of the visual world around them and have opportunities to create well-designed, meaningful visual messages using digital media.

The learning outcomes of the hybrid course include engaging with and thinking critically about visual culture, and applying design principles to create media and critique one another's work. This course requires a high level of student interaction, a supportive climate, and student engagement with challenging course material (theory, ethics, design principles, and production skills training). In matching affordances with outcomes, social media can enhance student engagement and interaction by providing a social layer to the learning environment.

Class blogs are useful tools for displaying and commenting on visual content, and for collective and individual engagement with visual texts. A class blog site provides a platform for both instructors and students to publish digital media, including images, videos, and slideshows (e.g., PowerPoint slides), giving students and instructors opportunities to create, talk back, and share within a learning community. Three primary outcomes encouraged by blogs include (1) technical literacy, (2) social interaction, and (3) critical reflection (transparency of posting in public and critically thinking about one's own work and the work of others). Microblogs help support these outcomes, as well as leverage the additional benefits of social presence and social influence, utilizing the network structure of both strong and weak ties.

Utilizing Blogs and Microblogs

Utilizing social media requires selecting appropriate tools and designing appropriate learning activities to benefit from them. A collective class blog

(Web log) was chosen to replace a traditional LMS, to provide an open interactive collaborative multimedia venue for assigning and publishing student work (http://msjviscom.wordrpess.com). As "blog owner" of the WordPress blog, I had administrator status giving me the ability to invite users and assign roles. Students were invited to join the blog first under the very limited default entry-level role of "contributor," which was quickly changed to the more participatory role of "author," granting users the ability to manage and publish blog posts. Students joined the collective blog by creating a WordPress account, which only requires an email address and password. After accepting their invitations to join the blog, students also had the option of creating their own personal blog, if they wished (only a few students created their own personal blogs).

All assignments were posted on the collective class blog, which included weekly journals and creative projects (both individual and group work). Student journals required individual responses to prompts that asked students to engage with theoretical concepts, design principles, and ethical implications of visual texts. Creative projects consisted of designing flyers and logos, and collaborating on creating digital photographic essays, Web videos, and the final project – a multimedia public service advertising campaign. Posting journal assignments and creative projects to the blog required students to upload and embed images and videos, add text, create hyperlinks, upload and embed portfolios (via SlideShare) – all requiring a fair degree of technical literacy.

The self-publishing multimedia environment enables vocal, textual, and visual content, or a multisensory approach to teaching that can more fully accommodate learning disabilities and various styles of learning (visual learners, social learners, auditory learners, etc.). The blog also enables students to combine critical reflection with visual texts. An important assignment for critical reflection is peer review of student work. Peering helps people learn from one another by requesting and receiving feedback and support described by Brown as "atelier learning" (Brown, 2006, as cited in Siemens & Tittenberger, 2009, p. 30). In this learning model, "the activities of all students can serve to guide, direct, and influence each individuals work" (Brown, 2006, as cited in Siemens & Tittenberger, 2009, p. 30). Student comments on each others' creative work help them learn from one another and the instructor, and to develop work at a more professional level based on constructive feedback.

The blog functioned as a more open LMS, but also as a shared venue for collaborative learning. Students were able to view each others' work, comment, and benefit from their collective knowledge. Those with diverse skills and talents, competencies and literacies, learned from one another in the practice of media production as well as technical literacy. Publishing to the blog requires social media literacy – opportunities to interact "meaningfully with tools to expand mental cognition" or "distributed cognition" (Jenkins, 2008, p. 4). Students with varying levels of comfort, experience, and "know how" turned to one another for support sharing literacy skills, tips, instructional videos, and knowledge. For example, when creating their logos, students took pleasure in sharing tips about how to screen capture their work on paid logo creation sites, without paying. I provided time and space for group interaction in class to help build a social foundation where students could be comfortable with each other and help one another in their collaborative projects.

Twitter, the micromessaging service, complemented the blog by providing a less formal social space to interact in ("hanging out" and "messing around"). I selected Twitter to achieve social presence, keep students interested in the class, and provide a model for effective use. Although I have my own Twitter account, I chose to set up a specific class account (http://twitter.com/MSJVISCOM) to provide a more focused hub that brings the class together around more targeted course-related Tweets (without sending Tweets to my own broader network of followers). I requested that students create their own Twitter accounts (only one student had already had one), and to follow the class Twitter account and each other in order to form a dense network of reciprocated following/ followers. I recommended to students who were active Twitter users to set up a student-related account to better target their academic Tweets. Another useful and easy way to follow a specific Twitter stream is to use hashtags (the symbol # before a key term like #viscom) to follow real-time results related to a particular conference, subject, or theme (also known as "live tweeting"). The ability to follow subjects via hashtags is recommended for students who have established Twitter accounts and are heavy users (my students were new to Twitter). I chose the reciprocal following/followed network model on Twitter to better establish a stronger connection between students and faculty that could last throughout the semester and beyond. By establishing a course/subject-specific user account, I was also able to build a learning community of followers more closely related to the topic of Visual Communication (as opposed to the Twitter network that follows my broader set of research interests); a network that now includes other visual communication students, artists, educators, and communication departments. Establishing a course-related Twitter network addresses the ecology of student's online lives, reaching students' Twitter streams more effectively than typing a class hashtag into a Twitter search window to initiate a subject-specific Twitter stream. Yet, hashtags were also utilized during class period to be used as a backchannel, though not always successfully as some students were distracted and annoyed by the multichannel environment.

The network structure enabled by Twitter, blogs, and other forms of social media encourage information to pass directly to others (strong ties), as well as circulate outward, from person to person (weak ties). Together, the network structure of strong and weak ties exposes students to new information, as well as extends their influence to a wider audience ("geeking out"). Twitter enables students to strengthen ties by sharing information about the course ("informers") and themselves ("meformers") within an articulated social network (Naaman et al., 2010). The social space of Twitter also enables students to gain and exercise social power. The micromessaging service provides a means to talk back to the class and contribute to a feedback loop, where the instructor can integrate student Tweets into lectures, assignments (often regarding due dates), and class discussion. Students also saw Twitter as another venue that could extend their audience base, tweeting about their final projects as a means of raising awareness about their campaigns.

Final Group Project

The final group project is a collaborative multimedia public service ad campaign that raises awareness about a local cause or service of the students' choice. Final projects require at least three different kinds of visual texts, and can include a flyer, logo, billboard, YouTube video, and Facebook fan page. Social media, including Facebook campaigns, Twitter accounts, and YouTube videos, were utilized as components of integrated social media ad campaigns. One final project, for example, "Forever in a Landfill," consisted of a logo, a billboard, a YouTube video, a tote bag, and a Facebook fan page. The Facebook page featured all of the components of the campaign as well as provided a space to post-related information and build a community around the issue (seehttp://bit.ly/aoYMS6). Although the video was uploaded to the video host site YouTube, Facebook (as well as WordPress) enables the embedding of YouTube videos into Facebook pages, demonstrating interoperability within the social media ecology. Facebook fan page status updates were syndicated or cross-posted to one of the student's Twitter accounts in order to reach new networks through savvy social marketing. Cross-posting became a favorite way for students to integrate the various spaces and components of their campaigns.

The ability to link various social media spaces through their interoperability resulted in sophisticated use of cross-posting. One way to leverage social media is through a multiplatform approach provided by cross-posting or the syndication of messages on various "Web platforms simultaneously" (Global Voices Advocacy, 2009). The "broad dissemination approach" can help organizations and campaigns connect to their core audience while reaching new niche members in new online spaces (Global Voices Advocacy, 2009). These interoperable affordances of social media were readily exploited by students in the creation of their campaigns ("learning by doing").

Following media across different spaces and media contexts, from Facebook to Twitter, and from Twitter to YouTube, involves "transmedia navigation" or "the ability to follow the flow of stories and information across multiple modalities" (Jenkins, 2008, p. 4). Initially, students may not identify transmedia activity as a skill because it is part of the normal information flow of their everyday online lives. Student activities that move from classroom to blog and from Twitter to YouTube enable transmedia patterns to emerge, demonstrating social learning through social media. Transmedia and cross-posting can engage learners in the course, and also encourage the migration of interests beyond the classroom. Additionally, social media helps connect students to a larger network that now includes other students, communication programs, and practicing professionals in the field.

DISCUSSION

A central theme in this work has been exploring the opportunities of using social media to support the social learning method. Social learning holds much appeal as a theory; students learn from interactions with one another and the instructor, and the instructor programs and structures opportunities for interaction. In practice, however, there are challenges and risks that need to be managed, reflecting some of the limitations of my application of social learning using social media.

Resistance to Social Media in the Classroom

In teaching Visual Communication there were moments of online inactivity, a week might go by between student posts, with a flurry of activity occurring just before face-to-face classroom meetings and due dates. There was also outright resistance to social learning, where some students voiced that it was ineffective. At times, student sharing was not always volunteered nor unfolded naturally. I often needed to explain my methods to encourage understanding and "buy in," as well as structure ways to stimulate students who preferred individual development and achievement over collaborative learning opportunities. I often adapted student desire for autonomous learning by suggesting modular ways of approaching and executing group assignments. I also accepted that some students (two students out of eighteen) were not going to work well in groups and let them work individually.

Power Sharing

With social media comes social power. Working with social media requires flexibility, the ability to adapt, relinquish some control, and work with the unexpected. Social media can alter "classroom power dynamics" signaling "to students that they're in control" (Young, 2010, p. 9). In some cases students may overreach, abuse opportunities to steer the class, and generally distract from meeting learning requirements. Not every instructor is comfortable with sharing control in class, and perhaps not all classes would benefit from power sharing. In an augmented class, I found that face-to-face meetings provide opportunities to restore a balance of authority and control by asserting class expectations and restoring a focus on learning outcomes. Possibly blended and online formats provide different approaches to managing the challenges of power sharing.

Student Privacy

An additional risk of using social media is the potential to invade student privacy. There is growing concern over privacy and security on social networks with data handling practices on SNS increasingly called into question, as well as the practice of government and employer surveillance of

SNS (Barriger, 2009). Educators that wish to meet students on familiar ground and utilize the online ecology of social media should not colonize students' established networks. Students should be empowered to create separate accounts, opt out of contexts of online sharing, and otherwise have their privacy respected so that open education does not compromise a student's right to manage and maintain their online presence and need for anonymity.

Discrepancies in Usage

Not all digital learners are alike and a greater understanding of their heterogeneity is needed to better account for the diversity of learners. Students demonstrate various levels of skills, competencies, and literacies, and varying rates of adoption of technologies based on their abilities and backgrounds. This must be taken into account when requiring student participation and designing the use of social media applications to promote social learning. When attempting to harness participatory culture for participatory learning, it is important to realize that students will have varying comfort levels, skills, and experiences. A way to address discrepancies in technological usage and know how is by providing various access points to participate and time for students to become comfortable to enjoy the benefits of peer mentoring. I provide training demonstrations throughout the semester, and I actively encourage a supportive environment based on sharing and not on competition.

When designing courses using social media, student participation can be enlisted by encouraging interest-driven opportunities. The microblog, for example, provided flexibility and freedom for students to follow their own interests related to the class and to their own day-to-day lives. Providing a comfortable climate also encourages social interaction. By explicitly creating less formal places of play and experimentation, students enjoy more autonomy to express themselves, which facilitates more open exchange and establishes social bonds. Students who form a sense of community may then be more inclined to share and provide peer-to-peer mentoring.

Although social learning can be enhanced by social media, it does not rely on it. The turn to the social in learning theories emphasizes social contexts for learning, yet online contexts are changing and underlying practices of "cultural competencies and social skills" can be overlooked (Jenkins, 2008, p. 4). Online spaces should be integrated with the educational goals of the class and not merely a distraction in order to provide a context for meaningful interactions.

Designing Meaningful Interactions

Providing social media tools is not enough. Maximizing opportunities for social learning requires creating meaningful and "grounded interactions" (Brown & Adler, 2008, p. 2). Activities that elicit interaction and participation or "grounded interactions" are typically structured as group work, which provide time and space for small groups to work on collaborative projects. It remains a challenge to educators to design content in ways that bring learning and socializing together in meaningful ways that best utilize the particular affordances of particular social media tools and face-to-face instruction. In my experience, the classroom returned as an important component in tying online and offline spaces of the class together. Augmented learning formats that maintain linkages and mutual contingencies between online and offline work help ground meaningful interactions for social learning.

Through a reflection on actual uses in my own Communication Studies class, I have suggested ways that participatory culture and social media can be harnessed to effectively achieve social learning, and augment and expand the time and space of the classroom, with important implications for teaching other disciplines. Insights into social learning styles and contexts help recognize effective ways that social media can reach students within the flow of their everyday online lives. Social learning through social media also helps accommodate disparities among traditional, nontraditional, and other unique learners in higher education. Blogs and microblogs used together provide a social foundation that helps encourage social learning and allows students to express strengths and to share in ways that may be otherwise stifled in a traditional learning environment. Blog sites can provide a dynamic, interactive, and supportive forum for students to respond to each other's work. And Twitter enables students to exercise a broader level of freedom and social influence. The affordances of social media cannot determine student engagement and learning, but social media combined with the design of diverse and meaningful group assignments and activities can bring students at various levels of talent, ability, and interest together to participate and learn from one another. Further research into pedagogical uses of emerging forms of networked and social media can help educators adapt to a diverse student body, to new styles of learning, and to a changing and challenging learning environment.

REFERENCES

Ambinder, M. (2009). The revolution will be Twittered [Electronic version]. *The Atlantic*, June. Available at http://politics.theatlantic.com/2009/06/its_too_easy_to_call.php

Barriger, J. (2009). *Social network site privacy: A comparative analysis of six sites*. Canada: The Office of the Privacy Commissioner of Canada.

boyd, d., & Ellison, N. (2007). Social network sites: Definition, history, and scholarship. *Journal of Computer Mediated Communication, 13*(1). Available at http://jcmc.indiana.edu/vol13/issue1/boyd.ellison.html

boyd, d., Golder, S., & Lotan, G. (2010). Tweet, tweet, retweet: Conversational aspects of retweeting on Twitter, January. Paper presented at the Proceedings of HICSS-43, Kauai, HI. Available at http://danah.org/papers/TweetTweetRetweet.pdf

Brown, J. S., & Adler, R. P. (2008). Minds on fire: Open education, the long tail, and learning 2.0. [Electronic version]. *EDUCAUSE Review, 43*(1), 16–34.

Castells, M. (2009). *Communication power*. Oxford: Oxford University Press.

Chickering, A. W., & Gamson, Z. (1987). Seven principles for good practice in undergraduate education. *AAHE Bulletin, 40*(7), 3–7.

Christakis, N. A., & Fowler, J. H. (2009). *Connected: The surprising power of our social networks and how they shape our lives*. New York: Little, Brown, and Company.

Dunlap, J. C., & Lowenthal, P. R. (2009). Tweeting the night away: Using Twitter to enhance social presence. *Journal of Information Systems Education, 20*(2). Available at http://jise.org/Issues/20/V20N2P129-abs.pdf

Gillman, D. (2006). *We the media: Grassroots journalism by the people, for the people*. Sebastopol, CA: O'Reilly Media.

Giridharadas, A. (2009). Democracy 2.0 awaits an upgrade. *New York Times*, September 12. Available at http://www.nytimes.com/2009/09/12/world/americas/12iht-currents.html?scp=1&sq=Giridharadas,%20A.%20%282009%29.%20Democracy%202.0%20awaits%20an%20upgrade.&st=cse

Global Voices Advocacy. (2009). Cross-posting for advocacy: An introduction to effective social media integration. Available at http://advocacy.globalvoicesonline.org/projects/advocacy-20-guide-tools-for-digital-advocacy/cross-posting-for-advocacy/

Hargittai, E. (2002). Second digital divide: Differences in people's online skills. *First Monday, 7*(4), 1–18. Available at http://chnm.gmu.edu/digitalhistory/links/pdf/introduction/0.26c.pdf

Hargittai, E. (2008). The role of expertise in navigating links of influence. In: J. Turow & L. Tsui (Eds), *The hyperlinked society: Questioning connections in the digital age* (pp. 85–103). Ann Arbor, MI: The University of Michigan Press and The University of Michigan Library.

Hickins, M. (2009). A new view of government 2.0. *Information Week*, August 19. Available at http://bit.ly/12IPHL

Ito, M., Baumer, S., Bittanti, M., boyd, d., Cody, R., Herr-Stephenson, B., Horst, H., et al. (2010). *Hanging out, messing around and geeking out: Kids living and learning with new media.* Cambridge, MA: MIT Press.

Ito, M., Horst, H., Bittani, M., boyd, d., Herr-Stephenson, B., Lange, P. G., Pascoe, C. J., & Robinson, L. (2008). *Living and learning with new media: Summary of findings from the digital youth project.* [White Paper]. Available at http://digitalyouth.ischool. berkeley.edu/files/report/digitalyouth-WhitePaper.pdf

Jenkins, H. (2008). *Convergence culture: Where old and new media collide.* New York: NYU Press.

Jones, S., & Fox, S. (2009). Generations online in 2009. Pew Internet and American Life Project. Available at http://www.pewinternet.org/Reports/2009/Generations-Online-in-2009.aspx

Lenhart, A., & Madden, M. (2005). Teen content creators and consumers. Pew Internet and American Life Project. Available at http://bit.ly/bAZWYI

Miller, P. (2001). Learning styles: The multimedia of the mind. *Educational Resources Information Center ED, 451,* 140.

Naaman, M., Boase, J., & Lai, C-H. (2010). Is it really about me? Message content in social awareness streams. Paper presented at the Proceedings of CSCW-2010. Available at http://infolab.stanford.edu/~mor/research/naamanCSCW10.pdf

Neo, T. K., & Neo, M. (2004). Classroom innovation: Engaging students in interactive multimedia learning. *Campus Wide Information Systems, 21*(3), 118–124. Available at http://www.emeraldinsight.com/10.1108/10650740410544018

Palfrey, J., & Gasser, U. (2008). *Born digital: Understanding the first generation of digital natives.* New York: Basic Books.

Perez, S. (2009). Study reveals high levels of Twitter use at conferences. Read/Write Web, July 27. Available at http://bit.ly/4miqX5

Rainie, L. (2010). How social users of social media have changed the ecology of information, February. Paper presented at the biennial meeting of the VALA Libraries, Melbourne, Australia. Available at http://bit.ly/alKWdm

Rogers, E. M. (2003). *Diffusion of innovations* (5th ed.). New York: Free Press.

Seitz, A. R., & Shams, L. (2008). Benefits of multisensory learning. *Trends in Cognitive Science, 12*(11), 411–417. Available at http://bit.ly/9Ppvh5

Shirky, C. (2008). *Here comes everybody: The power of organizing without organizations.* New York: Penguin Press.

Siemens, G., & Tittenberger, P. (2009). *Handbook of emerging technologies for learning* (Available at http://techcommittee.wikis.msad52.org/file/view/HETL.pdf). Manitoba: University of Manitoba.

Skog, D. (2005). Social interaction in virtual communities: The significance of technology. *International Journal of Web Based Communities, 1*(4), 464–474.

Tierney, W. G., & Hentschke, G. C. (2007). *New players, different game: Understanding the rise of for-profit colleges and universities.* Baltimore: The Johns Hopkins University Press.

Vygotsky, L. S. (1978). *Mind and society: The development of higher mental processes.* Cambridge, MA: Harvard University Press.

Wellman, B., Quan-Haase, A., Boase, J., Chen, W., Hampton, K., Diaz, I., & Miyata, K. (2003). The social affordances of the Internet for networked individualism. *Journal of Computer-Mediated Communication, 8*(3). Available at http://jcmc.indiana.edu/vol8/issue3/wellman.html

Young, J. R. (2008). Why professors ought to teach blogging and podcasting. [Electronic version]. *Chronicle of Higher Education*, *30*(25), 111–114. Available at http://chronicle.com/article/Why-Professors-Ought-to-Teach/9287/

Young, J. R. (2010). Teaching with Twitter: Not for the faint of heart. *Chronicle of Higher Education*, March. Available at http://chronicle.com/article/Teaching-With-Twitter-Not-for/49230/?sid = at

TEACHING SOCIAL MEDIA SKILLS TO JOURNALISM STUDENTS

Geoffrey Roth

ABSTRACT

Social media is rapidly become deeply ingrained as part of the journalistic process, and in some cases is replacing traditional journalism as a means for distributing news. To effectively teach journalism at the university level, we must incorporate social media as both a learning tool and a subject for examination in our classes. This chapter looks at three areas that should be incorporated in teaching journalism. The first is media literacy and social media. The chapter examines the tools and critical thinking needed to distinguish reliable from unreliable information before it is passed on to a news audience. The second is the use of social media as a tool for gathering information. The chapter looks at how social media can be used to make and maintain contacts, dig for unique and impactful stories, and use your social media contacts to improve and enhance your reporting. The third is how to effectively use social media to distribute information, and the pitfalls that can occur when your personal use of social media conflicts with your professional life as a journalist. Each section of the chapter ends with exercises teachers can use with students to hone their social media skills.

Teaching Arts and Science with the New Social Media
Cutting-edge Technologies in Higher Education, Volume 3, 127–140
Copyright © 2011 by Emerald Group Publishing Limited
All rights of reproduction in any form reserved
ISSN: 2044-9968/doi:10.1108/S2044-9968(2011)0000003010

SOCIAL MEDIA AND MEDIA LITERACY

A student of mine who is one of my Facebook friends and an active poster recently put up a Facebook posting that said ABC News was going to rebrand its *World News with Diane Sawyer* as *America's News with Diane Sawyer* and quoted ABC executives as saying it was to reflect the fact ABC News was going to shut down most of its overseas bureaus and concentrate on covering news in the United States. The post generated a lot of response. However, the date of the post was April 1. When I commented on the post saying that I thought this was an April Fool's joke, the student replied that he had seen the report on the popular and usually reliable news gossip site NewsBlues.com. I informed him that the author of the site was known for posting a fake April Fool's story every year. The website also posted a story that day saying ABC News had purchased several new news robots that could take the place of photojournalists. Embarrassed that he had fallen for a joke and passed the information on as legitimate news, he removed the post from Facebook. I used this story as an example to my students of both the pitfalls of using social media to distribute news and the pitfalls of information in this new age of fast, cheap, unedited, and unregulated information distribution. As I consoled the student, I told him that news organizations with far more experience than him have fallen victim to the same dissemination of false information.

On June 26, 2009, in the midst of the media hype over the death of Michael Jackson, reports began to surface, were tweeted around the world, and were reported by news organizations, that the actor Jeff Goldblum had also died, killed in a fall while shooting a movie in New Zealand. The problem was the reports were false, and were traced back to a website that intentionally puts out false stories about actors' deaths. In yet another, and justified, send-up of the foibles of the news business these days, on the next day, Stephen Colbert brought Jeff Goldblum onto his show to give his own eulogy (Colbert, 2009).

There has been much discussion in the professional journalism world about how to protect the basic tenets of truth and trust. From panel discussions to newly written codes of ethics, three concepts are constantly brought up; reliability, transparency, and information distribution. We shall look at all three of these areas, and what teachers need to know and teach students so they can become effective journalists in the 21st century.

RELIABILITY

There are two principal aspects to reliability that need to be communicated to journalism students. The first is the reliability and verification of information that is disseminated, and the second is the reliability of the sources of social media information. These two aspects can work separately and jointly to ensure that information obtained through social media is accurate before being reported.

RELIABILITY OF INFORMATION

Early in 2010, the Radio Television Digital News Association's (RTDNA) Ethics Committee along with the Poynter Institute developed and published Social Media and Blogging Guidelines (RTDNA, 2010). The second item in the list of guidelines states, "Information gleaned online should be confirmed just as you must confirm scanner traffic or phone tips before reporting them." Students need to be taught that no more reliability should be placed on unsourced or unchecked information obtained via social media outlets than is placed on other types of information that needs to be vetted before it is published. The reference in the RTDNA guidelines to scanner traffic offers up a perfect example of what can happen when information that is unchecked is disseminated. This example can easily be found online and used as a teaching example in class.

On September 11, 2009, the eighth anniversary of the 9/11 attacks, CNN reported breaking news that the Coast Guard had fired shots at a suspicious boat in the Potomac River, playing radio "scanner" transmissions of the incident. Local station WJLA put up a helicopter and began broadcasting live video of the boats in the water. Fox News and the Reuters news service started reporting the story as well (Kurtz & Duggan, 2009). The problem: it was a training exercise, and these news organizations went with the information before it could be confirmed. While there was an intense debate afterwards if the Coast Guard should have given media organizations the heads up about the exercise, the bottom line is that these local and national news organizations reported unreliable and unconfirmed information. The moral of these stories is that students (and journalists) need to be made aware that information derived from social media must be treated the same, or even more cautiously, than any other source of information. The end result of not substantiating information, be it from traditional sources or from Twitter, is the same.

RELIABILITY OF SOURCES

Whether it is information from sources gathered for breaking news or information from sources that constitute an entire news operation, the issue of trust is at the forefront. Do the news consumers trust your information? Do you trust the sources of the information you are gathering?

In February, 2010, The Paley Center for Media sponsored a panel on Social Media and Journalism that featured two panelists heavily involved in the use of social media and user-generated content. Rachel Sterne is the founder and CEO of the news website GroundReport.com. Robert Mackey is the editor and lead blogger for The Lede on the *New York Times* website, nytimes.com. Much of their discussion focused on the coverage of two major news stories that, in the early stages of each event, were driven by social media, the terrorist attack on Mumbai and the earthquake in Haiti (Mackey, Robins, & Sterne, 2010).

In both cases, a combination of few professional news resources that were in place and the general confusion created by either a manmade or natural disaster, led to early information about the events trickling out through sources such as Twitter and YouTube. The challenge in both those cases was to figure out how reliable the sources of that information were before publishing the information.

The first thing to teach students is to determine *who* is sending out the information.

In both the aforementioned stories, the journalists knew some of the people who were sending out tweets, either because they were professional colleagues or because they had received reliable information from them in the past. Past performance is a great barometer for measuring reliability. GroundReport.com has, over time, built up a group of citizen contributors who have gained the trust of both the editorial staff and the readers of the website.

If you do not know the sources of information there are ways of determining how reliable the information may be.

- Geographic confirmation – as you search for information through various means such as Twitter or YouTube (more on that in the next section) there are methods for determining if the information is actually being generated from the area where the news event is taking place. You can use IP addresses and geo-tracking from sites such as Twitter and YouTube to see where the content originated.

- Consistency of information – if you are receiving information from multiple sources and the details of the information from those sources are consistent with each other, you can start to believe that the information is reliable. As an example, last year a moderate earthquake struck the San Jose area shortly before noon PT. My newsroom in Fresno, CA began monitoring Twitter for tweets from the San Jose area. As more and more tweets began popping up describing similar observations, we felt comfortable enough to begin relaying that information to our viewers in our noon newscast.
- Comparisons of social media information to information flowing from traditional news sources – if the information you are receiving from social media sources is consistent with information you are seeing from other sources, you can have a higher confidence level of the accuracy of the information. Going back to the story I told earlier in this chapter about the student who passed on false information about ABC renaming its newscast, if the student had checked around to other sites that report on news industry news, he might have begun to become suspicious when he saw that no one else was reporting what would have been a major story if the information had been true.

Summary

While social media can be a powerful tool for gathering and disseminating information, the inherent immediacy and the common use of anonymity can make it an unreliable source for accurate information. Journalism students must be taught the methods for verifying and trusting information gathered through social media before that information is incorporated into the news product that is disseminated to your audience.

An Exercise

Over the course of a week, have students monitor breaking news stories on either a news website or a local or national TV news channel. Have them log information that is reported based on information from social media or so-called citizen journalists. Look at that information in class and discuss how reliable the information was, and/or if there was any clear examples of bad information that was passed along by the news organizations.

SOCIAL MEDIA AS A RESEARCH TOOL

The flip side of using social media as a way of putting out information, is using it to gather information. Sandy Malcolm (2010), the former Executive Producer for Video at CNN.com, talked about using social media at a recent seminar on social media and journalism held at the Broadcast Education Association convention.

The first thing she told the group was "The Twitters, the Facebooks, these platforms are more of a platform to help you in terms of leads, but then you have to take it from there." While she was focusing on the discussion about the reliability of social media information, she did so in the context that social media can function as a powerful source for finding out what is going on out there. Just like traditional sources such as contacts and press releases, social media can be a source for story ideas.

A REAL-TIME SEARCH FOR INFORMATION

The example I used earlier of the earthquake in San Jose is a perfect example of gathering real-time information in a breaking news situation. How did our news organization do it?

There are several Twitter search engines available such as Tweekdeck, Monitter, and Twitter Search that can do real-time searches of keywords, in which you can also narrow down search results from specific geographic areas.

Using Monitter, which gives you the power to search for Tweets within a specified area, we did a search for "San Jose earthquake" and within seconds the tweets started flooding in and within minutes we had enough information to supplement our coverage with real-time descriptions of what was happening. Call it the 21st century version of calling people on the phone to get eyewitness accounts, except this is faster and you can get information from more sources in a shorter amount of time.

TAPPING IN TO THE TWITTOSPHERE

Similarly, constantly monitoring social media sites can lead to great, exclusive stories.

At my same station in Fresno, CA, we routinely monitored social media using keywords that related to our viewing area to see what turned up. One day the search term "Clovis," which is a suburb of Fresno, turned up this tweet:

Preparing presentation to CUSD for at-home online learning program.

Using the Twitter profile of the person who sent this out, we were able to get his email address and sent him a message asking him for more information about the program. His initial reaction was shock, and he wanted to know how we had found out about this project he was working on, which had not been made public yet. (The privacy issues that people do not understand about social media is, in itself, a subject for another entire book). After we informed him that he had announced it to the world with his tweet, he checked with his superiors and eventually gave us a great, exclusive story about plans by the Clovis Unified School District to offer an online, at-home, education program for high school students who were at-risk and having difficulties learning in a traditional school environment.

Beyond the Twitter search engines are search aggregators such as whostalkin.com that allows you to mine dozens of social media and social video sites with keyword searches. There are so many out there that there is even an aggregator of social media monitoring tools: http://wiki. kenburbary.com/.

TWITTER AS A CONTACTS LIST

When you wanted to find out what an expert or a news source thought about something, you picked up the phone and called. When email came into its own, you sent an email to someone. Now, you can constantly monitor what people are thinking and doing by following them on Twitter. Many reporters will tell you that they follow dozens, if not hundreds, of people who are related to their beats to see what they are saying. Sports reporters follow sports figures, education reporters follow school board members, and on and on. The practice is becoming so common that athletic departments at universities are trying to play catch up to determine if they should place limits on what their student athletes can say on social media sites.

Social media can also be a powerful tool for tracking people down. Most people will include some type of contact information, be it an email address or phone number, on their social media site profiles. How were we able to track down the man who was building an online school for the Clovis school district? He included his email address on his Twitter profile.

WHERE DID YOU GET THAT PICTURE?

The reason I first joined Facebook and MySpace, back when those sites were mainly a domain for the younger crowd, I did it for one specific reason. Almost always, whenever we were doing a story about someone under 30, you could find a picture and information about that person on those social media sites. There continues to be ongoing discussions about the ethics of using material off these sites in news stories, but most news people would tend to agree that if the material is available to the general public, it is fair game.

WHO NEEDS A SATELLITE TRUCK?

When I was planning a class discussion on how media outside the United States viewed the concept of journalistic objectivity, I lucked out, in that a journalist for an Israeli newspaper was planning on being on campus that day and was available to be in my class. But, I also wanted a second viewpoint. I had a colleague who used to be a U.S. correspondent but was now working for a Russian cable news operation called Russia Today. He was based in Moscow. How was I able to have him participate in the class discussion? Through a very powerful online tool – Skype. By hooking up a webcam to the classroom computer and using the room's projection system, a simple Skype call brought my colleague in to the classroom to participate in our discussion.

Skype can be a great tool for doing face-to-face interviewing, and the conversations can even be recorded to include in video stories, either online or on the air. Even Oprah uses Skype to interview guests on her syndicated talk show.

CROWDSOURCING

One of the biggest buzz words to come out of the merging of journalism and social media is *crowdsourcing,* and it is a skill every journalism student should know and use. Put in its simplest terms, it is using your vast connection to people through social media to gather information and ideas or to have people review or contribute to your work.

People use crowdsourcing every day and probably do not even realize it. Not too long ago, a friend put out a simple question to her friends on Facebook, asking them if any of them had used a particular video camera and what they thought about it. She had several hundred Facebook friends, and she got dozens of answers that helped her decide if she was going to buy that particular camera. She may not have realized it, but she utilized crowdsourcing to get information. Several reporters I know constantly use crowdsourcing to come up with story ideas, with even a very honest "Hey, it's a slow news day. Anyone know of anything going on today?"

Summary

Social media can be an incredibly powerful tool for gathering information in the journalistic process. It can be used to monitor breaking news, to find exclusive stories, to track down and maintain relationships with sources, and to generate information from a large group of people.

An Exercise

Assign students to come up with a story idea involving the campus using crowdsourcing through their current social networks. Have them use social networking tools to find information and sources to interview as the basis for writing a story on the story idea they found. In class, review the stories and discuss how the idea and the information was gathered via their social networks.

SOCIAL MEDIA AND NEWS DISSEMINATION

The final part of the social media mix in the world of journalism is the output of information. So far we have focused on gathering information. Now it is time to talk about using it to reach our audiences. There are several sets and subsets of social media information dissemination. Some of its uses are direct, such as blogs and tweets. Some are a subset of a larger media, such as Twitter feeds embedded on a website. And some are a marketing tool to draw consumers to your product.

ANYONE CAN BE A NEWSPAPER OR A TELEVISION STATION THESE DAYS

Think about it. What caused the concentration of media into the hands of a relatively few number of people and corporations? Much of it was the cost of distribution. It did not cost much for one person to sit down and gather the information for a great story, but it cost millions of dollars to put that story into a newspaper that required printing presses, news print, ink, and a vast distribution system to get that story to readers. The same holds true for television news. A basic set of TV gear could cost nearly $100 thousand, and you needed millions of dollars to run a television station.

Fast forward to 2010. The students in my Journalism Tools course take $120 Flipcams use a $600 laptop computer and Windows Movie Maker software, and post stories that they email in to our class website set up on a free blogging program. Go to www.journalism10.posterous.com. Anyone across the world with internet access can see the stories written and produced by my students. All for under $1,000.

Sites such as YouTube turn can turn anyone into a television production company with instant worldwide distribution. Sites such as Justin.tv or UStream.tv, or even Skype, can allow anyone with a minimal amount of inexpensive equipment to broadcast live around the world.

While anyone can, and many try, to become their own media empire, the fact of the matter is that there are still a relatively few number of sources that people turn to for information. It is hard to keep an exact tract of the number of blogs out there, but the well respected website Technorati, which tracks and lists blogs, estimates that there are somewhere in the neighborhood of 113 million blogs, with nearly 175 thousand being created every day.

Despite that, one a few hundred have what anyone would consider a sizable enough number of readers to be considered a mass medium. Trust, respectability, and fame contribute to a large number of people turning to sites such as *The Huffington Post* or *The Drudge Report* , while only a couple of dozen people are reading my sister-in-law's blog (http://www. iwannatalkaboutbrooke.blogspot.com). Depressingly, that may be more than the number of people who read my blog (www.nomoredeadlines.com).

So, when it come to teaching journalism students about social media and news dissemination, the approach has to be twofold, looking at how both small scale and mass media operations use social media.

TRANSPARENCY

The RTDNA guidelines for social media states, "If you cannot independently confirm critical information, reveal your sources; tell the public how you know what you know and what you cannot confirm."

Guidelines that held true for older sources of information should hold true for the new ones. The Associated Press (AP) provides information to news organizations around the world and often news organizations will use this information without specifically attributing it to the AP. However, in some situations, such as the case of breaking news, it is often common practice to attribute information to the AP that cannot be independently confirmed. How many times have you heard a newscaster say "According to the Associated Press…" as a qualifier before giving out information.

In the case I referred to earlier, when my TV station reported what people were saying on Twitter about the earthquake in San Jose, we made it crystal clear on the air that Twitter was the source of this information.

If you use information from a blog or other social media site, you should always make clear who the information is coming from, and if the source of the information has a vested interest that your readers or viewers should know about.

And, if you make a mistake, correct it and make sure your audience knows about it. The RTDNA guidelines gives a great example of this:

For Discussion in your Newsroom: When an Army psychiatrist killed 13 people at Fort Hood, Twitter messages, supposedly from "inside the post" reported gunfire continued for a half hour and that there were multiple shooters. Journalists passed along the information naming Twitter writers as the sources. The information proved to be false and needed to be corrected. If one or multiple shooters had been at large, withholding that information could have caused some people to be in harm's way. The nature of live, breaking news frequently leads to reports of rumor, hearsay and other inaccurate information. Journalists must source information, correct mistakes quickly and prominently and remind the public that the information is fluid and could be unreliable.

Questions for the Newsroom: What protocols does your newsroom have to correct mistakes on social media sites such as Twitter and Facebook? -Does your newsroom have a process for copyediting and oversight of the content posted on social media sites? What decision-making process do you go through before you post? -What protocols do you have for checking the truthfulness of photographs or video that you find on Facebook, YouTube or photo-sharing sites? Have you contacted the photographer? Can you see the unedited video or raw photograph file? Does the image or video make sense when compared to the facts of the story? -Who in the newsroom is charged with confirming information gleaned from social media sites?

SPREADING THE WORD THROUGH SOCIAL MEDIA

Almost every major news outlet these days has a Twitter account and a Facebook page. Many of the reporters and news managers at these organizations also have their own accounts affiliated with their news organizations. Some have even developed apps for the iPhone. All are meant as new ways to reach out to their audiences. These channels of communication are meant to both immediately inform their audiences and to, in the hopes of the news organizations, draw their audience to their main channels of communication. A court reporter may tweet updates from a trial throughout the day to keep viewers informed, but also hope the interest generated by doing that will cause the viewers to want to watch the complete report on the trial later on the station's newscasts. A quick update on a story posted to Facebook will contain a link to a fuller version of the story on the news organization's website. An email alert can inform an audience of breaking news that will make them want to tune in or go to a website to get more information.

HOW PERSONAL IS TOO PERSONAL?

While social media is used as a means of distributing information in the mass media, it is also used to distribute information, and opinion, on a much smaller scale. People do have their personal blogs. They tweet information. They have Facebook pages. But, students must also be made aware that as part of a journalistic organization, you could corrupt and damage the public's perception of that organization by what you say and do personally in the social media world.

Again, we turn to the RTDNA guidelines for some food for thought:

> Remember that what's posted online is open to the public (even if you consider it to be private). Personal and professional lives merge online. Newsroom employees should recognize that even though their comments may seem to be in their "private space," their words become direct extensions of their news organizations. Search engines and social mapping sites can locate their posts and link the writers' names to their employers.

- There are journalistic reasons to connect with people online, even if you cover them, but consider whom you "friend" on sites like Facebook or "follow" on Twitter. You may believe that online "friends" are different from other friends in your life, but the public may not always see it that way. For example, be prepared to publicly explain why you show up as a "friend" on a politician's website. Inspect your "friends" list regularly to look for conflicts with those who become newsmakers.
- Be especially careful when registering for social network sites. Pay attention to how the public may interpret Facebook information that describes your relationship status, age, sexual preference and political or

religious views. These descriptors can hold loaded meanings and affect viewer perception.

- Avoid posting photos or any other content on any website, blog, social network or video/photo sharing website that might embarrass you or undermine your journalistic credibility. Keep this in mind, even if you are posting on what you believe to be a "private" or password-protected site. Consider this when allowing others to take pictures of you at social gatherings. When you work for a journalism organization, you represent that organization on and off the clock. The same standards apply for journalists who work on air or off air.
- Keep in mind that when you join an online group, the public may perceive that you support that group. Be prepared to justify your membership.

These guidelines are especially relevant for students who often do not think about the implications of what they post online or how easy it is for anyone to get this information.

In my classes, I do two things to point out the pitfalls. First, I relate to them stories about otherwise highly capable people I wound up not being able to hire because of photos and other material posted on their personal websites or Facebook that made my immediate supervisor nervous. Second, I conduct an in-class experiment. I log on to Facebook, and randomly pick two or three students in the class and type in their names in the search box. Almost always, I find their Facebook page, which usually does not have very much privacy protection, and start showing the class all of the information and pictures of the students, and of the students' Facebook friends, that I now have access to thanks to their lack of privacy protection. Inevitably, the next day when I go back to those students' Facebook pages, I find that they have now enabled most of the privacy protection functions.

Summary

Social Media can both help gather and distribute information. Reliability, transparency, and marketing can all help to make social media an effective way for news organizations to disseminate information.

An Exercise

Have each student create their own blog using WordPress of Blogspot. Have them populate it with some class projects and have them use their current social media outlets to market their websites. Using a tracking site such as

Google Analytics, have each student see how many people they can get to read their blogs over the course of a semester.

CONCLUSION

To be successful as journalists and communicators in the future, students must learn how social media can be used within the framework of journalistic endeavors. Students probably are already well versed in the functions of social media, and use these outlets for personal communication. The job of the journalism professor is to teach them how to harness that power.

REFERENCES

Colbert, S. (2009, June 29). *Jeff Goldblum will be missed.* Available at http://www.colbertnation.com/the-colbert-report-videos/220019/june-29-2009/jeff-goldblum-will-be-missed

Kurtz, H., & Duggan, P. (2009). CNN jumps the gun on Coast Guard story. *Washington Post*, September 12, A.7.

Mackey, R., Robins, M., & Sterne, R. (2010, February 12). *Social media and journalism.* Panel in: *A way forward: Solving the challenges of the news frontier.* New York: The Paley Center for Media.

Malcolm, S. (2010, April 15). Making the relationship work: Integrating social networking into daily news coverage. *Educating for Tomorrow's Media* conference. Las Vegas, Broadcast Education Association.

Radio Television Digital News Association (RTDNA). (2010). Social media and blogging guidelines. Available at http://www.rtdna.org/pages/media_items/social-media-and-blogging-guidelines1915.php [no longer accessible].

UNLOCKING THE VOICE OF THE UNIVERSITY: THE CONVERGENCE OF COURSE, CONTENT, DELIVERY, AND MARKETING THROUGH SOCIAL MEDIA

Ryan Busch

ABSTRACT

Social media ideology represents a missed opportunity of vital importance to colleges and universities. The core tenants of this ideology include wider and freer access to information through the use of emerging technologies. Colleges and universities should consider implementing social media ideology to improve efficiencies in the delivery of learning and organizational operations. As example, the chapter highlights two innovative companies founded on innovations representing a doctrine of convergence – socializing course, content, delivery, and marketing into a broader format, which not only educates the student, but also expresses the unique qualities of the organization itself. Examples include Tech University of America, eduFire, and an experimental course model developed as the result of an introduction of the leaders of these two organizations.

Teaching Arts and Science with the New Social Media
Cutting-edge Technologies in Higher Education, Volume 3, 141–166
Copyright © 2011 by Emerald Group Publishing Limited
All rights of reproduction in any form reserved
ISSN: 2044-9968/doi:10.1108/S2044-9968(2011)0000003011

This chapter, at its heart, is about innovation. It is an exploration of how concepts that take shape for one purpose can inspire wholly new uses; a convergence. Social media has become a facet of everyday life, but its use in education is still in its infancy. This chapter reveals how higher education institutions have found social media both inspiring and daunting. The reveal begins with an explanation of how to view social media – not as a set of tools, but as a developing philosophy. The exploration continues by considering how the general structures of higher education organizations both limit the embrace of social media and can find greater benefit from choosing to allow this embrace. To highlight this disconnect and to show the potential benefits to organizations, the chapter details the innovations of two educational organizations [eduFire and Tech University of America (TUA)]. The exploration continues by explaining a dynamic collaboration between Piccolo International University (PIU) and eduFire to create a new type of educational experience. From these examples, the exploration concludes by offering suggestions for prospective innovators who might seek to free the unique qualities of a higher education organization. The chapter explores how innovation results from the convergence of course, content, delivery, and marketing through social media can unlock the voice of the university.

SOCIAL MEDIA IDEOLOGY

Social media is not Facebook. Nor is it YouTube. Neither is it blogging or tweeting. Facebook, YouTube, blogs, and Twitter are merely platforms and channels; players on a much broader stage collectively referred to as social media. This distinction is important. Social media thought-leader, Brian Solis, suggests avoiding such a narrow focus on individual players and viewing the entire landscape itself as platform – such that: "Social media is, at its most basic sense, a shift in how people discover, read, and share news and information and content. It's a fusion of sociology and technology, transforming monologue (one-to-many) into dialog (many-to-many)" (Solis, 2007, para. 5). Thus, social media as an ideology focuses on reducing communication barriers, constraints on ideas, and hindrances to the transfer and exchange of knowledge. Education in general, and higher education in specific, mirrors this ideology. This connection is not new. John Dewey expressed the link between social constructs, communication, and education:

> Not only is social life identical with communication, but all communication (and hence all genuine social life) is educative. To be a recipient of a communication is to have an enlarged and changed experience... All communication is like art. It may fairly be said,

therefore, that any social arrangement that remains vitally social, or vitally shared, is educative to those who participate in it ... not only does social life demand teaching and learning for its own permanence, but the very process of living together educates ... A man really living alone (alone mentally as well as physically) would have little or no occasion to reflect upon his past experience to extract its net meaning. (Dewey, 1922)

But the advent of empowering technologies (like the Internet) and associated ideologies (like social media) have been little exploited to create improved communications exchanges between social and educational structures. In fact, there is a deep disconnect between the ways that educators, marketers, and administrators at colleges and universities currently utilize social media in communicating the voice of the institution.

Rather than creating connections between people and information (of all types), higher education has continued to segregate information and communications into neat categories. While some information is educational and some information is promotional; information is rarely communicated for both educational and promotional use. Preemptively evaluating information asserts the values of the people or groups who categorize and reduces the ability of information consumers to evaluate independently the usefulness of that information. Thus, if the categorizers presume that some information is educational and other information is promotional, then the categorizers approach communicating these different information parcels by channeling them according to category. This process of channeling produces two voices for a university: one voice that of educational quality and institutional mission; the other, the voice of marketing rhetoric. Social media ideology promotes information to flow freely, information consumers (not information producers) to determine the value to information, and creates dialog of free communication exchange to benefit all parties. But for now, colleges and universities have categorized social media itself as primarily a marketing platform.

Rachel Rueben relates an example of how Ohio State University (OSU) stopped using the Facebook *wall* because it was overwhelming for a single marketing manager to deal with more than 6,900 fans and a school population in excess of 50,000 students (2008, pp. 9–10). Feeding a demand for fresh content to a group of this size was too large a challenge for one person to manage. Rather than attempting to supply a vast audience with a single resource, this decision begs the question of why did OSU's leaders not think to integrate support from faculty members into the program. Surely the students would have preferred to communicate with faculty members rather than with an overwhelmed marketing staffer. A 2005 OSU report shows 1,017 Associate Professors as part of the faculty (Craft & Phillips, n.d., p. 5).

Adding the voices of more than 1,000 faculty members to OSU's Facebook effort would have improved OSU's ability to manage the communication effort. Furthermore, those interested in OSU would have had access to a greater variety of information, not just information categorized as marketing rhetoric. The results of this sort of engagement would, no doubt, have looked different from one executed for only marketing purposes – but there would still have been a definite marketing benefit from the effort.

Ruben's notes about OSU are made more fascinating by the fact that this story is conveyed as part of a best practices section in her writing. In truth, the social media program at OSU does represent a best practice example of a university's use of social media. As such – repeating this example is not meant to disparage the efforts of OSU's leaders, but rather, to indicate that this limited application highlights opportunities for greater innovation in the use of social media by members of the higher education community.

CATEGORIZATION OF INFORMATION

Higher education systems might generally be broken into four areas. These areas are course, content, delivery, and marketing. In essence, these four areas represent the general categorization of information by higher education institutions.

A potential student can encounter each category directly and further identify each as a distinct component of an institution's voice. But each category exists with some level of separation from its counterparts.

Course

Courses represent a basic framework for the transfer of knowledge between institutions and learners. Courses are composed of information (or content); degrees are composed of courses, and so forth. Course offerings might be generalized and divided into two categories. To differentiate, call them *commodity courses* and *distinct courses*.

Commodity courses represent required, foundational offerings that vary little from one degree to another. Furthermore, the content of commodity courses varies little from one institution to another. The lack of variation means that commodity courses are typically simple to transfer between institutions. The ubiquitous English 101 course represents a commodity course. Essentially, commodity courses are courses of mass consumption. They are fungible, unremarkable, and designed for the largest learning audience.

Distinct courses appeal to a more targeted audience. Whereas English 101 courses roll off assembly lines (and offer little distinction for an institution), the Massachusetts Institute of Technology (MIT) website (n.d.) lists a course, entitled *Out of Context: A Course on Computer Systems That Adapt To, and Learn From, Context* (Media Arts and Sciences, para. 29), which represents a hand-crafted, artisanal offering. Distinct courses can trigger remarkable reactions from prospective students because they can be found nowhere else but at a specific institution. Thus, distinct courses differentiate one institution from the next and form unique value propositions for institutions.

Content

Content is the elemental information from which courses are composed. Although instructors do create course-specific content (such as syllabi, handouts, and other types of supplements), content is most commonly derived from textbooks. As such, textbooks form the basis of courses. Most textbooks in use today come from major publishing companies, are heavily marketed, and are available at considerable cost to consumers. To further clarify, the information used to create courses is accessible to consumers only for a fee. The fee serves as a natural barrier between the information and the information consumer. Because the information is locked behind a price tag and courses are created from this content, the voice of the university is also restricted.

The MIT course noted above is part of an intriguing attempt to free these constraints. The "MIT OpenCourseWare (OCW) is a web-based publication of virtually all MIT course content. OCW is open and available to the world and is a permanent MIT activity" (Unlocking Knowledge, Empowering Minds, para. 1). Although meritorious, the OCW falls short of a true innovation in a couple of important ways.

Content showcased in the OCW is only semi-sharable. MIT's website offers these details about sharing: "If you download, copy, modify, reuse, remix, or redistribute 'all rights reserved' content without permission of the copyright owner, and if your use is found not to be 'fair use,' then you may be liable for copyright infringement" (FAQ: Fair Use, para. 6).

Even if there were no barriers to sharing OCW content, the content repository is just that; a catalog. Interactive learning engagements with MIT faculty members are absent from the OCW program: "solutions to homework assignments, quizzes, and exams are only discussed and presented in the classroom, and not made available in print or electronic format" (FAQ: Getting Started, para. 5). If course and content alone were enough to promote a true learning process, then libraries and correspondence programs would alone suffice. This indicates that there is greater value to be had through instructor involvement.

Delivery

Instructors deliver commentary and guidance on content. Instructors can deliver live lectures in-person, informally educate during office hours, or asynchronously online through learning management systems (LMS). Delivery in any format represents a key link between content and learning. Thus, the category of delivery includes two essential components: the instructor and the format.

The Instructor
Instructors represent unique voices for the university. Teachers inform the content in dialog with learners. Thus each teacher and course is unique. Even commodity courses are unique from this perspective alone. Thus, the voice of the school is only as special as the teachers there employed.

Although some notable instructors surface in commodity courses, most preeminent instructors are not found teaching commodity courses. Commodity courses are most often the domain of graduate teaching assistants and adjunct faculty members. Lower-level faculty members are in greater and can be scheduled with greater dexterity by administrators. This format serves the assembly line well; allowing the prized craftspeople to focus on the specialty work and the general laborers to manage the mass production elements.

Gaining access to prestigious faculty members is not easy; to study with an MIT professor, a student must first be admitted to MIT. Be the instructor's voice (or that of the institution) standard or distinct, schools contain these voices within the classroom walls. Only admitted students can learn from these voices and the voices do not extend beyond the campus.

The Format
The voice of a school can be contained geographically as with traditional, brick-and-mortar classrooms. These limitations are a distinct barrier to information. Technology provides a means to surmount this barrier.

Engaging an online format of some sort is one manner in which universities can extend learning beyond geographical. Whether online offerings are core elements of instructional format or secondary offerings associated with other delivery formats, delivering education online is a step toward incorporating the ideology of social media. Consider the success of online formats such as University of Phoenix. "As of February 28, 2010, 458,600 students were enrolled at University of Phoenix" (Apollo Group & Inc., n.d., *Just the facts: Enrollment*, para. 1). To further support this appeal, University of Phoenix documents indicate that by March 2010, there were 538,000 graduates of its programs (Apollo Group & Inc., n.d., *Just the facts: Alumni*, para. 1). Although University of Phoenix educates such a vast and geographically dispersed population, the transfer of knowledge from instructor to student is no more observable there than inside the buildings of MIT. Although the admissions standards differ and the courses are more widely syndicated, the instructor-influenced learning process still remains cloaked behind the veil of the classroom walls (virtual or otherwise) because password-protected LMS bar free access to dynamic instructor-led learning events.

Instructors offer a unique voice to the university and online formats represent an opportunity to syndicate that voice more widely. The free expression of the unique voices of most colleges and universities (whether in a physical classroom or virtual) is often restricted from engaging in a dialog with learners. There are technological tools that make free expression possible, but are most often focused on the category of marketing.

Marketing

Connecting the distinct course, content, and delivery categories with the imaginations of prospective students is the domain of the marketing department. Marketing promotion takes many forms: brochures, recruitment events, catalogs, view books, and (more recently) digital marketing. In all promotional formats, the marketer seeks to communicate the unique value proposition of the institution to the prospective student. Digital marketing promotions have made this communications effort more effective. The preeminent form of digital promotion in higher education has been lead generation. Lead generation is one of the most effective promotional formats employed by colleges and universities to recruit and enroll new students. But unnecessary inefficiencies and costs often temper the effectiveness of lead generation.

Lead Generation

Lead generation is a process whereby an organization targets a specific audience of interest and seeks to encourage members of that audience to take a desired action. Colleges and universities use digital networks to place advertisements that collect pertinent demographic information from the prospective students. Advertising entices students to fill-out information request for institutions. Once the form is completed, the information is delivered to an institution's enrollment team. Although the dynamics of the process vary, institutions will typically attempt to contact the potential student telephonically to secure an enrollment. Lead generation is an effective means for generating a large volume of information requests, but the process of gaining an information request is done through a tightly formatted communications process.

The inefficiency starts with the failure to communicate enough informa-tion at the point of the initial enticement. Prospects are shown highly engineered information in limited amounts. Providing too much informa-tion upfront reduces the number of potential information requests. Instead, marketers rely on charismatic enrollment teams to fill-in gaps in the prospect's knowledge about an institution. These gaps in knowledge often include vital information about the school itself, the programs offered, tuition and associated costs, and the delivery format. Filling these engineered knowledge gaps is costly in terms of both time and resources. The more blanks to fill-in, the longer the telephonic process lasts, the more expensive the recruitment process.

Whether they gain an enrollment or not, institutions are obliged to pay for the opportunities to have these enrollment conversations. This obligation escalates the inefficiency. Lee Kantz, founder and CEO of edufyimarketing.com (a new lead generation company) and former CMO of the University of Illinois Global Campus (UIGC), noted that depending upon the program offered, UIGC conversion rates approximate industry standards; ranging from roughly one to five percent and averaged out between two and two and a half percent (approximating industry standards) (personal communication, April 5, 2010).

Universities employ lead generation programs, with all their inefficiencies because these programs offer marketers and school leaders an established and proven model. These industry standards allow easy benchmarking and because marketing departments rely on outside agencies to produce the prospective students – institutions can implement lead generation programs in a relatively straightforward manner.

Implementing and measuring efforts relating to the use of social media ideology is not as simple. Few benchmarks exist and the simplicity of relating costs to revenues is not as direct as with lead generation. If "Many of today's prospective students are 'secret shoppers' who are gathering information about your institution on the Web and from their acquaintances using social media like Facebook, without making themselves known to your institution" (Noel-Levitz, Inc, 2009, p. 8), then some level of social engagement is vital.

When higher education employs social media concepts merely as an element of the marketing plan, it prevents broader planning and use of these programs by strategists who fully embrace this ideology. As such, the use of social media ideology in higher education marketing programs limits broader usage in the categories of course, content, and delivery. The limits of execution found in each category have a basis in similar operational silos inherent in higher education organizations.

SILOS DO NOT PROMOTE SHARING

Intraorganizational silos are dangerous walls created within an organization in an effort to optimize the performance of specialized job functions. Complex data streams and multipronged operational units have created similar challenges for non-educational businesses. The concept of enterprise resource planning (ERP) has been used to explain both why silos disrupt the focus on the customer and how rethinking the issue from the planning stage up resolves the problem: "Used in conjunction with efficient processes and a motivated work force, ERP can help a utility change from confusing, isolated silos to a transparent, joined up business which knows all about the condition of its assets, the deployment of its employees and the needs of its customers" ("Breaking Down Operational Silos through ERP", 2004, p. 2). Silos may be viewed by higher education institutions accordingly: marketing directors are not teachers, teachers are not administrators, and administrators are not researchers. This organizational structure mirrors the information categorization of educational content is not marketing rhetoric and instructors are not spokespeople.

Certainly these artificial boundaries are malleable in specific examples. But on the whole, marketing directors employed 40 hours per week to direct marketing efforts are rarely granted leave to spend 15 hours of that time teaching marketing courses. Instead, the education factory hires a specialist craftsperson, an instructor, to do this work. Likewise, if a specialist is occupied with students or research, he or she can offer little help to a marketing director.

Specialization itself is not inherently troubling. The silo walls, however, are troubling for the simple fact that walls create barriers to open communications between specialists in the same way that categories are barriers for the free exchange of information. The specialists working on one side of the wall cannot see the work being done on the other side, nor can they see in which direction the work is headed. This micro-view of the process is antithetical to an educational model that embraces the ideology of social media.

Organizations that embrace social media ideology will seek to optimize the sharing process in nearly every level. University leaders who wish to explore the real potential of social media to optimize learning programs must start by reducing silos within the organization first. Unifying course, content, delivery, and marketing categories will be difficult if those leading the efforts must do so from within segregated structures.

ORGANIZATIONAL EFFICIENCY REDUCES COSTS AND IMPROVES ACCESS

Siloed organizations are less efficient. Siloed outputs are also less efficient. Siloed organizations with siloed outputs are more inefficient than either issue. Inefficiency increases costs and passes these costs, in the form of higher prices, on to students.

Price is a fundamental barrier to access; the higher the price, the fewer who can afford it. Although the U.S. government has established compensating measures, this is a form of artificial life support. According to The College Board (2009), "Over the decade from 1998–99 through 2008–09, grant aid per undergraduate student increased at an average of 3.4% per year after adjusting for inflation" (Economic Challenges Lead to Lower Non-tuition Revenues and Higher Prices at Colleges and Universities, para. 10). The government has limits to how much and how long it can dole out support. With aid increasing at such a rate, it seems natural to assume that colleges and universities would attempt to curtail costs. But as of late 2009, the costs of attending college have continued to rise in excess of the rate of inflation (Pope, 2009). There exist only a few possible outcomes once governmental support reaches its limit; further reduction in college access to or a more assertive government intervention into higher education processes. Costs and fees charged to students must be curtailed to avoid either scenario.

The Effect of Cutting Costs versus Improving Efficiencies

Cost-cutting measures often reduce academic quality. This most frequently asserts itself in the forms of increasing class size or reducing faculty size. These effects are two sides of the same coin. If an institution wishes to keep student-to-teacher ratios steady, then culling the faculty means reducing the number of students and thus reduces access. If the institution wishes to keep access high, then cutting the faculty means increasing the number of students in a course. Consider the University of Hawaii. In 2009, budget troubles at the university demanded an increased class size for *Introduction to Accounting* from 60 students to 120 students (KITV.com, 2009). According to KITV.com (2009), Hamid Pourjalali, a professor at the university puts this in perspective: "When you increase the class size, it has a direct effect on the quality of the delivery of the course and I can sense it, and you can sense it in students that their interest level has dropped significantly" (Budget Cuts Force Larger Class Sizes at UH, para. 6).

Academic departments schedule courses based on enrollment projections. Universities set enrollment caps as a means of planning admission figures. Thus, administrators limit how many students the university will serve at a maximum. The methods for creating these caps vary, but the existence of caps at all is antithetical to the tenants of social media's emphasis on increased access to information. This structure differs from the world of business. Imagine Toyota capping the production of cars based not on market demands but on budget constraints. If the car buying market was clamoring for more cars, then Toyota (or any other business) would evaluate methods to fulfill the market demand. As Terence Chea writes:

> colleges are capping or cutting enrollment despite a surge in applications from high school seniors, community college students and unemployed workers returning to school ... Colleges that previously accepted all qualified students are becoming selective, while selective schools are becoming more so. Most community colleges have open-access policies, but demand for classes is so intense that many students can't get the courses they need. (2010, para. 2)

Reducing delivery is a tactic employed by businesses in trouble, not businesses that are thriving. With demand increasing, supply reductions from colleges and universities indicate desperate rather than conscientious measures. Conscientious measures would seek first to optimize efficiency.

CONVERGENCE

Embracing social media ideology mandates the establishment of a community of feeds; both into and out of a system, which blend the elements of course, content, delivery, and marketing into a more transparent design. A community of feeds might be envisioned as an audio speaker. Course, content, delivery, and marketing serve as feeds into the speaker and the speaker blends these feeds into a single, resonant output. Separately, the inputs are raw and unrefined, but gathered and polished by the college or organization, the signals are optimized. The process of optimization produces greater efficiency across the system.

Square Peg, Round Hole

This community-fed design creates a new model for higher education systems. Colleges and universities have been adapting non-technological methods to fit an increasingly advanced world, rather than creating new methods based on new ideas from a changed landscape. The situation is that of the quintessential square peg, round hole. Institutions can jam, hammer, and force the peg into the hole. They can carve, plane, and whittle the edges of the square until it moves into place; but all methods that begin with a square and target a circle leave gaps. From these gaps emerge the innovations. By starting with rounder shapes and more efficient designs, innovators are using social media as the basis for the development of new methodologies capable of freeing the voice of the university.

Privacy Concerns and Message Control

Creating a community of feeds can be a scary prospect for many higher education institutions. Many organizations realize a need to control information. Colleges and universities often restrict the community of feeds concept from worry over the loss of control over the intuitional message and the privacy rights of students. But both concerns might be somewhat misplaced.

Services like Facebook offer various methods for users to control personal privacy. Ralph Gross and Alessandro Acquisti considered the concern about student privacy in 2005. They concluded in their study of over 4,000 Carnegie Mellon University students using Facebook that, while these students presented substantial amounts of personal information on their Facebook profiles, the students made personal decisions to present this information by not employing Facebook's mechanisms for restricting access (2005, p. 10, para. 1). This is not to say that privacy is not an important concern, but rather to note that even though tools are available to control privacy, privacy is the user's personal prerogative.

Personal prerogative is also at the heart of an educational institution's attempts to control its message. Control is something of an illusion as it assumes that only the institution has access to information syndication. The democratic nature of social media allows anyone to talk about anything at anytime. Childs Walker notes "College officials have to accept that once they establish a Facebook destination for students, they forfeit control of what's said in the space. Students are as apt to trade dirt on poorly managed programs or boring social scenes as they are to promote a college's virtues" (Walker, 2009). But this conversation can occur whether the college or university has its own page or not. Avoiding the concerns does not make them go away. The next section focuses on innovators who do not avoid the conversation, but embrace it as a competitive advantage. The innovators are the round pegs who employ social media ideology.

ROUND PEGS: INNOVATIONS EMPLOYING SOCIAL MEDIA IDEOLOGY

Two round pegs represent this movement: TUA and eduFire. The third, PIU, represents a proof-of-concept test that organized the use of tools into a new type of model; a *blended, social-learning model*. PIU's collaboration with eduFire provides an example of the advantages offered to a traditional asynchronous online university in embracing a new type of learning experience.[1]

Tech University of America

Steve Cooper is adamant, "A social networking site should not just enable a classroom-it should be the classroom" (Wendt, n.d., para. 3). Cooper is the founder of TUA. He conceived the idea for TUA in early 2008 and the website suggests that:

> The rationale for this paradigm shift in higher education is three-fold: rising cost of tuition, increasing demand of end-user innovation by faculty and students, and the right of everyone to access knowledge ... By teaching our courses in social networking sites we can deliver distance learning courses in the most dynamic, interactive and current medium of online communications. This also allows us to significantly reduce our cost of delivering our programs and those savings are passed on to our students. (Welcome to Tech University of America, 2009, para. 1)

The basic message is this: rather than integrating with social networks, a college should *be* the social network. Building a university directly within Facebook offers two distinct advantages: reducing delivery costs and improving market visibility.

Many universities licensing the most popular LMS can expect to pay hundreds of thousands of dollars per year (Natel, 2009), but Cooper's use of Facebook circumvents the need for licensing fees. This does not mean that adopting Facebook is a free endeavor. Costs include the efforts of developers to customize the look TUA in the Facebook environment. But similar costs, in addition to LMS licensing fees, exist for universities employing popular LMSs.

Beyond just reducing costs, TUA has the added advantage of access to more than 100 million U.S.-based Facebook users – a reflection of Facebook's growth of 144.9% between 2009 and 2010 (Corbett, 2010, para. 2). Institutions like University of Phoenix value this audience, but the university's efforts to date still fall short of complete integration. The student population at University of Phoenix is much larger than the more than 39,000 people who *like* its page on Facebook (Facebook, n.d.). This might be due, in part, to the use of Facebook as a marketing extension rather than a platform itself. Were University of Phoenix to adopt Facebook directly as a delivery platform, the results might be very different. That difference is what drives Cooper's interest. Rather than identifying students in one place and moving them to another, Cooper's innovation keeps the connection between TUA and Facebook users constant. Essentially, not all University of Phoenix students are Facebook users, but all Facebook users can be TUA students.

Delivering wider access to students and increasing marketing opportunities are not the only innovations in the TUA model. Cooper's TUA integrates the feeds of course and content into the system. Course offerings can be quickly adapted to keep pace with the speed at which new information emerges, according to the TUA website: "Our approach affords our faculty and students the opportunity to actively engage in learning by being able to change and update course materials in a wiki-like manner in order to make them more relevant" (End-User Innovation, para. 5). Given particular student and faculty interest new courses can emerge quickly and include more relevant materials.

The content used in a course can come from a creative variety of open-source solutions. Using open-source content further reduces resistance in the program and enhances access. The TUA website indicates YouTubeEDU (Open Source Content, para. 3) as an example of such open-source content offering. Moving a step further, TUA students will use offerings from textbook companies like FlatWorld Knowledge or Textbook Media which provide engaging and interactive textbook solutions at no charge.

Because Cooper looks at revenue from a different perspective TUA can continue to operate regardless of the free elements. In Cooper's concept, TUA charges for a desired result rather than the learning process; as the motto on its website suggests, TUA offers "Open access courses in social networking sites – free for everyone, only $99/month if you want to earn a degree" (Web site Banner, n.d.). The course, content, and delivery efficiencies created at TUA allow revenue to follow a *freemium* format.

Fred Wilson, a venture capitalist involved in technology-based businesses, conceptualized that freemium means: "Give your service away for free, possibly ad supported but maybe not, acquire a lot of customers very efficiently through word of mouth, referral networks, organic search marketing, etc, then offer premium priced value added services or an enhanced version of your service to your customer base" (Wilson, 2006, para. 1). This model is becoming far more ubiquitous as social media philosophies continue to spread, and Cooper sees the application to education as well. The concept of *free* has some magic to it. One of its most magical qualities is that it allows many people to investigate and explore new ideas without doing a specific cost-benefit analysis. Facebook's exceptional growth well represents Wilson's suggestion that a *freemium* company "acquire a lot of customers very efficiently." Facebook added over 400 million active users worldwide between its founding in 2004 and February 2010 (Facebook, 2010).

Cooper hopes to realize similar benefits from the *freemium* revenue model for TUA. But TUA is still at an early level of innovation. As of 2010, TUA has not graduated any students from the program. This is not atypical for a new university, especially one not much more than two years old. Students have become savvier about accreditation, thus accreditation is one of the key factors for students when selecting a school. The Distance Education and Training Council (DETC), a key player in accrediting online and distance learning programs, notes that institutions seeking accreditation must "have been enrolling students for two consecutive years under the present ownership and with the current programs" (Eligibility & Standards, para. 3). Although Cooper's TUA was conceived of in 2008, according to the TUA website it did not offer its first course until March 2009 (Tech University of America Timeline, para. 1). Although currently unaccredited, Cooper is planning for the future. Cooper's plans include gaining national accreditation from the DETC once eligible and a few years later pursuing regional accreditation from the Higher Learning Commission (HLC) (Gonzales, 2009).

Although TUA is still seeking students, still seeking accreditation, and still in its early years, the premise itself represents an innovative adoption of social media ideology. Cooper is establishing a unique blend of courses, content, and delivery elements and the marketing of the university rests squarely on the shoulders of this blend.

eduFire

Jon Bischke, like Cooper, has foisted much on his shoulders. Bischke is the founder of eduFire, a community-driven platform for delivering live, audio/visual learning events. eduFire currently hosts a community of users dispersed throughout the world in more than 200 countries with 65% of the more than 70,000 member learning community within the United States (J. Bischke, personal communication, April 27, 2010). eduFire does not define the roles of community members, any user can be either student or teacher. About 10,000 users classify themselves as teachers, around 70,000 classify themselves as students, and about 5% consider themselves both students and teachers (J. Bischke, personal communication, April 27, 2010). Rather than registering as one or the other, a user simply joins the community and can decide which roles he or she wishes to play after joining. This democratic philosophy is evident from eduFire's website statement that the mission of eduFire is to: "Revolutionize education. Our goal is to create a platform to allow live learning to take place over the Internet anytime from anywhere. Most importantly ... for anyone. We're the first people (that we know of) to create something that's totally open and community-driven (rather than closed and transaction-driven)" (About eduFire, para. 1). Bischke's eduFire and Cooper's TUA concepts have several strong similarities; but most important, a *community-first philosophy*.

This community-first concept is gaining wider consideration. Steve Gilfus, one of the founders of the Blackboard LMS, and Frank Ganis, an early member of the Blackboard team, suggest that one of the fundamental challenges for delivering online educational interactions is that popular LMSs emphasize an approach of "content first, context second and community third" (2010, p. 7, para. 1). In an interesting whitepaper, *Promise of Community Changes Education Technology Paradigms*, they suggest that:

> Social networking sites like MySpace, Facebook, Twitter, Delicious and many other 'Web 2.0' technologies created new definitions and paradigms for technology based communities. In these technologies the focus is on building community first, providing context within that community second, and then sharing content within the community third, a direct opposite approach to existing Learning Management Systems. By nature of their design, new social technologies provide a more accurate construct for building effective learning communities and enabling a set of richer community learning experiences (Ganis & Gifus, 2010, p. 7, para. 2).

This is true of eduFire as well. Although the company employs a widely available software technology (Adobe Connect) to facilitate the learning experience, the primary revolution that eduFire's primary innovation is establishment and optimization of a large and active learning community. The community itself determines how learning occurs. Users within the community have vast options for sharing information. Users can freely exchange content (both within the platform and via external syndication channels), they can review and rate courses and teachers and establish interconnected relationships with other users.

User-driven relationships have another revolutionary aspect when considered against a more traditional view of formal education: quality control and course creation are a community function, not an administrative one. The market assigns value to courses and instructors. In this case, the market is primarily the eduFire user community. Because eduFire offers users the ability to rank the quality of teachers, the cream rises to the top. Likewise, if enrollment is an indicator of interest, those courses with the greatest appeal will be the most highly attended.

Enrollment potential is self-regulating through another eduFire innovation: teachers and students set the prices for the courses. eduFire does not directly charge specific fee to members of its learning community. Teachers set the price for courses; without limits to how much or how little they can charge: "the better ... you are the more likely you will be to attract new students and charge a higher price for your services" (eduFire, 2007–2009, about teaching on eduFire). Consider this concept on an institutional level. Students and schools make similar decisions on a programmatic level.

Ivy League schools charge more than community colleges ostensibly because they offer access to instructors and resources of a higher caliber. eduFire drills this concept down to the individual course level. An eduFire student could conceivably (if such instructors were teaching in the eduFire community) construct an education experience composed entirely of Ivy League instructors (from various universities) and composed of the same courses taught at those institutions. Given that scenario, there appears to be little difference between learning experience at such an institution and study conducted through eduFire. Steve Cooper and Jon Bischke share similar perspectives. Their common perspective provided the basis for PIU's recent concept test.

Piccolo International University

PIU is a small, state-licensed, but unaccredited university in Scottsdale, Arizona. It has a storied history. PIU comes from merging of two schools; The Institute of Construction Management & Technology (ICMT) and Nouveau University (NU) – both began operations in the still-vibrant real estate market in 2005. In 2007, the first students were admitted to ICMT and by October 2008 NU had acquired ICMT (Piccolo International University, 2009). Before the end of that same year, the new ownership (led by Laura Palmer Noone, the former long-time president of University of Phoenix) had rechristened the blended organizations as a single university under the new name: PIU. Steve Cooper, the founder of TUA, was the original chief executive officer of ICMT and led it through the acquisition by NU. Cooper's involvement with PIU would have ended there were it not for a shift in strategy by Noone's organization. Approximately a year after the acquisition of ICMT, Cooper again found himself across from Noone at the negotiating table; buying back what had become PIU. As fate would have it, Cooper first sold and then regained the much-prized PIU and resumed operations of the university by late 2009.

As the owner of two yet-to-be-accredited schools, Steve Cooper had a challenge on his hands. Accreditation is a central requirement for schools seeking to offer federal student loans. A lack of accreditation presented a two-fold challenge for enrollment processes: knowledgeable students looked for accreditation, as proof of PIU's credibility, and these same students sought student loans to finance the cost of education. Lacking the ability to address these needs directly, schools like PIU must become more innovative in the pursuit of new enrollments. Steve Cooper embraced this need as an opportunity rather than as a challenge. The opportunity Cooper sought came in the form of a simple introduction.

PIU AND EDUFIRE: CREATING THE BLENDED, SOCIAL-LEARNING MODEL

As PIU was changing ownership in 2009, Higher Ed Gadfly (a digital marketing, social media, and e-learning innovations consultancy) introduced Steve Cooper to Jon Bischke. Cooper and Bischke, under the guidance of Higher Ed Gadfly, developed fast ties through common perspectives on both the challenges to higher education and opportunities from social media ideology. Jon Bischke had not developed a formal higher education channel within the eduFire community. Higher Ed Gadfly and Steve Cooper suggested that it was time eduFire tried an experiment.

The experiment would have PIU offering a blended-course concept. PIU's regular online program is a standard, asynchronous modality, delivered through the Angel LMS. PIU courses are eight weeks in duration. Under the auspices of this experiment, PIU would enroll current PIU students in a standard section of the course *Digital Marketing 201 (DMK201)* (Busch & Cooper, 2010). In addition to this course delivered through the Angel LMS, the instructor of the course would also offer a series of live lectures delivered in the eduFire community platform; one lecture per week for each week of the regular course. The lectures would be one-hour long, and each lecture would introduce the topic of study in accordance with the syllabus for the course. Members of the eduFire community would be able to participate in the live lectures. Students enrolled at PIU would be able to participate in both the live lectures and the asynchronous course. Members of both groups could engage the course's instructor and each other during the live lectures. eduFire-only users would be able to attend these lectures at no cost, but would receive no credit for the course.

Structure of the Proof-of-Concept

The goal of this experiment was to establish proof-of-concept that the course, content, and delivery of a college course could unlock the voice of a university. By unlocking the voice, through open delivery and enrollment, the concept would serve as a more effective introduction to the programs offered at a university and better prepare potential students for possible formal enrollment; thus supplanting the need for traditional, high-cost marketing tactics like lead generation marketing.

The experimental course entered the eduFire community about two weeks prior to the official start of the PIU-based cohort. The course syllabus was loaded onto the platform as part of the process for setting-up a new course within the eduFire community. This syllabus supplied the basis for organic search results on major search engines because the eduFire system is optimized to improve search engine results. Furthermore, each weekly lecture was accompanied by a new slide show presentation specifically related to the topic for the week. Slide shows of lecture notes were presented to attendees during the live lecture and then posted to several social channels designed to accommodate the sharing of this type of content. These platforms included: SlideShare, Scribd, and eduFire's own content section (Busch & Cooper, 2010). The program received a twofold benefit from the posting of notes: students had archival access to these notes for continuing study and search engines found this new content in relation to the course syllabus further optimizing organic search engine rankings.

In addition to the postings of these information-rich content resources, several other social tools were used to promote the course. Twitter was used to notify large audiences of both pending lectures and the posting of new notes. Both Digg and StumbleUpon, two social bookmarking tools (designed to showcase valuable content both within the individual user communities associated with these services and to the larger user population of the Worldwide Web at large through enhanced search engine optimization) were also employed.

Value Revealed

Few students from eduFire had enrolled in the lecture series during the two weeks between the establishment of the course in the eduFire community and the official start of the course. The lack of fast enrollments was not a concern. PIU, including all variations of its composition, was a small school: enrollments from all iterations top to about 300 and it has graduated about only 16 students since ICMT's 2007 inaugural year. But course enrollments were about to grow.

The first lecture of the series occurred on December 3, 2009. The first lecture of the series introduced 17 unaffiliated students (20 students registered, but three of the first 20 were affiliated with PIU) to PIU. Interest grew steadily from that point.

Week two of the series added 12 additional attendees. Week three saw three more students. By the fourth week of lecture, the population had grown from zero students, before the start of lectures, to 40 students not previously affiliated with PIU. By the final lecture, the cohort had grown to 75 unaffiliated students. This count indicates an average weekly growth of 8.29 new students per week (the data in Table 1 indicate the enrollment growth throughout the eduFire/PIU course experiment and account for enrollments with previous relationship to PIU). PIU in its various iterations had been offering courses for about 104 weeks to achieve enrollments of about 300 students. This means that PIU had enrolled an average of about 2.88 students per week across its operational life span. The short, eight-week proof-of-concept experiment resulted in a 288% improvement over the PIU's historical weekly student acquisition rate.

Similar success was achieved in search engine optimization from the previously described methodologies. During the run of the experiment, PIU's lecture notes from the eduFire lectures earned first page organic search rankings (for the phrase "Digital Marketing Lecture") on these search engines: Google, Yahoo, MSN/Bing, and Ask.com (Busch & Cooper, 2010). In many cases, the notes generated multiple results on page-one searches and on secondary and tertiary pages, and ranked as high as first and second positions in a Google search (Busch & Cooper, 2010).

Table 1. Enrollment Growth by Week for the edFire/PIU Course Experiment.

Lecture Date	Week in Course	eduFire Community Enrollments	Less Those Affiliated with PIU	Unaffiliated Students	Weekly Enrollment Increase
December 3, 2009	1	20	(4)	24	
December 10, 2009	2	30	(5)	35	11
December 17, 2009	3	33	(4)	37	2
January 17, 2010	4	44	(5)	49	12
January 14, 2010	5	52	(4)	56	7
January 21, 2010	6	59	(4)	63	7
January 28, 2010	7	71	(4)	75	12
February 4, 2010	8	78	(4)	82	7

Additionally, the social media channels mentioned previously further increased the reach of PIU's voice during this experiment. Combined views if the notes posted to SlideShare, Scribd, and eduFire's content section exceeded more than 1,400 views (Busch & Cooper, 2010). These results are further enhanced because no marketing budget was allocated to this experiment nor were any funds spent on marketing.

Evaluating the Value

Although these results are impressive, it is important to consider a few simple facts about the outcomes of this experiment. The experiment was a proof-of-concept test rather than a formal research project. Rather, eduFire, PIU, and Higher Ed Gadfly sought to understand the potential for the concept in reality. Further research would better inform these results and confirm this model's potential for use as an ongoing instructional modality. Notwithstanding, the concept does appear to indicate strong potential and further explorations would optimize the results.

Further optimization includes transforming enrollments in a freemium lecture series into a revenue generating opportunity for PIU. Although nearly 80 students attended the lecture series, and the cost for acquiring these students was virtually nil, the experiment did not generate actual revenue for PIU or eduFire. But the full impact of the experiment remains to be seen. The posted notes continue to garner search engine attention and view counts of the notes were still growing even after the last lecture on February 4, 2010. The effort has observable value even if the balance sheet shows that the PIU/eduFire experiment was considered a wash in terms of costs expended and revenues earned. This is especially evident when weighed against countless examples of failed marketing tests which included higher cost than the revenues gained.

HOW TO EMPLOY SIMILAR TECHNIQUES AND PRINCIPLES TO TRANSFORM OTHER UNIVERSITIES

The examples from innovators provide lessons for all educators and marketers in considering how to use social media ideology more effectively at higher education institutions. Those who search for methods of using social media as a means for freeing the voice of the university should first assume the perspective of the people and companies profiled in this chapter. The core tenant that must be adopted is a willingness to break from the mold of established higher educational practices. Creating new models built at a foundational-level. Integration alone (between new tools and older models) is not enough; only through true convergence can educational leaders unlock the voices of other institutions.

Convergence through the Course

Innovators should consider revising how the institution views courses. The PIU/eduFire experiment highlights the value of courses as marketing tools. Digital marketing is an appealing subject to teach. PIU's digital marketing course represents a distinct course because it is the introductory course to a degree track at PIU. The course was more appealing to audiences than a commodity course. Distinct courses free the voice of the university; they are of distinct value and can therefore carry the voice of the institution farther than the commodity course.

Embracing social media ideology can lead to the example of TUA. The TUA model allows the creation of new distinct courses at the behest of the members of a learning community. Rather than creating all courses within the silos of the academic departments, innovative educators and marketers should seek to include the members of associated learning communities and to listen and respond their input. This practice provides direct opportunities for the hosts of these courses to entice new enrollments and to inspire greater interest and dialog between the members of an institution and the members of an associated learning community.

Convergence through the Content

Those seeking true innovation might avoid the tendency to rely on restricted content (such as textbooks) and embrace opportunities to promote content offered through sharable formats (such as YouTube). Social media ideology promotes greater opportunities for sharing and syndicating content; unlocking the voice of the institution requires the adoption of this philosophy.

Innovators might move a step further by engaging faculty members and instructors to both create and freely distribute customized content. The PIU/EduFire experiment effectively used this technique to attract a larger audience. Rather than promoting the voice of a major textbook publisher, the content revealed a unique perspective from the instructor of the course and insights into the university itself.

Convergence through the Delivery

Unlock the voice of the university by freely offering course and content through unrestricted delivery methods. Rather than hiding the instructional process behind closed doors and firewalls, promote the process in full view of the learning community. Channel community involvement through an amplifier (like eduFire) to release the value of the university to the benefit of the members of the learning community at large.

Give prospective students the ability to observe directly and comment on the quality of the instructor, the quality of the content, and the technologies employed by the institution. Prospective students can then make more informed decisions before seeking admission to the institution. In essence, filling student knowledge gaps before interacting with admissions teams increases the efficiency of the process because prospective students can better anticipate the experience. Setting expectations before the admissions process can also improve student retention and success rates. Shorter, more successful enrollment cycles, fewer drop-outs, and greater numbers of successful students translate into greater financial gains and more efficient expenditures. Thus, the open delivery format can serve both the educational and the marketing goals.

CONCLUSION: CONVERGENCE IN WHOLE

Set a goal to reduce or eliminate organizational silos. Improve efficiencies between operational teams before looking for costs to cut. By optimizing core strengths before reducing service level, institutions can maintain access to quality learning experiences and better serve associated learning communities. eduFire, TUA, and PIU are all small, but nimble organizations that must innovate to survive. Consider how they have put the convergence of social media ideology and higher education into practice and imagine the possibilities for respected institutions of higher learning with more established access to learning communities. By adopting convergence as a principle, and based on Social media ideology as a philosophy, higher education leaders can renew the promise of higher learning to a wider community of learners.

NOTES

1. In June 2010, after this chapter was written, Camelback Education Group, Inc. (the holding company led by Steve Cooper which owns Piccolo International University) acquired eduFire and Ryan Busch has joined with Cooper to further develop Camelback Education Group's properties.

REFERENCES

Apollo Group, Inc. (n.d.). *Just the facts: Enrollment.* Available at http://www.phoenix.edu/about_us/media_relations/just-the-facts.html. Retrieved on April 10, 2010.

Apollo Group, Inc. (n.d.). *Just the facts: Alumni.* Available at http://www.phoenix.edu/about_us/media_relations/just-the-facts.html. Retrieved on April 10, 2010.

Breaking Down Operational Silos through ERP. (2004). *Utility Week, 22*(18), 24. Retrieved on April 9, 2010 from Business Source Complete database.

Busch, R., & Cooper, S. (2010, March 2). Successfully using social media to lower delivery and enrollment costs, extend reach, monitor quality, and add value to the diploma. Presented live at the 12th Annual Education Industry Investment Forum. Available at http://www.iirusa.com/education/day-two.xml

Chea, T. (2010, January 14). College applications rise, but budget cuts cap enrollment. *USA Today.* Available at http://www.usatoday.com/news/education/2010-01-14-college-admissions_N.htm. Retrieved on April 12, 2010.

Corbett, P. (2010, January 4). Facebook demographics and statistics report 2010-145% growth in 1 year [Blog Entry]. Available at http://www.istrategylabs.com/2010/01/facebook-demographics-and-statistics-report-2010-145-growth-in-1-year/

Craft, S., & Phillips, J. (n.d.). *Associate professors at OSU: Results of the career enhancement survey.* The Ohio State University, Institutional Research and Planning.

Dewey, J. (1922). Democracy and education: An introduction to the philosophy of education. Available at http://etext.lib.virginia.edu/etcbin/toccer-new2?id=DewDemo.sgm&images=images/modeng&data=/texts/english/modeng/parsed&tag=public∂=2&division=div2. Retrieved on April 19, 2010.

eduFire. (2007–2009). About eduFire. Available at http://edufire.com/about. Retrieved on April 12, 2010.

eduFire. (2007–2009). About teaching on eduFire. Available at http://edufire.com/faq. Retrieved on April 12, 2010.

Facebook. (2010). Press room. Available at http://www.facebook.com/press/info.php?statistics#!/press/info.php?timeline. Retrieved on April 12, 2010.

Facebook. (n.d.). University of Phoenix. Available at http://www.facebook.com/universityofphoenix. Retrieved on April 12, 2010.

Ganis, F., & Gifus, S. (2010) *Promise of community changes education technology paradigms: Community, context, content.* The Gilfus Education Group. Washington, DC. Available at http://www.gilfuseducationgroup.com/white-papers/community-platform-for-education. Retrieved on March 22, 2010.

Gonzales, A. (2009, March 27). Tech University of America offers classes via social networking model. *Phoenix Business Journal.* Available at http://phoenix.bizjournals.com/phoenix/stories/2009/03/30/story9.html?surround = etf&b = 1238385600^1801467. Retrieved on April 1, 2010.

Gross, R., & Acquisti, A. (2005). Information revelation and privacy in online social networks. *Proceedings of WPES'05* (pp. 71–80). Alexandria, VA: ACM. Available at http://www.heinz. cmu.edu/~acquisti/papers/privacy-facebook-gross-acquisti.pdf. Retrieved on July 14, 2010.

KITV.com. (2009). Honolulu News. Available at http://www.kitv.com/news/20678226/ detail.html. Retrieved on April 29, 2010.

Massachusetts Institute of Technology. (n.d.). *Courses.* Available at http://ocw.mit.edu/OcwWeb/ web/courses/courses/index.htm#MediaArtsandSciences. Retrieved on March 30, 2010.

Massachusetts Institute of Technology. (n.d.). *About OCW: Unlocking knowledge, empowering minds.* Available at http://ocw.mit.edu/OcwWeb/web/about/about/index.htm. Retrieved on March 30, 2010.

Massachusetts Institute of Technology. (n.d.). *Help: FAQ: Fair use.* Available at http://ocw.mit.edu/OcwWeb/web/help/faq7/index.htm#3. Retrieved on March 30, 2010.

Massachusetts Institute of Technology. (n.d.). *Help: FAQ: Getting started.* Available at http://ocw.mit.edu/OcwWeb/web/help/faq1/index.htm#3

Natel, R. (2009, June 1). Price ranges for learning management systems in 2009 [Blog Entry]. Available at http://brandon-hall.com/richardnantel/2009/06/01/price-ranges-for-learning-management-systems-in-2009/

Noel-Levitz, Inc. (2009). Noel-Levitz whitepaper: Connection enrollment and fiscal management. Coralville, Iowa. Available at https://www.noellevitz.com/NR/ rdonlyres/9A41D0EE-47D2-45AA-9798-E9C95AD9830B/0/ConnectingEnrollmentand-FiscalManagement09.pdf

Piccolo International University. (2009). About Piccolo International University (PIU). Available at http://onlinepiu.com/about-piccolo-international-university. Retrieved on March 29, 2010.

Pope, J. (2009, October 20). College tuition costs rise again [blog entry]. Available at http://www.huffingtonpost.com/2009/10/20/college-tuition-costs-ris_n_327398.html

Reuben, R. (2008, August 19). The use of social media in higher education for marketing and communications: A guide for professionals in higher education [downloadable pdf document from blog]. Available at http://doteduguru.com/wp-content/uploads/2008/08/ social-media-in-higher-education.pdf

Solis, B. (2007, June 29). The definition of social media [blog entry]. Available at http://www.webpronews.com/blogtalk/2007/06/29/the-definition-of-social-media

Tech University of America. (2009). About us and contacts. Available at http:// www.techuofa.com/index-5.html#pageContent. Retrieved on April 2, 2010.

The College Board. (2009). Press releases. Available at http://www.collegeboard.com/press/ releases/208962.html. Retrieved on April 15, 2010.

The Distance Education and Training Council. (n.d.). *About the distance education and training council's process of accreditation.* Available at http://www.detc.org/ accreditationprocess.html#eligibility. Retrieved on April 9, 2010.

Walker, C. (2009, September 30). Colleges learn to live with social media. *The Baltimore Sun.* Available at http://www.buffalo.edu/news/pdf/September09/BatlSunSocialMedia.pdf. Retrieved on July 14, 2010.

Wendt, J. (n.d.). Q&A with Jeff Wendt: Steve Cooper, founder, Tech University of America. *Today's Campus Online.* Available at http://todayscampus.com/minute/load.aspx?art= 1756. Retrieved on April 2, 2010.

Wilson, F. (2006, March). My favorite business model [blog entry]. Available at http:// www.avc.com/a_vc/2006/03/my_favorite_bus.html

PART III
LEARNING ARTS AND SCIENCE
IN THREE-DIMENSIONAL
VIRTUAL WORLDS

UNDERSTANDING COMMUNICATION PROCESSES IN A 3D ONLINE SOCIAL VIRTUAL WORLD

Aimee deNoyelles and Kay Kyeongju Seo

ABSTRACT

The aim for this chapter is to better understand the dynamics of social communication processes within Second Life®. Understanding communication processes in 3D online social virtual worlds is vital in embracing contemporary social issues and improving interpersonal and organizational relationships as these environments are rapidly growing in popularity in the education sector. In this chapter, we observed an undergraduate communication class and discussed four powerful interrelated forces behind the students' communication processes: (1) gamer status; (2) avatar appearance; (3) physical proximity; and (4) virtual proximity. Our findings can inform Arts and Science educators in general and Communication instructors in particular about how learners socially communicate and interact within a 3D online social virtual world and how teachers can foster students' communication and collaboration in this environment and support their content creation and collective knowledge building.

Teaching Arts and Science with the New Social Media
Cutting-edge Technologies in Higher Education, Volume 3, 169–187
Copyright © 2011 by Emerald Group Publishing Limited
All rights of reproduction in any form reserved
ISSN: 2044-9968/doi:10.1108/S2044-9968(2011)0000003012

Today's modern technology has added a whole new dimension to the academic discipline of communication. With the rapid expansion of user-centered social media such as blogs, wikis, and virtual worlds, the landscape of communication education is being dramatically transformed; its focus is reaching beyond face-to-face dialogue among people and flourishing well into online virtual interactions on a global scale. Three-dimensional online social virtual worlds such as Second Life® (SL) represent a unique example of the emerging social media, immersing users in a graphical environment where multiple people can log on, communicate, and create content in real time. These virtual worlds are rapidly growing in popularity in the education sector. In 2007, the number of educational institutions that formed an official group or owned virtual space within SL was around 170 (Jennings & Collins, 2007). Three years later, the number of institutions owning virtual space had nearly quadrupled (McGowan, 2010). While these environments afford communication and collaboration with others from around the world, there are also technically complex and cyber safety issues present. Therefore, understanding communication processes within these environments is vital in embracing contemporary social issues and improving interpersonal and organizational relationships.

Three-dimensional online social virtual worlds offer the ability to communicate and collaborate with others from a distance, both verbally (through text or voice) and nonverbally (avatar appearance, gestures, and content creation). Such environments "provide opportunities for real-time simulation, experiential learning, and collaboration in a virtual environment that do not require participants to be in the same physical location" (Jennings & Collins, 2007, p. 180). However, users may also sometimes share the same physical space. For example, some classes may meet face to face, with students communicating with others in both the "real world" computer lab and virtual world contexts. A primary reason for incorporating 3D online social virtual worlds in education is their distinctive ability to simulate authentic learning settings that are impractical or impossible to provide in traditional classroom settings (Gee, 2007). Participating in a simulation of a realistic situation can encourage the user to relate the experience to the overall learning objectives, thus easing learning transfer (Jones, Morales, & Knezek, 2005).

The focus of this chapter concerns a better understanding of how learners socially communicate and interact within 3D online social virtual worlds, particularly in SL, with consideration of the overall educational context. The underlying conceptual theory employed for this study is social constructivism. This concept is largely based in the writing of Vygotsky (1978), who emphasizes the crucial interplay of social interaction, language, and culture on learning.

Social interaction ensures that each student will take control of their own learning path, rendering a more personalized and active knowledge building process. Learners actively construct knowledge as they attempt to make sense of their experiences, influenced by previous knowledge and through sharing various perspectives with others (Woo & Reeves, 2007). The act of negotiating and solving problems with others leads to a shared understanding called intersubjectivity, and to enculturation, in which a learner learns the accepted norms and values of the currently established culture. The individual construction of meaning occurs when intersubjectivity exists within the enculturalized zone of proximal development, which is defined as the distance between the individual's developmental level of problem solving and the developmental level with expert guidance (Dimitriadis & Kamberelis, 2006). Social interaction from all participants is an essential facet toward creating a community of learners.

From this viewpoint, learning cannot be divorced from the social contexts in which it is situated. Therefore, social constructivist researchers strive to understand the process of learning by investigating interactions between learners and contexts (Wertsch, 1998). Within the context are tools that people use to perform activities. While face-to-face discussions are mediated through direct verbal and nonverbal communication, different tools are used to mediate communication in a traditional online environment, most notably text-based asynchronous discussion boards and chats. Physical appearances are typically not seen, voices are typically not heard, and face-to-face conversational cues are absent. In contrast, SL allows for synchronous verbal and nonverbal communication mediated by avatars, along with a shared visual space, potentially enabling more meaningful social interaction and collaboration from a physical distance. In the case of accessing SL in a shared physical space such as a computer lab, learners negotiate the dual social contexts of the "real" and virtual.

From an educational perspective, it is beneficial if learning contexts closely resemble future contexts, so learning is more easily transferred across situations. For instance, a communication student will learn more meaningfully about the nature of public relations within a practicum in an authentic context rather than simply reading about best practices. A practicum presents a real-life task, one that is ill-defined and is examined from multiple perspectives (Woo & Reeves, 2007). Designing authentic tasks in typical online learning environments (threaded discussion boards, for instance) is a challenge. Due to the rich 3D design affordances, SL allows for more realistic settings for learning transfer to occur. Certainly, each student will approach this environment with a specific knowledge base and technical skill, which will affect how students understand each other, integrate in the virtual culture, and construct individual and group knowledge.

Therefore, understanding the dynamics of social communication processes is truly important in achieving meaningful learning in SL. This inquiry will inform not only Communication instructors but Arts and Science educators in general about how to foster students' communication and collaboration in this environment and support their content creation and collective knowledge building.

THE UNIQUENESS OF SECOND LIFE

SL, regarded as being "in a social medium class of its own" (Levinson, 2009, p. 4), holds a lot of promise to support educational activities grounded in social constructivism. There are three kinds of 3D virtual worlds: game-based, educational, and social (Schultze & Rennecker, 2007). Game-based virtual environments such as Worlds of Warcraft are characterized by a "highly scripted, typically quest-driven narrative" (p. 338), with clearly defined objectives and outcomes. Similarly, educational virtual worlds such as River City are specifically designed and developed with the intention to support learners in meeting specified objectives. In contrast, social virtual worlds enable users to coproduce the design and content of the environment.

SL is an example of a social virtual world, which provides an open-ended experience for the user, as emphasis is placed on collaborations of a social nature. Publicly released in 2003, SL grew rapidly; its membership increased from 500 users to 6.9 million in 2007, to more than 15 million in 2009. SL is unique in supporting three modern trends: "increasing focus on people as organizing principle of the network ... improving ability of our computing devices to represent information visually across three dimensions ... [and] allowing users to generate content, fostering sense of ownership and pride among participants that fuels the growth of the community" (Johnson, 2008, p. 7). Unlike video games that contain pre-embedded objectives, in SL, the design, the content, and the objectives of the environment are perpetually shaped directly by its users. This allows a rich opportunity for learning, with learners actively creating and displaying their own knowledge to the online community in multiple ways, such as exhibits, embedded media, discussions, and performances.

Simulated bodies called avatars represent the user and serve as the visual point of view in the environment. Users can customize their own avatars from scratch or search through the world for ready-made features, such as the

Fig. 1. Customized Avatars. Avatars Generate a Shared Immersion in the Environment.

massive Kool-Aid man and Darth Vader with a light saber in hand that we observed in a communication class (see Fig. 1). An avatar appearance can change at any time, as each user has a complex and organizationally segmented inventory of possessions which they can use or add to. Some items in the inventory typically include body features such as a hairstyle, pieces of clothing, personal adornments, and personal possessions that may be acquired as gifts or giveaways in the environment. Possessions other than what was created or presented to them may be purchased using the SL currency, the Linden dollar.

Although users usually remain physically distanced, interactions in this environment are unique from other web tools in that they are multisensory and mainly synchronous, which stimulates a feeling of immersion in the environment, strengthening the feeling of a social community. Levinson (2009) shares:

> When you're in SL, you feel, much more than on other non-simulation sites on the Web, as if you are really in this community. The combination of the moving graphics and voices, the way you can move your avatar through this environment, creates a powerful illusion that you are actually in, rather than looking at, listening to or reading it. (p. 152)

Instead of relying on static text and images to represent users, avatars serve as the primary social cue, affording opportunities for both verbal and nonverbal language. When avatars interact, they inhabit the same visual space, reducing the social and psychological distance between them and eliciting a heightened shared experience (Boellstorff, 2008).

In order to navigate through the environment and customize content, the learner must take control and make decisions, leading to immersion in the learning process and positively affecting motivation (Dieterle & Clarke, 2009; Malone & Lepper, 1987). SL enables activities such as role playing through avatars and scenario building, allowing learners to appreciate multiple perspectives and temporarily assume the responsibilities of a professional without incurring consequences (NMC, 2007). For instance, Edirisingha, Nie, Pluciennik, and Young (2009) studied distanced archeology students in SL as they used their avatars to explore spaces resembling real-life villages, observing social norms such as rules about gender and age. The authentic and immersive space of the villages contributed to students' improved understanding of the subject matter by reinforcing the concepts that were learned in more traditional ways. In addition, the synchronous element of exploration increased the generation and diversity of group construction knowledge. This example displays how SL can effectively enhance student communication, due to its unique ability to deliver on key educational trends: user-centered creation of customizable content, social expression, and construction of collective and collaborative global knowledge (Dieterle & Clarke, 2009).

SOCIAL STRATEGIES IN SOCIAL VIRTUAL WORLDS

Our extensive review of the literature concerning social virtual worlds revealed some important social strategies that students used in order to communicate with others, revealing implications to the unique environment of SL. In general, students report feeling more engaged and connected to other students in a 3D social virtual world when compared to traditional online settings (Annetta, Murray, Laird, Bohr, & Park, 2008). One possible explanation for this perception is the degree of immersion one experiences in the 3D space, with learners sharing the same visual area and interacting in real time. Although these synchronous affordances more closely resemble a traditional face-to-face learning environment than an asynchronous text-based online environment, socialization in a 3D virtual world bears many differences. According to Edirisingha et al. (2009), one gains access to the virtual environment through online socialization, which happens when "participants establish their online identities, find others with whom to interact online, understand the nature of the online environment and how it is used for learning, and develop trust and mutual respect to work together at common tasks online" (p. 460). This online socialization must be present in order for more critical discourse to emerge (Garrison & Cleveland-Innes, 2005; Stein et al., 2007).

Aspects of "real world" identity such as gender and personality carry over into the online environment and blend with technical affordances, shaping an online identity that is not entirely virtual (McGerty, 2003). A user achieves a high level of social presence in an online setting if he or she feels that personal identity is being accurately projected to other users, thereby being perceived as "real" in the virtual space (Garrison & Arbaugh, 2007). In SL, this projection of identity is primarily achieved through customization of the avatar. When asked why they chose certain looks for their avatars, middle-school aged learners in Feldon and Kafai's study (2008) all mentioned basic aesthetic considerations, but some also voiced the desire to make the avatar more like their "real" selves. In an observational study on social behavior in SL, Yee, Bailenson, Urbanek, Chang, and Merget (2007) found that "real world" norms such as eye gaze and interpersonal distance are realized in the virtual environment. With this in mind, it is important that learners possess the technical competence to customize and direct their avatar in the desired preferences.

Avatar customization not only allows learners to establish personal identity in the virtual world, but also can be employed for purposes of social interaction. Taylor (2003) stresses the importance of the avatars, defining them as bodies that "not only facilitate the production of identities, but also social relationships and communication. They are not neutral and indeed their power lies in the very fact that they cannot be" (p. 38). Studies by Feldon and Kafai (2008), Kafai (2008), and Edirisingha et al. (2009) examined students' relationships to their avatars and found that "much avatar activity primarily served as a focus for social interaction and conversation" (Feldon & Kafai, p. 590) within the virtual environment. In Feldon and Kafai's study, participants who shared a computer lab also engaged in face-to-face conversation such as critiquing others' avatars and trading inventory items. This informal social communication, while not academic content-related, did serve to strengthen both online and offline group cohesion, which contributed toward feelings of community. Since avatar customization is important in establishing online identity and supporting social interaction, and socialization must be felt before meaningful learning can occur, time must be allotted to allow for this activity. In addition to time, those with less experience will need technical support to achieve this. They may find more technical difficulty customizing their avatar in the manner they wish, compromising the feeling of personal identity and "realness" in the virtual space.

It is important to note that most learners prefer socially interactive and collaborative activities over individual work in 3D virtual settings (see Delwiche, 2006; Kao, Galas, & Kafai, 2005; Mayrath, Sanchez, Traphagan, Heikes, & Trivedi, 2007). Participating in social activities such as contests and scavenger hunts can promote feelings of community and cohesion (Sanchez, 2007). The degree of feeling real in this virtual space also depends on the characteristics of the environment which surround the avatar. Proximal social spaces such as residences should be designed to foster both formal and informal learning opportunities and reduce perceptions of social isolation (DeLucia, Francese, Passero, & Tortora, 2009; Dickey, 2005; Jarmon & Sanchez, 2008).

Although the studies mentioned above reveal initial promising findings concerning socialization in online social virtual worlds, a closer examination is needed to investigate how discussions held within SL are developed into constructive communication and what drives and affects this process. To provide an insight into this, we present four factors shaping student communication in this chapter.

FOUR FACTORS SHAPING COMMUNICATION PROCESSES

To understand how students interact in SL in more depth, we observed an environmental communication class with 13 female and 8 male undergraduate students. This course was designed to explore the communicative structure and functions of virtual worlds through the lens of Environmental Communication, using SL as the virtual platform. For instance, in one activity, the students explored an ecologically conscious island in SL and discussed the depictions of nature found there, relating them to Cronon's typology. The class met in the computer lab on a weekly basis to access SL. In order to capture the dynamic interactions between learners, the computer lab context, and the SL context, multiple sources of data were collected. Five "real world" observations were conducted in the computer lab, and four students were interviewed at the end of the course. A total of 82 individual SL-related blog entries were collected over the quarter, which comprised of prompted written reflections and virtual snapshots the students had taken. Finally, two surveys were distributed, including both closed and open-ended questions, along with demographic information such as age, gender, and previous experience with multiuser games and SL.

To understand the ways in which the data sources converge, we employed grounded theory techniques proposed by Corbin and Strauss (2008). During the initial phase of analysis, concepts were identified from the multiple sources of data and then coded and subcoded, eventually being organized into categories. In the next phase, data was put back together and connections were drawn between codes and categories. We then returned to the data and searched for additional evidence to support the developing themes. Through this thematic analysis, we identified four powerful interrelated forces behind the students' communication processes: (1) gamer status; (2) avatar appearance; (3) physical proximity; and (4) virtual proximity.

Gamer Status

We found that gamer status was the core concept that explained the majority of interaction patterns and knowledge construction within SL. Learners who had previous sustained personal experiences with 3D virtual worlds (game-based or social) were more easily enculturated into the SL environment, were more comfortable with establishing an online identity and communicating with others, and possessed more advanced technical skill. Those who were less experienced with regards to virtual worlds faced problems with technical skill, which inhibited their participation and self-expression in the environment. They also were not as aware (or occasionally wary) of the nature of SL, which led to diminished social interaction and participation.

Previous experience with gaming, however, does not automatically make one a "gamer." "Gamer status" was determined by reviewing the sources for evidence of identification with gaming and technical ability. For instance, Mary begins her blog entry with, "Due to the fact that I do not consider myself a gamer ..." Britney, who indicates that she has played online games and SL before, also says "I'm not very technologically savvy at all so that could be a serious problem for me in [SL]." This stands in contrast to Jon, who shares in an interview that "I play a lot of video games ... usually, someone that plays a lot of video games, they can pretty much pick up a game ... I can usually get it and not have to think about it too much." From this classification, two major groups emerged: those with gamer status (like Jon) deemed "gamers" in this study, and those with less gamer status (like Mary and Britney) deemed "nongamers." In the class, there were 10 students who were classified as gamers and 11 students as nongamers.

One's gamer status was closely related to his/her level of technical skill and world view. Level of technical skill, such as ease with navigation and communication, and world view, the way one interprets the nature of SL and interactions within the environment, both influence individual communication and participation. To determine one's level of technical skill, we examined the multiple sources of data and noted any signs of technical difficulty. The technical skill spectrum ranged from "I can't do anything in here" to "I caught on and got a grasp on the basics pretty quickly." Utterances like these provide a clue as to how people are getting around (or perceive they are getting around). "World view" was also defined by considering the multiple sources of data. Some nongamers held a somewhat disconnected world view, that 3D social virtual worlds like SL are "a fantasy world for people to retreat to when their real lives aren't satisfying," while gamers tended to have a more accepting world view, such as "It is just cool to see what SL has to offer to people and the way that it helps you learn." Gamers had strong technical skill and more of a willingness to take risks in the environment, but displayed less serious connection to the world for academic purposes. All of the nongamers exhibited inadequate technical skill, but there was variation in their conceptions of the nature of the environment and subsequent involvement in this environment.

Avatar Appearance

The avatar is a powerful vehicle to establish online identity since it affords rich verbal communication (text/voice) and nonverbal communication (appearance and gestures). First, there are nearly limitless ways that an avatar's appearance can be customized, including body shape, skin/hair/eye color, clothing, and accessories. In addition, users may assume a completely different look for their avatar, such as nonhuman forms. Considering the data from these students, avatar customization is influenced by gamer status.

Nongamers tended to customize their avatars to assume a typical human young adult appearance. The majority of these students dressed their avatars in real-life clothing and adopted traditionally "attractive" features (see Fig. 2(a) and (b)). There are several possible explanations for these avatar customization choices. First, nongamers are likely to be unfamiliar with virtual world norms due to inexperience, while gamers regularly interact with

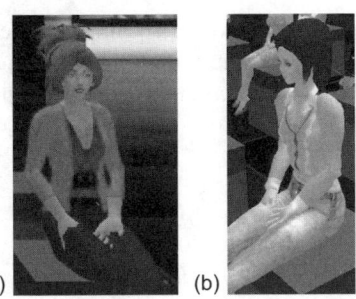

Fig. 2. (a) Elena and (b) Georgina. Most Non-Gamers Designed Avatars that Resembled their "Real-Life" Selves, Influenced by Cultural Expectations of Beauty.

"unusual" looking avatars or digital agents. Since nongamers do not have experience with this, they may heavily rely on real world norms instead, one being the social pressure for women to be physically attractive in real life. When we asked Mary about her avatar, she describes:

> In [another course], we were learning the characteristics of what, as a culture, we thought was beautiful, and it was like the same things that I had applied to my avatar. Like she had, like a smaller, like childlike face but, with the bigger baby eyes and long hair, which as a society, something we value as well. I think that's also a reflection of our cultural values and the things that we see as beautiful.

Mary recognizes the link between real world and in-world conceptions of beauty. Georgina, who dons her avatar in the most typical real-life outfit, describes in her blog that "I project my identity through my avatar because I made her look as closely to myself as possible ... but I did make her a little thinner than my [real life] physique." This stands in contrast to the gamers, which were more experimental in their avatar choices (such as Darth Vader and Kool-Aid Man).

In addition, nongamers tended to be more preoccupied by avatar customization than gamers and tended to focus communication on customization. For instance, when instructed to explore an environmental island and take a group snapshot, Uma, Tina, and Harriet form a group and travel to the same place. As they prepare to take the snapshot, some comments are voiced out loud, like "We look cute," "Look at my hair," and "Look at how pretty she is." A minute later, Uma rescinds the last remark and exclaims, "I've got to change my appearance – I look like I got run over by a train," then adds a white miniskirt over her hot pink pants and changes her hairstyle. While the nongamers tended to be more focused on physical appearance, the gamers were more interested in exploring the site and immersing in the environment surrounding them. For example, Kevin's

Kool-Aid man went swimming with a whale, while Robert's Darth Vader located a helicopter to fly. Communication among gamers usually revolved around their actions. At least once a class, a gamer would ask out loud to a fellow gamer, "Did you see that?"

Physical Proximity

In this class, the students were physically together in the computer lab when they accessed SL. This simultaneous context greatly affected online interactions for both gamers and nongamers. The most obvious effect was that little verbal (instant messaging) and nonverbal (gestures) communication was mediated through avatars; instead, the majority of communication occurred in the computer lab among people. In our interview, we asked Chrissy, a nongamer, if she used the SL communication features and she answered, "We did occasionally, but a lot of times, people I communicated with were like right next to me, so we would talk out loud and then type things … it's harder when they're sitting right next to you." Jon agreed in his interview when asked if he used SL to communicate, "Not to my classmates, just because you know, it's so much easier to actually speak than type."

When students are together in the same room, there is less of a need to use avatars to mediate communication, for the primary community exists in the physical world. Uma, a nongamer, agreed in her blog, "I never felt like I was really there. I always had my classmates right next to me who I was in full conversation with and it seemed to keep me out of the virtual world." The real-life context of the computer lab directly hindered the development of the online learning community in SL, and diminished communication with distanced users as well. There is less need to express self in an online environment when one is visibly seen and heard in real life. Being distanced forces the community members to express themselves online in order to reduce the psychological and emotional distance between members. When members share the same physical space, there is much less necessity for online expression. Therefore, the educational promises of a 3D online social virtual world (sharing the same visual space with others; simulating authentic settings, enabling distanced communication) are somewhat diminished in this configuration.

Of course, there were benefits to being in the same classroom. Those that needed more technical help could receive it in a face-to-face manner. Students could see others' screens, which prompted questions; being in the same space "allowed for learning things that one did not know enough to ask about" (Fields & Kafai, 2007, p. 9). However, since gamers and nongamers were usually mutually exclusive, nongamers were often reluctant to ask gamers for technical help, and often were not able to fully help other nongamers due to lack of expertise. In addition, in our observations, around half of the class rarely spoke out loud. This is a concern since the same half rarely spoke in SL. For this class, sharing a physical space diminished online communication, and being present in SL together may have diminished real-life communication.

Virtual Proximity

In addition to consideration of physical proximity, virtual proximity is also a powerful influence on social interaction. When users' avatars share the same space at the same time in a comfortable setting, feelings of immersion and social connection are enhanced. For this reason, proximal social spaces should be designed within SL in order to foster both formal and informal learning opportunities (Dickey, 2005; Jarmon & Sanchez, 2008). In our course of study, two virtual spaces were designed and implemented: the social space of a dormitory building and the formal learning space of a conference hall. Both were located on the university's island, which largely resembled the actual campus. Most students reported taking comfort in having a "home" base that resembled the real world. In his blog entry, Deon wrote, "The university campus is considered ideal because it makes us feel at home when we are exploring SL. We don't feel like an outsider." Some wished they had more choice in the construction of the university space. Kevin recommends, "If we are allowed to create what we feel the environment is, then maybe we can express ourselves better. We all view the environment differently and we have all experienced different things in our lives." This comment is especially pertinent in this communication course which focused on environmental communication. Applying academic content, such as theories of environmental communication, toward design of virtual space can stimulate meaningful learning and collaboration in a virtual setting.

Dorm Rooms

Each student was assigned a room in a dormitory residence tower on the university campus. Similar to the comments about the comfort of the "real world" university atmosphere, the rooms were regarded as a safe home base and a place to congregate (see Fig. 3). Mary explains, "You could teleport home whenever you didn't know where you were, and I knew that other students would be there. And it was almost like a second home in a sense, because it *was*

Fig. 3.   Virtual Dorm Room. Most Resembled "Real-Life" Rooms.

like the dorms my freshmen year ..." Being given the opportunity to furnish their rooms, the students were offered a space to nonverbally express themselves. For instance, Nora shares in her blog that "I really wanted a picture of Angela Davis to put in my room so I am going back to buy that later." While the rooms afforded a permanent social meeting place, they also were tied to more formal learning elements. For instance, Oliver applied the classroom discussion of nonverbal communication to his design of dorm room, writing "This was a time for us to get creative. The more inanimate objects brought into SL, the more realistic it becomes. These aspects of nonverbal communication serve as a bridge between SL and real life."

However, they need to be technically skilled in collecting objects and placing them in their dorm rooms, in order to make them more socially engaging. Mary shared, "I had trouble building things so my room was really boring, so I never really liked to go back to it." When asked how the dorms contributed to the community, Jon says, "I don't know if it did ... I think if people tried to make their apartments more fun, I can see that being good for the community. People might want to hang out there more. Doesn't seem anyone else really has anything ..." Jon's room, complete with a pool table, hot tub, and fully functioning trampoline, attracted the most avatar visitors and even prompted an informal dance party to emerge. While not exactly educational, it did succeed in bringing students together to communicate online.

Although these spaces afforded social opportunities, students reported wanting more time to simply socialize in the dorm rooms and visit each other. Chrissy explained that "Everyone was separated in their own dorm room ... a real dorm has a multipurpose room. So if they had something like that, to hang out in, that would be fun." In this case, communication would not be academic content-related, but it would serve to strengthen group cohesion and increase social presence, supporting future collaboration. To broaden students' experiences beyond their immediate peers, each student was assigned to share the dorm room with a student from another university class that was using SL. The intention of this arrangement was that these distanced roommates could leave notes for one another, furnish the room, and occasionally meet at the same time from different physical locations. However, the students in our class were not explicitly aware of this plan, and communication was compromised. Annie recalled:

> I was assigned a room and the next time I came back, it was all decorated. Once the names were put up in the rooms, I noticed some guy named Dave was living in there from another class. I thought, well I can't move his stuff out.

Assigning distanced roommates is ideal when both are made aware of the other's virtual and "real-life" identities and can interact at the same time.

Conference Hall
Just as "real-life" dorm rooms encourage social communication in the virtual environment, "real-life" classrooms do as well. On the final computer lab session, students filed their avatars into a conference hall on the university island and listened to a distanced guest speaker give a presentation about environmental communication. Although it may seem uninspired to hold class in a virtual conference hall given all the opportunities available, the most positive statements concerning SL were for this event, from gamers and nongamers alike. The authenticity of the setting encouraged them to feel immersed as an online community. Sam said, "I did feel like I was there because everyone's avatars were actually sitting in the lecture hall, as if it were a real class." Uma liked this factor of group cohesion, explaining that "it was neat to see our entire class in the room while listening to the lecture through SL." Oliver added, "I was actually taken aback by how real the SL presentation felt. I felt as though we were connected as a class to what [guest speaker] was talking about."

A major factor that contributed toward communication was the guest speaker's use of his own voice to deliver the presentation instead of text. His use of voice appeared to positively influence communication. Mary wrote,

"Having attended a group presentation in SL, I see how the richness of the interaction is increased with voice. I felt as if [guest speaker] was indeed a functioning human and not just a computer generated image." Harriett, usually hesitant to interact with others, shared, "Hearing his actual voice made me feel more connected than I have ever felt in-world. This was the first time that I honestly felt like I was present." At the conclusion of the presentation, the class traveled to a virtual island the speaker had mentioned. Patricia wrote, "I thought that it was cool that we could all go and see the places that he was actually talking about." In this session, SL allowed the class to interact with a distanced presenter and immediately take a field trip to an authentic space with him, something they could not have done in the traditional classroom.

CONCLUSION

Educators of Arts and Science, particularly Communication instructors, are encouraged now more than ever to embrace the ever-increasing array of social media that is available through the Internet. Although 3D online social virtual worlds like SL support content creation and collective knowledge building, the limitation is that most were not specifically developed to support learners. The challenge for educators is to determine how to foster and assess learning experiences in these environments. The findings of this chapter suggest the following important instructional considerations. First, the teacher should make sure the class knows each other's names and appearances in SL as nongamers may be more reluctant to speak to someone they do not know in the virtual world, even if they are in the same classroom. Second, it is important to make the design of the virtual social and learning spaces more "real world" for inexperienced students. When the instructor encourages students to act like they should and gives them clues on how to act, the students can see more connections between real and online. We found that the computer lab context derailed virtual communication and diminished the need for online identity. Therefore it is helpful to have at least some of the sessions held at a distance. Consistent technical support is an absolute necessity for a class with nongamers. Support should cover not only technical skills, but also talk of the norms of this online culture as students would need both of these facets to gain more confidence in this environment. A class going into the virtual world will have markedly different experiences depending on their gamer status, avatar appearance, physical proximity, and virtual proximity. Therefore it is important that the teacher closely monitor how students' communication processes are shaped and foster successful learning experiences in this environment.

As this environment is rapidly growing in popularity, more research should explore its educational potential and find more diverse and innovative approaches. For example, future research can directly compare SL interactions with traditional online asynchronous interactions. Are there differences in social and cognitive indicators? How does the synchronous nature of SL interaction compare with the synchronous nature of other types of social media? In addition, more research is needed to consider diverse populations, settings, and learning endeavors.

REFERENCES

Annetta, L., Murray, M., Laird, S., Bohr, S., & Park, J. (2008). Investigating student attitudes toward a synchronous, online graduate course in a multi-user virtual learning environment. *Journal of Technology and Teacher Education, 16*, 5–34.

Boellstorff, T. (2008). *Coming of age in Second Life: An anthropologist explores the virtually human*. Princeton, NJ: Princeton University Press.

Corbin, J. M., & Strauss, A. L. (2008). *Basics of qualitative research: Techniques and procedures for developing grounded theory*. Thousand Oaks, CA: Sage.

DeLucia, A., Francese, R., Passero, I., & Tortora, G. (2009). Development and evaluation of a virtual campus on Second Life: The case of SecondDMI. *Computers & Education, 52*, 220–233.

Delwiche, A. (2006). Massively multiplayer online games (MMOs) in the new media classroom. *Educational Technology & Society, 9*(3), 160–172.

Dickey, M. (2005). Engaging by design: How engagement strategies in popular computer and video games can inform instructional design. *Educational Technology Research and Development, 53*(2), 67–83.

Dieterle, E., & Clarke, J. (2009). Multi-user virtual environments for teaching and learning. In: M. Pagani (Ed.), *Encyclopedia of multimedia technology and networking* (pp. 1033–1041). Hershey, PA: Information Science Reference.

Dimitriadis, G., & Kamberelis, G. (2006). *Theory for education*. New York: Routledge.

Edirisingha, P., Nie, M., Pluciennik, M., & Young, R. (2009). Socialisation for learning at a distance in a 3-D multi-user virtual environment. *British Journal of Educational Technology, 40*, 458–479.

Feldon, D. F., & Kafai, Y. B. (2008). Mixed methods for mixed reality: Understanding users' avatar activities in virtual worlds. *Educational Technology Research and Development, 56*, 575–593.

Fields, D. A., & Kafai, Y. B. (2007). Tracing insider knowledge across time and spaces: A connective ethnography in a teen online game world. Paper presented at the 2007 Computer Supported Collaborative Learning Conference, New Brunswick, NJ.

Garrison, D. R., & Arbaugh, J. B. (2007). Researching the community of inquiry framework: Review, issues, and future directions. *Internet and Higher Education, 10*, 157–172.

Garrison, D. R., & Cleveland-Innes, M. (2005). Facilitating cognitive presence in online learning: Interaction is not enough. *American Journal of Distance Education, 19,* 133–148.

Gee, J. P. (2007). *What video games have to teach us about learning and literacy.* New York: Palgrave Macmillan.

Jarmon, L., & Sanchez, J. (2008). The Educators Coop: A virtual world model for real world collaboration. *Journal of the Research Center for Educational Technology, 4*(2), 66–81. Retrieved from http://research.educatorscoop.org/ASIS&T08_Jarmon_and_SanchezREVISED.pdf.

Jennings, N., & Collins, C. (2007). Virtual or virtually U: Educational institutions in Second Life. *International Journal of Social Science, 2*(3), 180–186.

Johnson, L. (2008, April). Thru the looking glass: Why virtual worlds matter, where they are heading, and why we are all here. Address presented at the Federal Consortium on Virtual Worlds, Washington, DC. Retrieved on March 21, 2009 from http://www.ndu.edu/irmc/ThruLookingGlass.pdf

Jones, J. G., Morales, C., & Knezek, G. A. (2005). 3-dimensional online learning environments: Examining attitudes toward information technology between students in Internet-based 3-dimensional and face-to-face classroom instruction. *Educational Media International, 42,* 219–236.

Kafai, Y. B. (2008). Gender play in a tween gaming club. In: Y. Kafai, C. Heeter, J. Denner & J. Sun (Eds), *Beyond Barbie and Mortal Kombat: New perspectives on gender and computer games* (pp. 111–124). Cambridge, MA: MIT Press.

Kao, L., Galas, C., & Kafai, Y. (2005). "A totally different world": Playing and learning in multi-user virtual environments. Paper presented at the Digital Games Research Association Conference, Vancouver, Canada. Retrieved from http://www.digra.org/dl/db/06275.00211.pdf

Levinson, P. (2009). *New new media.* Boston: Allyn & Bacon.

Malone, T. W., & Lepper, M. R. (1987). Making learning fun: A taxonomy of intrinsic motivations for learning. In: R. E. Snow & M. J. Farr (Eds), *Aptitude, learning and instruction. Volume 3: Conative and affective process analyses* (pp. 223–253). Hillsdale, NJ: Erlbaum.

Mayrath, M., Sanchez, J., Traphagan, T., Heikes, J., & Trivedi, A. (2007). Using Second Life in an English course: Designing class activities to address learning objectives. Paper presented at the World Conference on Educational Multimedia, Hypermedia and Telecommunications 2007, Chesapeake, VA. Retrieved on January 21, 2009 from http://research.educatorscoop.org/EDMEDIA07.proceeding.pdf

McGerty, L. (2003). Nobody lives only in cyberspace: Gendered subjectivities and domestic use of the Internet. In: J. Turow & A. Kavanaugh (Eds), *The wired homestead* (pp. 337–346). Boston: Massachusetts Institute of Technology.

McGowan, M. (2010, February 6). New virtual world could revolutionize education. *Lubbock Avalanche-Journal.* Retrieved from http://lubbockonline.com/stories/020610/fea_559349620.shtml

New Media Consortium (NMC). (2007). *The Horizon Report.* Austin, TX. Retrieved on April 5, 2009, from http://www.nmc.org/pdf/2007_Horizon_Report.pdf

Sanchez, J. (2007). A sociotechnical analysis of Second Life in an undergraduate English course. Paper presented at the World Conference on Educational Multimedia, Hypermedia and Telecommunications 2007, Chesapeake, VA. Retrieved on January 21, 2009, from http://research.educatorscoop.org/ed-media.htm

Schultze, U., & Rennecker, J. (2007). Reframing online games: Synthetic worlds as media for organizational communication. In: K. Crowston & S. Wynn (Eds), *IFIP international federation for information processing* (Vol. 236, pp. 335–351). Boston: Springer.

Stein, D. S., Wanstreet, C. E., Glazer, H. R., Engle, C., Harris, R., Johnston, S. M., et al. (2007). Creating shared understanding through chats in a community of inquiry. *Internet and Higher Education, 10*, 103–115.

Taylor, T. (2003). Multiple pleasures: Women and online gaming. *Convergence, 9*, 21–46.

Vygotsky, L. S. (1978). *Mind in society*. Cambridge, MA: Harvard University Press.

Wertsch, J. V. (1998). *Mind as action*. New York: Oxford University Press.

Woo, Y., & Reeves, T. C. (2007). Meaningful interaction in web-based learning: A social constructivist interpretation. *Internet and Higher Education, 10*, 15–25.

Yee, N., Bailenson, J., Urbanek, M., Chang, F., & Merget, D. (2007). The unbearable likeness of being digital: The persistence of nonverbal social norms in online virtual environments. *CyberPsychology and Behavior, 10*(1), 115–121.

MULTI-USER VIRTUAL ENVIRONMENTS FOR INTERNATIONAL CLASSROOM COLLABORATION: PRACTICAL APPROACHES FOR TEACHING AND LEARNING IN SECOND LIFE

Annie Jeffery, Scott Grant and Howard M. Gregory

ABSTRACT

With an increasing trend toward the use of participatory culture and networked learning in education, opportunities to explore real examples of participatory culture are invaluable. Interwoven into seemingly simple collaborations are pedagogical, cultural, knowledge management, social, temporal, technical, as well as legal issues. A further layer of complexity is added when considering international networks and collaborations. However, such issues add a level of understanding important to participatory cultures. Enabled by communities of practice, and social constructivist learning, a range of bricoleur skills are developed from technical to higher level cognitive skills amongst students. These skills map many aspects of Jenkins' Participatory Culture, and the skills essential to our 21st century students.

Teaching Arts and Science with the New Social Media
Cutting-edge Technologies in Higher Education, Volume 3, 189–210
ISSN: 2044-9968/doi:10.1108/S2044-9968(2011)0000003013

In this chapter, we review an empirical study where the 3D technology, the virtual social world Second Life, supported learning for 21st century digital learners and how social networking and scaffolding contributed to international educational collaboration.

INTRODUCTION

Virtual online and social learning is increasingly used worldwide at all educational levels. The cost of laptop computers, smart phones, and broadband Internet connections have significantly dropped over the past few years (Kharif, 2009), becoming affordable for an increased number of students. Easily available tools such as Web 2.0 and Web 3.0 tools have also developed from their infancy, become more user-friendly and are often freely available on the Internet. These tools contribute to a new kind of learning that is engaging, interactive, and real-time. Technology today contributes to learning by granting opportunities unavailable to students in the past. As Driver and Driver (2008) suggest the new 3D Web represents the combination of social media, digital media, new technologies, and the maturing Internet user. This new Internet is immersive and "promises to uncover previously unheard-of dimensions in engagement, which will in turn increase [educational and] workforce collaboration, effectiveness, and retention" (Driver & Driver, 2008).

Virtual worlds (VW) today are in the same position as the World Wide Web was 15 years ago, confusing and brimming with unrealized potential. "If the telephone, radio, film, and TV helped to define life in the twentieth century, the VW is the one true new medium of the twenty-first century. The VW combines aspects of all of these earlier technologies, creating something novel in human experience" (Damer, 2008). Foreshadowing the current state of the Internet today and the likely move to the 3D Web, Murray (1998) observed that an ever growing amount of human culture, education, entertainment, business, and communication is now contained within and accessed through the networked computer. "The digital domain is assimilating greater powers of representation all the time, as researchers try to build within it a virtual reality that is as deep and rich as reality itself" (Murray, 1998).

THE ART OF THE SOCIAL BRICOLEUR

There are many definitions of VW, but for our purposes it is the potential of networks through which collaboration occurs, that is of primary importance. Bell's redefinition in 2008 highlights both the networks of people and of machines as a "synchronous, persistent network of people, represented as avatars, facilitated by networked computers" (Bell, 2008). It is these networks that offer new opportunities for online education in general, and for VW in particular.

Bricolage, originally developed by Lévi-Strauss (1990), has proved useful to a wide variety of disciplines. The term's origins lie in the French concept of the bricoleur, one who engages in 'Do-It-Yourself' to repair and build using the tools and materials that come to hand. Within education, Turkle and Papert discuss bricolage as a style of 'soft' programming, whereby the learner engagement with the programming process is more 'painterly' that computer engineering. Rather than detailed planning and development by section, the learner might create and expand her code through experimentation (Turkle & Papert, 1990).

> Anne's case makes it clear that the difference between planners and bricoleurs is not in quality of product, it is in the process of creating it. As in the case of Alex, Anne does not write her program in "sections" that are assembled into a product. She makes a simple working program and shapes it gradually by successive modifications. She starts with a single black bird. She makes it fly. She gives it color. Each step is a small modification to a working program that she has in hand. If a change does not work, she undoes it with another small change. She "sculpts." At each stage of the process, she has a fully working program, not a part but a version of the final. (Turkle & Papert, 1990)

In their work on constructionism, Papert and Harel discuss styles for solving problems, they suggest that "contrary to the analytical style of solving problems bricolage [could be used] as a way to learn and solve problems by trying, testing, [and] playing around" (Papert & Harel, 1991).

At a more organizational level, Weick (2001) suggests that provision for the following should be made if bricolage is to be successful:

- intimate knowledge of resources
- careful observation and listening
- trusting one's ideas
- self-correcting structures, with feedback

A learner in a VW may enlist the help of surrounding communities for materials and expertise, to quickly master skills and achieve objectives. It is these found objects, advisors, and skills that form the toolkit and materials for the social bricoleur, and it is these skills that the social network learner employs within VW and more importantly within experimental learning scenarios. There is an element of "Do-It-Yourself" to much of the teaching and learning currently taking place within VWs such as Second Life (hereafter referred to as SL). In 2009, before the implementation of SL's viewer 2.0, the support for social media was limited or non-existent. Both teachers and learners were making do with whatever they found around them, rather than having access to dedicated educational and social media resources. As we see from the case study, this form of social bricolage was very much in evidence through this BSU/Monash project.

LITERATURE REVIEW

Communities of Practice (CoP) were originally identified through the working practices of Xerox repair representatives in a 1991 study examining "the variance between a major organization's formal descriptions of work (...) and the actual work practices performed by its members" (Brown & Duguid, 2000). Wenger, McDermott, and Snyder (2002) described how CoP are successful over time, because they are made up of volunteers who generate excitement, relevance, and value with regards to a common passion. Wenger et al. call this spark "aliveness," something that cannot easily be designed. Jones and Bronack (2007) support the idea that CoPs are not social per se but form around accomplishing tasks that matter to those involved. The "aliveness" of the CoP comes from sharing, inviting, transferring the enthusiasm of the group for the task at hand – making others excited about being part of the community and its purpose. Since the publication of Brown and Duguid's (2000) original work, CoPs have benefitted from both "Web 1.0" and "Web 2.0" technologies, and it is assumed that these communities will continue to benefit and grow within 3D environments.

Participatory culture as defined by Jenkins, Clinton, Purushotma, Robison, and Weigel (2006) is a culture where strong support exists for creating and sharing one's creations, and passing knowledge from the more experienced to the novice. Members of this culture believe their contributions matter and feel socially connected to one another. SL provides an excellent environment for participation, learners may collaboratively build, script, train each other, and hold discourse in real-time as well as take part in events, groups, communities, and role-play activities.

Many universities have used and are using VW for educational purposes. For instance, educators at Appalachian State University were early adopters of the technology and developed the concept of Presence Pedagogy. This pedagogy is based on social constructivism, wherein learners construct learning together fostering a collaborative environment or CoP. In this environment, each learner becomes a potential instructor, peer, expert, and novice (Bronack et al., 2008).

The border between who is a teacher and who is a learner is blurred because each can contribute their specific knowledge to the collaborative learning community.

Jones and Bronack (2007) discuss the importance for an instructor to support the development of learning communities in a 3D environment to allow students opportunities to communicate with others, from both inside and outside the class. The assigned instructor takes on the role of facilitator and helps students with interaction and feedback and to scaffold learning. SL enables synchronous and asynchronous discourse in many ways; in-world private instant messaging, text-chat, building, interaction, touch, sound, and voice. Indeed voice has been a welcome addition to SL, and in 2009 52% of people in SL used voice (Sharma as cited in Wakefield, 2009). Voice also played an essential element in the Monash University film projects, although the team needed to be mindful of the associated accessibility and technical issues. VW CoPs benefit from the range of communication options, and VW educators from the wealth of opportunities available to support learner preferences. (Hardaker, Jeffery, & Sabki, 2010). This variety of communication methods enhances the sense of presence a learner might feel. In Brown and Cairns' study (2004) users described this sense of presence as a feeling of actually being there while fully engaging in the activities, oblivious of time and reality with full attention to the activities in the world. In this sense, presence equates to the sense of Flow as described by Csikszentmihalyi (1990).

Research from Portugal (Esteves, Antunes, Fonseca, Morgado, & Martins, 2008) outlines the successful use of a SL-based CoP for teaching computer programming. Using an action research methodology, the researchers tried to alleviate the high failure rate students had experienced due to difficulties contextualizing programming lessons. Using SL to create and compile programs made the process engaging, whilst giving the students access to a CoP to help understand and fix programs, as well as allay any feelings of isolation or disconnection during the programming process (Gregory, 2009).

A study carried out in 2009 at Monash University in Melbourne, Australia, found that "collaborative language activities in an immersive VW improved students' self-efficacy beliefs about their capacity to use Chinese language in a variety of real-life contexts" (Henderson, Huang, Grant, & Henderson, 2009). The study involved 100 undergraduate students engaged in a beginners' level Chinese language and culture course and was based on a collaborative activity in a Chinese restaurant in SL to identify and order food in Mandarin.

At Boise State University (BSU), the Educational Technology Department (EdTech) started teaching courses in SL in January 2007. In her
research there, Dawley (2009), identified that 3D VW environments, such as
SL, lend themselves particularly well as frameworks for the ways people
interact, teach, and learn collaboratively. Dawley found that social network
technologies impact on learners' thinking processes and the development of
consciousness and refers to this as social network knowledge construction
(Dawley, 2009). Twenty-first century learners are no longer passive
receptacles of spoon-fed learning, but seek out learning, building knowledge
together in a more social way. These networks and communities are able to
share and collaborate in increasingly easy ways across national and cultural
divides.

From these studies, it can be seen that SL lends itself very well to the
culture of participatory education (Jenkins et al., 2006), social network
knowledge construction (Dawley, 2009), and presence pedagogies (Bronack
et al., 2008). It is a widespread view that collaboration and communication
are essential in the 21st century classroom. Learners need transferable skills
that will remain relevant over the course of their professional careers. This
makes collaboration, networking, and communication essential skills for
educators to teach and students to learn (Brown & Adler, 2008). In this
chapter, we focus on a learning community in SL.

CASE STUDY

Background

In 2009, two instructors, one from BSU in the United States, a second from
Monash University in Australia, explored the potential for a joint project
between two seemingly disparate courses: Social Network Learning in
Virtual Worlds and Chinese Media Studies. To envision the combination of
skills and potential benefits, they created the relationship diagram outlining
both instructor and student relationships (Fig. 1).

BSU's graduate level course, Social Network Learning in Virtual Worlds,
was developed as part of BSU's pioneering program of classes in VW, which
are primarily offered in SL and attract learners from many backgrounds and
nationalities. Offered through the Department of Educational Technology,
the course explored social media and networks in combination with emerging
technologies and pedagogies in VW. At Monash University undergraduate
students, enrolled in Mandarin Chinese language and media literacy have

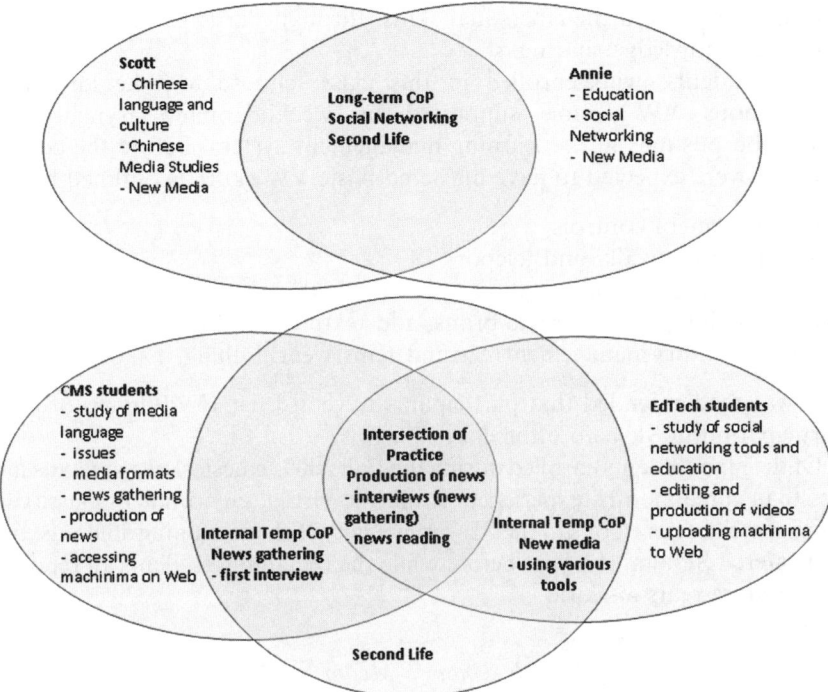

Fig. 1. Project Intersections of Practice.

been using SL to engage in interactive learning tutorials and simulation for the past two years.

EdTech 597: Social Networking in Virtual Worlds

In 2008, the BSU instructor developed a course for BSU that had never been taught in this environment before – EdTech 597: Social Network Learning in Virtual Worlds. The course, offered as an elective in the Master of Educational Technology Program, attracted nine students in 2009. The course consisted of weekly class activities, mandatory two-hour in-world meetings, reflective action including weekly readings and discussion, peer teaching, social networking participation, and a final synthesis project to wrap up the class (Jeffery, 2009). The students taking this class were expected to become active participants, expending time and effort to build the community and their personal skills, through participation, collaboration, and knowledge sharing.

The students, who enrolled in this class, choose to take an online synchronous VW course supported by weekly online asynchronous discussion postings in the learning management system. Before the course, students were expected to have mastered basic VW skills as outlined below:

- Use of camera controls
- Ability to fly, walk, and teleport
- Knowledge of communication tools (IM and chat)
- Basic building skills (create prims, add textures, and content)
- Basic inventory management (can find items, wear clothing, make a notecard)

It was recommended that participants have at least 15–20 hours of prior participation in SL before the course begins.

Of the nine students enrolled during the Fall 2009 semester, two students had less than three months experience within the virtual environment classifying them as novices or noobs in the SL vernacular. Of the remaining students, two had entered SL almost a year before while the remaining students all reported two or more years in-world.

Monash: Chinese Media Studies

At Monash, the instructor had engaged in Chinese language literacy for 10 years. The Chinese Media Studies class of fall 2009 had 28 students enrolled. Students in the Chinese language learning classes were engaged in synchronous interviews with specially invited Mandarin Chinese speaking guests. Recording students during in-world interviews and news reports, then posting the resulting videos to the Internet, helps both the students and the instructor to review the effectiveness of the language learning. This provided the students with the opportunity to enhance and review passive learning in an active, creative, and purposeful way. Thus, the students and instructor were able to follow the development of the language learning process in both visual and aural modes and fill in any knowledge gaps. This also provides the opportunity for peers, friends, and family to view the resulting interviews online.

The Project Proposal to BSU Students

Early in the Fall semester, the BSU graduate students were offered an opportunity to take over the entire production process of a VW video project including locations, set-up, machinima capture (creation of animated video by the screen capture of 3D environments), production, and editing of the videos. They would facilitate the entire process providing support to the Monash undergraduate students. The idea for the collaboration with Monash initially arose during an informal conversation between the BSU and the Monash instructor. It became evident from the discussion that some form of collaboration could be beneficial for both groups of students and instructors. The BSU instructor proposed that she offer the collaborative filming as a potential final project to her students. The BSU course was a unique and experimental graduate course, and any students wishing to take up the offer would get the chance to participate in a real life learning scenario with educators and students around the world. The project would give the BSU students a number of benefits and opportunities; to form wider social networks and expanded CoPs centered on a real life project; to use social media in a genuine learning situation, and to utilize a number of communication tools and strategies taught as part of the course. Additionally, an increasing interest in Digital Storytelling as a form of pedagogy had made machinima a useful skill set for students.

Overall, the BSU and the Monash instructors thought that transferring control of the filming to students would enable the BSU students to collaborate on an authentic Social Network Learning project for their final assessment, and for the Monash instructor to concentrate on teaching and learning. The Monash students would benefit by the ability to concentrate on their learning, rather than dividing their focus mastering the skills needed to create machinima projects.

The project was on a voluntary basis, and involved two separate rounds of filming, which would count as the capstone project for the BSU class. Several groups of Monash students interviewed native Chinese speakers as part of their assessment, which was recorded as part of the first filming session. During a follow up session, the Monash students would present a news desk report, focused on their group interviews, which needed to include a breakaway to an on-site reporter. For the BSU students, a successful film project would entail;

- Choosing a studio location and adapting it for use during the interview and news desk sessions.
- Coordinating film crews responsible for recording the interviews on multiple dates.
- Creating studio sets.
- Finding locations for on-the-spot reporter section of News Desk reports.

- Editing the resulting video files adding titles and end credits.
- Uploading the edited video files to YouTube by the deadline.

BSU Student Acceptance

Eight BSU students expressed interest in the project. However, an informal IM poll by the BSU course instructor discovered two major concerns; the amount of work needed to complete such a project and the novice skill levels of some participants. As one student put it "I think this project is going to stretch my skill set, I worry that the noobies (new users of SL) are going to be overwhelmed." However, the most pressing concern for the BSU students was the timing; the first scheduled Monash filming (Table 1) was to take place in the third week of the BSU class, allowing for only eight days of preparation. This was caused by unexpected differences between the US and Australian academic years, whilst it was the beginning of the fall semester in the United States, Monash students were nearing the end of their academic year and getting ready for the summer vacation. The expectation of the BSU instructor and the two graduate students who took the lead preparing for the first round of filming was that eventually all the eight BSU students who had expressed an interest would participate.

Table 1. Chat Show Interviews Schedule.

Date/Day	Group	Interviewee	Time (Australia)	Time (China/ Singapore)	Time (USA-SLT)	Status
17.9.09 Thursday	Group 3	Leo Pey	0 11:00:00	09:00:00	Wednesday 18:00:00	Confirmed
17.9.09 Thursday	Group 4	Greennote	0 12:30:00	10:30 am (20 mins)	Wednesday 7:30 pm	Confirmed
24.9.09 Thursday	Group 5	Jay	0 11:00:00	09:00:00	Wednesday 18:00:00	Confirmed
24.9.09 Thursday	Group 1	Ling	0 11:00:00	09:00:00	Wednesday 18:00:00	Confirmed
24.9.09 Thursday	Group 6 & 2	Prof. Tim Xie + Suzhou Uni Student	14:00:00	12 noon	Wednesday 9:00 pm	TBC
24.9.09 Thursday	Group 6 & 2	Suzhou Uni Student(s)	14:00:00	12 noon	Wednesday 9:00 pm	TBC

BSU Community Communications

The BSU students, facing a deadline arranged to meet the following Saturday, five days before the anticipated filming. Owing to the short notice and other commitments only two of the students attended, and as a result they took the initiative to move the project forward. They decided to use the Monash Lecture Hall as the studio location for the interview sessions, contacting the Monash instructor to request building permission for the location, inquire about filming times and access rights for the filming location. In addition, they choose to assume the roles that best suited their experiences and skill. The first student assumed the role of producer, assuming responsibility for all filming and editing, as she was familiar with advanced video editing tools and the Fraps screen capture software. The second student assumed the role of production coordinator due to his studies in the field of knowledge management. He created a project wiki to disseminate information regarding the meeting, the project schedule, student roles and the ever expanding tasks which needed to be completed in order to finish the project successfully. The growth of Web 2.0 has made the use of a wiki invaluable to enabling a CoP. Because the BSU class members had found the WikiSpaces (which they initially used for their course interactions) user interface to be cumbersome the production coordinator moved to PBWorks, a platform in which he was more comfortable and familiar. PBWorks has a more user friendly, intuitive interface which the experienced Web 2.0 student found easy to use when paired with the WYSIWYG (what you see is what you get) type interface. The students and instructors who used the new platform appeared to agree with this assessment.

Project Wiki

The project Wiki's first page was a draft for a scope note outlining the project requirements including sections outlining the opportunity, background, success criteria, risk, and risk reduction necessary for the project to complete with ease. These requirements were gathered from the in world meeting between the two course instructors and the BSU students. The production coordinator who created the document assumed that his instructors and fellow students would review his draft document and collectively correct mistaken assumptions. Indeed, the use of a wiki for review and editing was an invitation to collaborate which, unfortunately, did not occur. With participation limited to a few enthusiasts, several assumptions persisted until after the first filming.

The BSU students scheduled a second project meeting three days prior to the first scheduled filming. This meeting was attended by both course instructors as well as the students who had assumed the producer and production coordinator roles. Four more of the BSU students attended, including one who assumed the role of set designer and one who accepted the role of location scout for the planned news desk on-the-spot report sessions. During this meeting it was confirmed that Fraps would be the video capture software and that both the producer and production coordinator (acting as back up camera) had the software installed. Location decisions that had been made at the first meeting were shared and explained.

During this second meeting the project scope, outlined when the project was offered to the BSU students, began to be tweaked and expanded. During this meeting the Monash instructor shared expectations that added to the project scope: besides a project title within the title and end credits, each video needed to include student and guest names and English sub-titles (a request that proved too labored and time intensive when considering the projects time frame). The students' analysis phase of the project was rushed, and the Production Coordinator's requirements document had not been edited to include additional requirements (by either students or instructors). This lack of refinement of project requirements can be seen as the first contributor to project scope creep.

Owing to time commitments only the producer and production coordinator proved available to prepare for and film the first interview sessions. The production coordinator, therefore, accepted the role of set designer and assembled the required daytime talk show set capable of seating five to six avatars. Filling vacancies in project roles became a common event for the producer and the production coordinator. The Monash student schedule did not allow for either set approval or rehearsals before the live events. Indeed, the Monash and BSU students first met during the first filming session. Before the filming Monash students were given a typical interview shooting schedule. The shooting schedule was given in English and included a Chinese translation (Table 2).

First Monash Filming

Filming the interviews proved hectic. Owing to international time zones, the interview sessions took place during the morning in Australia, and evening for the producer and production coordinator located in the central and eastern United States. Monash students, still unfamiliar with the VW environment,

Table 2.   Chinese Media Studies TV Chat Show Shooting Schedule.

Start time − 5 mins	Sound check
Start time	Audience takes their seat; audience must sit down quickly
Start time + 5 mins	Guest arrives to be greeted by the interviewing group
Start time + 10 mins	Interviewing group and guest take their places on the sofa on stage
Start time + 15 mins	Interviewing group introduces themselves and their guest to the audience (both those present and the future viewing audience)
Start time + 17 mins	Interviewing group begins interview, asking both pre-prepared and spontaneous questions
Start time + 27 mins	Interviewing group winds up questions and prepares to thank guest
Start time + 30 mins	Interviewing group formally thanks guest and the audience and says goodbye; filming stops

Notes: Interviewing group must thank guest again off camera, and if there is a following interview, invite them to stay and watch if they would like. If there is a second interview, it would commence 5 minutes after the completion of the first.

needed additional time to find their way around the studio. The BSU students were still in the process of creating checklists for managing the filming process, whilst simultaneously learning to master the Fraps screen capture software. The first real project challenge occurred when the interview recording times averaged more than twice the expected 10-minute duration. Though, switching between student groups also created timing challenges.

An unexpected minor cultural issue occurred when the Monash instructor viewed the specially designed interview set shortly before the first filming. The set designer had purchased a single chair as a place of honor for the interviewee. However, the Monash instructor informed the BSU students that isolating a "guest" would be seen as impolite. The chair was therefore, replaced hastily with an additional sofa so that a student could sit next to the guest (Fig. 2). This illustrates that students needed to be aware of cultural differences, the necessity of project planning and value of dress rehearsals.

Interview Post-Production

While it had initially been the producer's intent to do all video recording and post production, with other students providing back-up, the size of the project

Fig. 2. In the SL Studio.

made this goal unattainable. Thus the producer and the production coordinator decided to share responsibilities to complete the video projects in a timely manner. The production coordinator had limited experience with the Fraps software and video editing, so he focused on using editing solutions within his experience for video post production, these included Microsoft Movie Maker, the shareware program Audacity (for audio editing) and Microsoft Photo Story 3 for the creation of title and end credits. The producer, focusing on video recording, used Adobe Premier, Adobe PhotoShop, SwishMax, PhotoStory3, GraphEdit, and Audacity. The producer found YouTube tutorials to increase their familiarity with Fraps and both students shared and developed tips and tricks as they became more proficient. The ability to use the same software throughout the process would have helped this sharing of ideas. However with such a strict filming schedule, it was decided that mastering a relatively new software solution would prove too time consuming and risk project delays or possible failure.

Once the initial interview filming and post production session had been completed, the student producer and production coordinator were able to take a moment to relax and regroup. With the assistance of their instructor, they began to explore ways to encourage more student participation in the hope of gaining assistance with the work load. A number of factors were identified as contributing to lack of participation from other students. The three most pressing issues were identified as:

- time issues (international time zones)
- skill level required
- role uncertainty

Filming times were dependent on the Monash class schedule and could not to be adjusted due to the demands on lab resources and inflexible student schedules. The other barriers were addressed by creating pages for new user resources, and possible project roles in the wiki. The BSU instructor developed a list of project roles based on skill levels and possible areas of interest, which related to the reportage theme. The project team also offered their fellow BSU students weekly progress reports and posted links to the completed videos to stimulate interest. No further students stepped forward to offer help with video production. However, students did express an interest in acquiring some of the skills demonstrated in the project and these were subsequently added to the BSU class.

Structure within the Monash Course

With the need to learn technology in addition to SL removed, the Monash students were able to strengthen their engagement and focus on language learning. Working in groups, the undergraduate students were able to build their own CoP as they discussed and planned the interviews, the way these were to be conducted, what questions would be asked, and what specific role each member would take. Most of this was done outside of the virtual environment, either face-to-face or through email. In their respective cohorts, the more experienced learners shared their knowledge, passing it on to those with less expertise than themselves. An added dimension for these students was having to interact with the BSU students only through the virtual environment. The Monash students also had to negotiate what they felt was needed to achieve the goals set for the chat show interviews, and the practical considerations set by the producer and production coordinator for the whole shooting process. In a sense, the Monash students conformed to many aspects of participatory culture with people working collaboratively in the day-to-day physical world as well as a networked VW. They collaborated not only within their own group, but to some extent within broader groups; together with the BSU students. The two groups' common goals and interests were very much intertwined, contributing to each group's success in the participatory learning experience. Ideas on how this shared experience can be further explored and built on in the future are considered further in the discussion section.

News Desk Report Filming

The first filming session left the producer and projection coordinator more prepared and comfortable with the assignment even without additional assistance. The second round of filming revolved around each group creating a news desk report relating to the interview they had performed earlier. The production coordinator worked with the set designer to find a location for the news desk studio and the set creation while the producer edited the interview videos. The set designer donated the use of her sky sandbox for the news desk studio. Three sets, to accommodate the three different sized groups, were created. These second sessions took place on a single day over the course of four hours (Table 3).

PROJECT ISSUES

As stated before, several erroneous project assumptions were discovered during the post production process. The average length of the interview

Table 3. A Project Filming Schedule.

Date/Day (Australia)	Group	Time (Australia)	Time (USA-SLT)	Location	Status
15.10.09 Thursday	Group 1	11:00–11:20	Wednesday 14.10.09 6:00–6:20 pm	To be decided	
	Group 3	11:25–11:45	Wednesday 14.10.09 6:25–6:45 pm	To be decided	
	Group 4	11:50–12:10	Wednesday 14.10.09 6:50–7:10 pm	To be decided	
	Group 5	12:15–12:35	Wednesday 14.10.09 7:15–7:35 pm	To be decided	
	Group 2	2:00–2:20	Wednesday 14.10.09 9:00–9:20 pm	To be decided	
	Group 6	2:25–2:45	Wednesday 14.10.09 9:25–9:45 pm	To be decided	

Source: SLT Wednesday, October 14, 2009, News Desk Reports.

sessions turned out to be closer to 25 minutes rather than the expected 10, with the longest of the 10 produced videos being almost 33 minutes. This substantially increased the time required to film and edit the videos. Closely related to this, BSU students were originally planned to post the videos on YouTube, as had been proved successful in the past. YouTube, however, limits uploaded files to a maximum of 2GB in size and 10 minutes in duration. The BSU Producer chose to invest in a personal Vimeo account so that files could be uploaded in their entirety.

There was also an issue with the requirement that the Monash student names (both real and SL) be included in the credits. After the completion of the project, it was discovered that while there was implied consent there was no written consent for student names to have been included in the videos. Later it was pointed out that the same held true for those being interviewed, however prior to filming, each interviewee had been given a written description of the project and a list of questions that would be asked, which could also be seen as implied consent. This issue lead to a lengthy debate within the core group to which there was no clear resolution. The point became moot when the Vimeo account was closed at the end of the semester. When planning video productions such as this, release forms need to be obtained and kept on file.

The two BSU graduate students who filled the roles of producer and production coordinator were very satisfied with their work and their contributions. They sensed pride, and relief, to have been able to successfully take ownership of the entire production process and complete on time. However, the two students also sensed disappointment that more students had not been as fully engaged and involved in the project. For an online group to be successful, the basic requirement of investing in active participation, in the creating and consuming of content, needs to be in place for the group to be sustainable. People who seek to manage CoPs find that they spend considerable time and attention on interacting, controlling, and encouraging members' behavior and that this takes time (Butler, Sproull, Kiesler, & Kraut, 2002). The BSU graduate students had expected that all their classmates would choose to take part in this project voluntarily. However, ultimately two students took time from their other activities to complete a project for they felt responsible, to which they had been entrusted by the Monash students.

However, it should be noted, if we use Nielsen's (2006) 90-9-1 model as our metric for participation we exceeded expectations. We can expect 90% of the participants in online networks (CoPs, listserve, etc.) to observe but not take part, 9% participate in activities occasionally, and 1% to be very involved as leaders and facilitators (D. Bedford, personal communication, July 16, 2010). Viewed using this model, our collaboration can be considered very successful; we had two members of the group totally committed, two others offering varying degrees of assistance and the remaining four acting as observers.

LESSONS LEARNED

Logistical Issues

The timing of meetings and activities was a key concern to project participants as members were located in the United Kingdom, Oman, the United States, Taiwan, Australia, Singapore, and China. This meant that activities could occur anywhere over a 24-hour period and lead to participants experiencing schedule conflicts with school, work and sleep. For example, Australia is nearly 12 hours ahead of the United Kingdom, whilst the United Kingdom is five or six hours ahead of central and eastern America. Additionally, differences in the academic calendars meant that the Australian academic year was ending at a time when the US academic year was starting. It was these conflicting schedules which limited the time between project conception, filming and production. It should be noted that time-related issues are a common feature of all VW collaborations.

The flow of information between the BSU project team and the Monash student groups, outside of the studio filming sessions, was conducted through the Monash instructor. This filtering through the instructor created an information bottleneck, which slowed communication and put undue stress on him. A wide range of information needed to be conveyed including, links to tutorials, a widget for time zone conversion, potential filming locations, set-up and sound check instructions. If Monash students had been able to access the project wiki, they may have been better prepared for filming. Contributing to the wiki may have allowed them to share their expectations, make suggestions, and gain understanding of the filming process before the first session. This would also have alleviated the information bottleneck between the two groups.

The limited time between project conception and first filming may have been a factor for those BSU students who ultimately declined to take part in the film sessions. In less experimental scenarios, it would be beneficial to offer participants enough time to prepare physically and mentally when offering a skill-sharing project of this type. With more time, an instructor would expect to be able to provide the means for relevant skills acquisition.

Reducing the Risks of Failure

The BSU students found that the tight filming schedule allowed only one opportunity to capture the Monash student sessions on video. This meant that it was crucial to create a plan that allowed for redundancy in filming by creating backups in case of technical or connectivity issues.

To reduce the risks of failure, the production coordinator posted a project needs assessment to the wiki. Unfortunately, this assessment was never completed and many assumptions were only corrected after they had become problematic. A project of this size and scope requires the documentation of expectations to ensure both smooth collaboration and success. "Over" documenting the scope and requirements of the project will help catch problems before they happen. Detailed analysis of requirements helps prevent scope creep. Scope creep can best be defined as the addition of unplanned for requirements discovered over the course of the project. Scope creep can cause frustration and may threaten the success of the project. Thus the need for planning, documenting, and monitoring of the project reveals the need for students to develop basic project management skills and for instructors to share clearly documented expectations.

The BSU course instructor incorporated many of the skills for creating machinima into class activities after the filming had taken place; for instance camera control tutorials, an introduction to screen capture and digital storytelling. In future courses, there would be an investment in essential machinima skills early on in the course to encourage participation and prepare students for the filming sessions. Good machinima skills, like all skills, come with intense practice and a commitment to learning. Collaboration between cohorts could be made better with the use of the wiki. Greater wiki contributions could be generated by introducing an element of assessment into the activity. Additionally, the opportunity for careful project analysis and design in the early stages would alleviate many of the issues encountered during this experimental project.

FINAL THOUGHTS

VW educators often devote a great deal more time and energy to VW teaching and learning than they might deal more traditional delivery methods. However, the benefits for learners would seem to make this effort more than worthwhile in terms of collaboration, communication, social presence and the potential range of pedagogies available for exploration. Successful community building, as we have seen here, stems from people who are motivated and engaged. Butler et al. (2002) suggests that leader(s) perform community building expecting to get something in return from the experience. The two BSU graduate students in charge of the Monash filming found the prospect of working with an international group of students, on a project that would stretch and expand their skills, to be both challenging and exciting. As they reviewed the project they realized that while experiencing the stress, frustration, pressure and ultimate successes of the collaboration, they also enjoyed themselves. They relayed that they felt the filming sessions were fun, and that the completion and posting of the Monash videos was exciting.

The interplay and transfer of knowledge between contexts is a key aspect of both late 20th and 21st education. External knowledge provided by communities outside the project proved invaluable. For instance, BSU students increased their mastery of the video software used for filming and production with online videos, manuals and detailed tutorials. In addition, to the acquiring and developing new skills during the SL collaboration, students were able to bring their own external knowledge and abilities into the project. For instance, The producer was able to bring her existing video skills, whilst the production coordinator used the graphics skills learned making SL textures and audio editing skills gained editing the videos to create multi-media PDFs for work. This blurring of the realms of work and play has been documented in research within VW games (Yee, 2006). Re-examining this blurring of work and play could form the basis of further studies of student engagement, and the way in which VWs may enable that engagement.

By the end of the project, four out of the nine BSU students had taken part in some capacity. All nine had expressed interest in learning associated skills and acquired some level of these same skills. This outcome supports the view that such projects are worthwhile forms of learning for students even in experimental scenarios; good planning and co-ordination would enhance these benefits.

REFERENCES

Bell, M. (2008). Towards a definition of virtual worlds [think piece]. *Journal of Virtual Worlds Research*, *1*(1), 1–5. Available at journals.tdl.org/jvwr/article/view/283/237.

Bronack, S., Sanders, R., Cheney, A., Riedl, R., Tashner, J., & Matzen, N. (2008). Presence pedagogy: Teaching and learning in a 3D virtual immersive world. *International Journal of Teaching and Learning in Higher Education*, *20*(1), 59–69.

Brown, E., & Cairns, P. (2004). A grounded investigation of game immersion. Available at http://delivery.acm.org/

Brown, J. S., & Adler, R. (2008). Minds on fire. *Educause.* Available at http://www.johnseelybrown.com/mindsonfire.pdf

Brown, J. S., & Duguid, P. (2000). *The social life of information.* Boston, MA: Harvard Business Press.

Butler, B., Sproull, L., Kiesler, S., & Kraut, R. (2002). Community efforts in online groups: Who does the work and why? Available at http://repository.cmu.edu/hcii/90

Csikszentmihalyi, M. (1990). *Flow: The psychology of optimal experience.* New York: Harper and Row.

Damer, B. (2008). Meeting in the ether: A brief history of virtual worlds as a medium for user-created events. *Journal of Virtual Worlds Research, 1*(1). Available at journals.tdl.org/jvwr/article/download/285/239.

Dawley, L. (2009). Social network knowledge construction: Emerging virtual world pedagogy. *On the Horizon, 17*(2), 109–121.

Driver, E., & Driver, S. (2008). The immersive Internet make tactical moves today for strategic advantage tomorrow. Available at http://www.thinkbalm.com/wp-content/uploads/2008/11/thinkbalm-immersive-internet-report-nov-20084.pdf

Esteves, M., Antunes, R., Fonseca, B., Morgado, L., & Martins, P. (2008). Using Second Life in programming's communities of practice. Universidade de Tras-os-Montes e Alto Douro. Available at home.utad.pt/~leonelm/papers/CRIWG/MicaelaCRIWG.pdf

Gregory, H. (2009). Play vs. work in virtual worlds: An examination a community of practice in Second Life. Knowledge management research proposal. Unpublished manuscript, Kent State University.

Hardaker, G., Jeffery, A., & Sabki, A. (2010). Learning styles and personal pedagogy in the virtual worlds of learning. In: S. Rayner & E. Cools (Eds), *Style differences in cognition, learning, and management: Theory, research, and practice.* London: Routledge.

Henderson, M., Huang, H., Grant, S., & Henderson, L. (2009). Language acquisition in second Life: improving self-efficacy beliefs. Paper presented at the ASCILITE 2009 New Zealand. http://www.ascilite.org.au/conferences/auckland09/?m=Call-for-proposals&ss=Proceedings-publication.php

Jeffery, A. (2009). EdTech 597: Social network learning in virtual worlds. Available at http://bit.ly/1ZmpBR

Jenkins, H., Clinton, K., Purushotma, R., Robison, A. J., & Weigel, M. (2006). Confronting the challenges of participatory culture: Media education for the 21st century. Building a field of digital media and learning. Available at http://digitallearning.macfound.org

Jones, J. G., & Bronack, S. C. (2007). Rethinking cognition, representation, and processes in 3D online social learning environments. In: D. Gibson, C. Aldrich & M. Prensky (Eds), *Games and simulations in online learning: Research and development frameworks* (pp. 89–114). Hersey, PA: Information Science Publishing.

Kharif, O. (2009). How low can PC prices go? Business Weekly. Available at http://www.businessweek.com/technology/content/mar2009/tc20090310_258460.htm

Lévi-Strauss, C. (1990). *The savage mind.* Chicago: University of Chicago Press.

Murray, J. H. (1998). *Hamlet on the holodeck: The future of narrative in cyberspace.* London: The MIT Press.

Nielsen, J. (2006, October 9). "90-9-1" rule for participation inequality: Encouraging more users to contribute. Available at http://www.useit.com/alertbox/participation_inequality.html

Papert, S., & Harel, I. (1991). *Constructionism.* New York: Ablex Publishing.

Turkle, S., & Papert, S. (1990). Epistemological pluralism: styles and voices within the computer culture. *Signs: Journal of Women in Culture and Society, 16*(1), 128–157.

Wakefield, J. S. (2009). VOIP: Integrated voice in two massive multi-user online games. Available at http://bit.ly/bGPlcb

Weick, K. E. (2001). *Making sense of the organization*. Oxford: Blackwell Business.

Wenger, E., McDermott, R., & Snyder, W. (2002). Seven principles for cultivating a community of practice. Available at http://www.askmecorp.com/pdf/7Principles_CoP.pdf

Yee, N. (2006). The labor of fun: How video games blur the boundaries of work and play. *Games and Culture, 1*(1), 68–71.

USING MULTI-USER VIRTUAL ENVIRONMENTS IN TERTIARY TEACHING: LESSONS LEARNED THROUGH THE UQ RELIGION BAZAAR PROJECT

Helen Farley

ABSTRACT

Second Life, as a three-dimensional social medium, provides an unparalleled opportunity for people to interact with each other and their surroundings in unfamiliar and innovative ways. After a brief introduction to the discipline of Studies in Religion at the University of Queensland (UQ), this chapter will examine some of the key characteristics of MUVEs in general and of Second Life in particular, with a view to assessing its suitability as an environment for learning based on andragogical and constructivist methodologies. Further, it will explore the original conception and development of the UQ Religion Bazaar project within Second Life.

The UQ Religion Bazaar project was originally conceived in 2007 and developed through 2008. It consists of a Second Life island situated in the New Media Consortium educational precinct and boasts a number of religious builds including a church, a mosque, a synagogue, an ancient

Teaching Arts and Science with the New Social Media
Cutting-edge Technologies in Higher Education, Volume 3, 211–237
Copyright © 2011 by Emerald Group Publishing Limited
All rights of reproduction in any form reserved
ISSN: 2044-9968/doi:10.1108/S2044-9968(2011)0000003014

Greek temple, a Freemasons' lodge, a Zen Buddhist temple and a Hindu temple to Ganesha. The island was used in two large first-year classes and for supervising distance postgraduate research students.

INTRODUCTION

Second Life is an internet-based, three-dimensional social world which can be accessed by individuals via a software client which runs on a personal computer (Linden Lab, 2010). Since 2003, this virtual world has captured the imagination and ire of the general public, on the one hand concerned at the implications and complications for a first life, and on the other intrigued by the possibilities that such a flexible environment affords. Many educators fall into this latter category; in higher education institutions a vanguard of adventurous educators have been quick to spot the possibilities for innovative teaching and learning in such environments, providing an unparalleled opportunity for people to interact with each other and their surroundings in unfamiliar and innovative ways (Conklin, 2007). Second Life is the foremost Multi-User Virtual Environment (MUVE) exploited by educators, with hundreds of tertiary institutions offering classes partly or entirely within Second Life (NMC, 2010). The UQ is also represented there, with the Studies in Religion discipline purchasing a Second Life island in 2007 and constructing a number of religious builds for use in two large first-year courses. After a brief introduction to the discipline at the university, this chapter will examine some of the key characteristics of MUVEs in general and of Second Life in particular, with a view to assessing its suitability as an environment for learning based on andragogical and constructivist methodologies. Further, this will be conducted with particular reference to the first-year Studies in Religion curriculum at the UQ. The chapter will conclude with some recommendations educators should keep in mind when thinking about teaching and learning in an MUVE, formulated in light of the UQ Religion Bazaar project (Fig. 1).

STUDIES IN RELIGION AT THE UNIVERSITY OF QUEENSLAND

The Studies in Religion programme at the UQ seeks to educate under-graduates and postgraduates about various ethical, cultural and religious systems in alignment with the graduate attributes espoused by this university. Similar attributes are promulgated by most tertiary-level institutions and

Fig. 1. The UQ Religion Bazaar Island in Second Life.

include the acquisition of knowledge about other cultures, the fostering of intercultural communication, an appreciation of cultural diversity, historical consciousness and a global perspective. Traditionally, content has been delivered via a tutorial and lecture programme, supplemented by assessment that necessitates field trips to religious spaces situated in South East Queensland.

The main focus of this study will be the introductory level courses offered within the discipline, namely *Meditation and Soul Journeys: Eastern and Western Spiritual Experience* and *Introduction to World Religions.* These courses form the core of the Studies in Religion major within the Bachelor of Arts degree and are compulsory for students undertaking a major or a double major in this discipline. The combined enrolment of these classes is between 150 and 200 students per semester. Although these courses are coded as introductory, they are popular electives for students in a variety of programmes including psychology, science and engineering, as well as other majors within arts. Consequently, students bring with them a wide range of prior knowledge and experience; for example, recently a PhD candidate enrolled in a medical science programme sought special permission to participate in the *Meditation and Soul Journeys* course.

The greater proportion of enrolments is comprised of first-year students – 18 or 19 years old – new to tertiary education; even so, mature-age students ranging in age from their early 20s through to the late 50s, form a significant proportion of this cohort. Though figures are not available for the numbers of these students enrolled in these courses, McInnis reports that nearly half of students commencing university studies in Australia are mature age (2001, p. 106). In addition, these courses have no prerequisites so there is no assumed knowledge, making it challenging to cater to all learners' needs. This increasing diversity of first-year students has long been recognized as a feature of Australian university programmes; for example, see McInnis and James (1996).

In order to give students a broad experience of other religious traditions, assessment in both of the large, first-year courses in Studies in Religion, required that students go to various places of worship around South East Queensland in order to observe and investigate religious practice. There were a number of issues with this approach. Firstly, Australia is a predominantly western Christian country. Even though there are significant populations from other cultures resident in the area, there are still relatively small numbers of Taoist temples, Hindu temples, Sikh temples and so on. Secondly, given that significant proportions of students are completing other courses, working to supplement their income, are sometimes parents themselves and are otherwise juggling the competing demands placed upon them, it is often prohibitively time-consuming to complete the assessment tasks, especially the off campus visits, to a satisfactory degree. And because of the public liability insurance implications, the institution is unwilling to streamline the process by providing transport and appropriate supervision. Finally, there are ethical implications when sending students to observe worship and other religious practices. Though most religious adherents are happy to let students observe them, some are disturbed by this. I discovered through informal discussions with practitioners that some felt like 'animals in a zoo' when students observed their worship. Though this exercise was considered to be invaluable in exposing students to an array of religious practices, the significant disadvantages rendered it desirable to find an alternative means of exposing students to alternative cultures and religions. As the course co-ordinator of both *Introduction to World Religion* and *Meditation and Soul Journeys: Spiritual Experience East and West*, I believed that perhaps Second Life could furnish a solution while still being consistent with my own beliefs about andragogical and constructivist learning.

THE RELEVANCE OF ANDRAGOGY AND CONSTRUCTIVIST METHODOLOGIES

Students enrolled in first-year Studies in Religion courses necessarily bring with them a wide range of experiences and knowledge. In order to design effective learning for all, it becomes necessary to consider appropriate approaches to teaching and learning taking into account this variety of experiences. Traditionally, when thinking of education and how educators participate in that process, it is discussed in terms of 'pedagogy'. But an analysis of what that term implies indicates that its use in the context of tertiary education is inappropriate. 'Pedagogy' literally means 'the art and science of helping children learn' (Knowles, 1980, p. 43). Though not the first to make the distinction, American educator Malcolm Knowles pioneered the development of adult learning theory, adopting the label 'andragogy' to distinguish it from pre-adult learning (Merriam, 2001, pp. 4–5).

Certain assumptions about the adult learner underlie the application of andragogy. As summarized by Merriam (2001), the adult learner is characterized as follows: (1) possesses an independent self-concept in addition to directing his or her own learning, (2) has accrued a reservoir of experience which potentially is a valuable resource for learning, (3) has learning needs which correlate to changing social roles, (4) is problem-centred and prefers knowledge that is immediately applicable and (5) is inspired to learn by internal factors rather than external ones (p. 5). Knowles urges educators of adults to take these characteristics into account when designing, implementing and evaluating educational experiences (Knowles, 1984, p. 5). Though this study is concerned with the education of adults, it should be noted that there has been much debate as to whether or not children and adults learn differently; for example Cyril Houle, mentor to Knowles, stated that 'education is fundamentally the same wherever and whenever it occurs' (Houle cited in Merriam, 2001, p. 6). In addition, he asserted that the concept of andragogy alerted educators to 'involve learners in as many aspects of their education as possible and in the creation of a climate in which they can most fruitfully learn' (Houle cited in Merriam, 2001, p. 6).

The aims of andragogy, to some extent, are consistent with the idea of constructivist learning, which is the conviction that knowledge is constructed by learners, not merely transmitted by a teacher or lecturer (Brown & King, 2000, p. 245). As with andragogical approaches to learning, constructivism places the learner at the centre of the process. Advocates of both constructivism and andragogy assert that learners should play an active role in learning. In order to facilitate this process, learners should be encouraged to explore and manipulate the learning environment (Dickey, 2003, p. 106). Central to the philosophy of constructivism, is the belief that meaning is made within the context of a situation (Brown & King, 2000, p. 245) which correlates with the idea of experience being a reservoir for learning in andragogical contexts whereby meaning is constructed through relation to prior experience (Merriam, 2001, p. 5). Though the desirability of collaboration is frequently not made explicit in the literature relating to andragogy (Schapiro, 2003, p. 151), constructivism advocates the provision of opportunities for collaboration between learners. This is enhanced by discourse and supports social negotiation between learners; information is shared, allowing learners to reflect on their learning. Three-dimensional virtual environments, such as MUVEs, are showing some promise as constructional learning environments (Dickey, 2003, p. 106) and a significant literature is building around this topic (e.g. see Nelson & Ketelhut, 2007; Sardone & Devlin-Scherer, 2008).

MULTI-USER VIRTUAL ENVIRONMENTS, THEIR CHARACTERISTICS AND AFFORDANCES

An MUVE is a computer-, server- or internet-based virtual environment that allows participants to move around and use various forms of communication (text chat, voice chat or instant messaging). It allows participants to create a virtual identity which persists beyond the initial session (Maher, 1999, p. 322; Ritzema & Harris, 2008, p. 110). The term was coined by Chip Morningstar and F. Randall Farmer in 1990 (see Morningstar & Farmer, 1991, p. 273) and is often used interchangeably with 'Virtual World' (VW) (see Castranova, 2001, pp. 4–5). Second Life is one of the most well-known MUVEs probably due to the intense media scrutiny it has attracted. Though it boasts nearly 16 million user accounts, only about 13,000 of these are based in Australia or New Zealand (Linden Lab, 2010b). Though Massively Multiplayer Online Role-Playing Games (MMORPGs) such as World of Warcraft resemble MUVEs in many ways, including users sharing the same virtual space and persistence of characters, they differ in important ways too. In MUVEs there are no 'levels' to be worked through or embedded fiction that directs the activities of participants; instead the content and experiences are shaped by users (Ondrejka, 2008, p. 231).

MUVEs are populated by motional 'avatars'; the term is derived from Sanskrit and used in Hindu mythology to denote the earthly form adopted by a deity, commonly Vishnu (Leeming, 2001). In MUVEs, this term denotes the representation of a character, controlled either by an individual or a software agent in the case of a 'bot', which acts somewhat like a virtual automaton (Duridanov & Simoff, 2007, p. 4). The choice of avatar can reflect a player's personality, gender or ethnicity. It is also possible for a learner to assume a completely different identity which in itself may constitute a significant learning experience, particularly important in role-playing scenarios (Annetta, Klesath, & Holmes, 2008, p. 2). Avatars are able to interact with and modify the virtual environment and are even able to interact beyond the confines of the MUVE if objects are linked to web pages (called 'web on a prim' in Second Life) (Tashner, Riedl, & Bronack, 2005b, p. 6).

There are many educators eager to exploit the unique affordances of MUVES, endeavouring to give their students the most authentic learning experiences possible. What better way to train an architect than to let him or her design and construct a building; walk around in it when completed and then go back and correct any deficiencies or experiment with alternatives? A prospective surgeon will learn best by performing surgery on a patient that cannot die, and a student of history will appreciate and more fully understand historical events if for just an hour or two they could take on a role and wander around a battleground or participate in a significant legal trial. For some disciplines, the educational affordances of a virtual environment such as Second Life are obvious (Salmon, 2009, p. 529). The simple presentation of information is arguably not as valid as engaging students in interacting with that information as becomes possible in an immersive virtual environment (Tashner et al., 2005b). The student has a firsthand experience, resembling real life situations that are too dangerous, too expensive or too difficult to replicate in real life. True to the principles of constructivist learning, the student remains at the centre of the learning experience with the teacher or instructor acting as the guide or facilitator.

Engaged students who are responsible for their own learning through an active approach tend to experience a deeper level of learning compared to those who are merely passive recipients of information. Problem-solving, authentic learning experiences, virtual learning, online collaboration and other active methods will usurp more conventional didactic approaches to learning. Further, Curtis Bonk and Ke Zhang (2006) also flag a greater emphasis on reflection for students to 'internalize and expand upon their learning pursuits' (Bonk & Zhang, 2006; Sanders & McKeown, 2008, p. 51) and this can be readily facilitated through interaction in and with an immersive virtual environment.

With the widespread availability and accessibility of MUVEs, educators are no longer restricted to the physical and geographical limitations of their school, community, state or even country. Just about anything that can be conceived can be created in a virtual environment. MUVEs enable educators to leverage social connections and learning methodologies in order to transform basic approaches to learning and communication

(Ondrejka, 2008, p. 229). Generational, professional, historical or gender gaps become obsolete in an environment where users cooperate to create knowledge and experiment with identity in collaborative spaces (Ondrejka, 2008, p. 229). Users, via their avatars, learn how to solve problems in the design by means of creation and modification of their own content. This intrinsic culture of participation suffused with pervasive learning makes MUVEs dynamic and stimulating learning environments (Ondrejka, 2008, p. 229).

Ideally, learning designs using these environments would imbed more authentic learning through collaboration, teamwork, problem-based and active learning (Bonk & Zhang, 2006, p. 251). The increasing importance of hands-on learning in the near future has already been glimpsed in the rising prevalence of realistic and complex simulations, interactive scenarios and commutative news stories (Bonk & Zhang, 2006, p. 251). This in part could be achieved in MUVEs through content creation in accordance with the learner's own ideas, learning goals and interests. This approach necessitates the acquisition of certain requisite skills which could be incorporated into educational designs favouring collaboration, peer-to-peer teaching and the creation of new types of 'learning communities' for both students and educators, underpinned by mediated immersion (Ondrejka, 2008, pp. 229–230; Clarke & Dede, 2005, p. 1; Tashner, Bronack, & Riedl, 2005a, p. 2117).

It is this ability of MUVEs to facilitate immersion that will contribute most to the engendering of these strategies. Immersion is achieved by actively engaging at least one of the senses, typically sight. The successful cultivation of immersion is characterized by the learner's impression of actually 'being there' in the virtual world and is a necessary condition for presence. This refers to a decreased awareness of one's existence in the actual physical space at the computer in a room or computer lab and an increased experience of being in the virtual world or MUVE (Witmer & Singer, 1998, p. 225). The sense of immersion can be enhanced by collaboration in the environment, leading to a sense of flow as well as presence resulting in an enhanced involvement and commitment to learning (McKerlich & Anderson, 2007, pp. 35–37).

In practical terms, this sense of immersion and presence can be leveraged to create quite realistic simulations and learning contexts. Within a MUVE, an object can be examined from many angles; it can be picked up, rotated, moved, stretched and otherwise manipulated. Similarly, a scene can be viewed from a number of angles, altered as the viewer's perspective changes. Experiments can be undertaken without the expense and inconvenience of real world consequences. In a physically safe and pliable environment, learners can undertake the hands-on learning identified as being highly desirable in contemporary learning contexts (Dickey, 2003, p. 106). This potential was also flagged by Bricken and Byrne after their early work in Virtual Reality (VR) (Bricken & Byrne, 1992) and identified by Winn as utilizing the same psychological processes as those evinced by participating in equivalent activities in the real world (Winn, 1993, p. 1).

Winn (1993, p. 4) further explains that knowledge is acquired in two ways. Firstly, it is constructed through direct interaction with objects and situations, such that the learner is unaware of its presence. This form of knowledge is direct, subjective and personal and is called 'first-person knowledge'. The other way knowledge is acquired is through description by someone else and is consequently objective, vicarious, communal and explicit. This Winn calls 'third-person knowledge' (Winn, 1993, p. 4). He further makes the point that 'first-person knowledge' is non-symbolic in contrast to 'third-person' knowledge which is symbolic and frequently codified. Transparency of the user interface in VR affords the acquisition of first-person knowledge and is consistent with constructivist approaches to learning in MUVEs (Winn, 1993, p. 4; Dickey, 2003, p. 106). Though the user interface is not so transparent in MUVEs – actions are mediated through a computer keyboard and mouse – participants are only minimally aware of its presence; this awareness decreases as the participant becomes more experienced in the environment.

Through a careful examination of the general features of MUVEs, a number of characteristics have been identified, that when used in conjunction with good educational design, would support andragogical and constructivist approaches to learning. A number of MUVEs are available to educators; these include IMVU, Activeworlds, Croquet Project, Project Wonderland and Second Life. Croquet Project and Activeworlds have been used by some K-12 and tertiary institutions (e.g. see Tashner et al., 2005a), but by far the most popular MUVE for educators is Second Life. Second Life has been used as the setting for a diverse variety of educational projects including the Island of Svarga that functions as a self-contained ecosystem, Roma: Ancient Rome (a classical role-playing build) and the International Spaceflight Museum complete with a functional planetarium and rocket rides (Kay & FitzGerald, 2008). Prominent within education circles in Second Life, is the New Media Consortium (NMC), an international, not-for-profit conglomerate comprised of some 260 learning-focused institutions committed to exploring and utilizing new technologies.

NMC has a considerable presence in Second Life; their educational precinct housing the Second Life islands of a number of universities including Princeton, MIT and Harvard University (New Media Consortium, 2010). The UQ joined the NMC in 2008, also sharing this space.

The popularity of Second Life can be attributed to several factors. First and foremost is the fact that a basic account is free while still allowing the user to customize his or her avatar which persists between sessions, the user can buy or sell objects and explore the myriad of available regions (Calleja, 2008, pp. 24–25). Users through their avatars are also able to build in open spaces called 'sandboxes', facilitating collaboration, sharing of resources and encouraging creativity (Calongne & Hiles, 2007, p. 72). A premium account is required if an individual wants to buy 'land' that costs from US$72 per annum (Linden Lab, 2010). In addition there are significant discounts on land purchase prices for educators and an active community of educators willing to lend their help, inworld resources and expertise. This is partially mediated through the SLED (Second Life Educators) email mailing list (Linden Lab, 2010c). In addition, there are large numbers of inworld classes for all residents providing information about negotiating Second Life covering basic and advanced topics, from building to avatar customization. For those educators preferring a more formal setting and theoretical focus to their learning, a number of US-based universities offer graduate courses in teaching in Second Life.

With many using Second Life purely for recreational purposes – shopping, socializing, listening to live music, taking informal classes, fantasy role-playing – it's not difficult to imagine the potential benefits and enjoyment possible using this MUVE for education. Though it does not necessarily resemble conventional 'education', education in an MUVE does become more attractive and often less intimidating (Stevens, 2006, p. 4). There are a number of characteristics of the environment and user interface that promote communication, collaboration and content creation. Video can be streamed into the environment, web pages can be linked to objects and three-dimensional objects can be coded and scripted to interact with avatars or other objects. Voice chat, utilizing VoIP, was added to the platform in August 2007 but individuals can also communicate through text chat and instant messaging (Ritzema & Harris, 2008, p. 100). This variety of tools and features makes it possible to create learning designs that cater for learners of all styles.

Though the potential advantages of Second Life as an MUVE cannot be overestimated, it is prudent to be mindful of the potential disadvantages, often overlooked in the literature relating to education in this environment. Probably the most significant problems encountered by users of Second Life relate to the technology. The system requirements of the programme can be prohibitive, particularly on those computers not configured for gaming or other pursuits requiring high quality graphics. This problem is exacerbated by the frequent software updates required by the increasingly sophisticated software servers of Second Life. In addition, many users may not have access to the broadband internet that is required to even access the environment (McKerlich & Anderson, 2007, p. 39). There is also a learning curve associated with using Second Life. Avatar controls can be challenging to master, locations may be difficult to find and a general unfamiliarity with how things work can diminish the quality of the experience leading to disengagement. One way to ameliorate this problem would be to provide a blended learning experience where regular classes would be supplemented by excursions into Second Life to apply the knowledge gleaned from a face-to-face session (McKerlich & Anderson, 2007, p. 39). It is crucial that both the advantages and disadvantages of this platform be considered when designing education for learners in Second Life. Some of these will be considered in detail in the next section.

BACK TO STUDIES IN RELIGION AT THE UNIVERSITY OF QUEENSLAND

Studies in Religion discipline convenor, Dr Rick Strelan teamed with me to successfully apply for a competitive Strategic and Learning Grant of AU$30,000 from our institution in order to establish the UQ's Studies in Religion presence in Second Life. The project is called 'UQ Religion Bazaar', a reference to the wide assortment of religions to be represented in the space (see Robinson, 2008, p. 15). A number of religious structures were constructed on the island including a mosque, a church, a synagogue, a Hindu temple, a Buddhist temple, a classical Greek temple, Freemasons' lodge and provision for ritual magic and neo-pagan rituals. These buildings were constructed in such a way as to allow ritual reconstructions and role-playing. In addition, there are a number of non-religious constructions including an amphitheatre, an office with a conference table and screen, numerous informal meeting spaces and a reconstruction of the UQ's iconic Great Court. The main use of this environment is for the large first-year Studies in Religion courses *Meditation and Soul Journeys: Spiritual Experience East and West* and *Introduction to World Religions*, both of which consider practices and rituals in a wide variety of religions and religious contexts (Fig. 2).

Fig. 2. The Mosque on UQ Religion Bazaar.

We believed that Second Life provided an immersive experience for students. Students could experiment with identity and perhaps experience empathy by taking on avatar of a different gender, culture or race. For example, a group of students gained a unique insight into the discrimination experienced by Muslim women wearing the burqa in western society, by dressing their avatars in that way and going to a public place in Second Life. In this way prior experiences could be reanalysed and reinterpreted. And students, working together in an environment where distinctions obvious in the physical world become obsolete, construct knowledge through firsthand experience. Through their avatars, students could explore every inch of a religious space with just-in-time information provided by way of notecards and triggered messages displayed in text chat at the various places. In addition, they could closely examine religious artefacts and outfits which would be very difficult, if not impossible for them to access in real life.

In each course, students were divided into tutorial classes of up to 20. Each class was taken at a separate time to a computer lab equipped with computers with the Second Life client installed. Initially, it was decided to have the students enter Second Life under supervision so that the lecturer and teaching assistant could render any help to the students if they needed it. Also, we were unsure as to how students would react in the environment. We wanted to be able to talk to them and observe their responses. Students self-selected into pairs and each pair was assigned an avatar that had already been created. We decided to do this so that students could work together to overcome any difficulties they encountered with the interface or with the assigned task. We saw benefit in students talking and working together to complete the tasks, building knowledge through experience and conse-quently confidence through collaboration. The students were of diverse ages from 18 years old to mature adults in their late 50s. Though the younger students could be said to be of the 'Net Generation' (Kennedy et al., 2009) or to be representative of Prensky's 'Digital Natives', the older students were less comfortable with the technology and could be considered to be 'Digital Immigrants' (Prensky, 2003). During the course, the older students became more confident with the technical aspects of Second Life under the guidance of support of their younger peers with whom they were partnered.

Each avatar had a certain number of religious outfits and an identical collection of religious objects readily accessible through the avatar's inventory. It was decided to provide the students with avatars as Second Life training was not available to the students outside of class and there was insufficient time in class to make up the deficit. Normally the library would take the responsibility of running courses for students in appropriate software and they do run courses in Microsoft Word, Endnote and so on. They declined to run courses about Second Life as it was not used at the university in any other academic courses. However, the provision of avatars significantly decreased the learning curve associated with entering the environment. On the downside, the research shows that students who build and customize their own avatars are more likely to feel a sense of identity with those avatars (Kafai, Fields, & Cook, 2007). By providing readymade avatars, we were depriving students of this experience. This was further exacerbated by students sharing an avatar. The individual student could not 'bond' with an avatar because he or she did not exercise sole control over it. Also, the interactions between members of the pair and with the larger group, including the instructors, made the students acutely aware of the

Fig. 3. A UQ Religion Bazaar Avatar Showing the Inventory.

physical environment, at the same time decreasing their immersion in the virtual one. Though we realised that these factors would significantly hinder the engendering of immersion during these initial activities, we felt we had little choice (Fig. 3).

In the initial session, each student pair was encouraged to explore religious spaces on the UQ Religion Bazaar island or other religious landmarks that had been provided (these places were previously investigated and permission to use them was gained from each build owner). Sometimes the build owner would make assistants, knowledgeable in the religious tradition being investigated, available to assist the students. Builds that were used included the Anglican cathedral on Epiphany Island and the Taoist Cijian Temple. The students completed a scavenger hunt which required they complete a number of tasks and provide proof of doing so. This might include providing a snapshot of their avatar in a certain pose at a particular religious build or bringing back an object that they found at the site. Not all of the class would go to the same site. Two or three pairs would go to each site and then move on to another. At the end of the expedition, each pair would report back to the tutorial class about what they had learned. A whole of class discussion ensued and students would share solutions for technical problems or issues that had arisen. This exercise was designed to help students learn the skills they needed to successfully negotiate their way around the Second Life environment while still learning about various religious traditions.

The next activity to be completed in Second Life was much more complicated and required considerably more planning and preparation. Students left the pairs they had formed; instead working in groups that they had already been allocated into. There was a major piece of group assessment, unrelated to the Second Life activities, to be submitted at the end of the semester. In order to get students used to working in a group, students were asked to self-select into groups of five or six at the beginning of the semester. For the next activity in Second Life, students were asked to work in these groups. To assist communication between the group members, a private page with a discussion board and a file sharing function was set up for each group on the course Blackboard site. The students could determine if and when they used the page.

For the next activity in Second Life, students in their groups were asked to research part of a religious ritual or historical event of significance to a religion, with a view to re-enacting it in the virtual environment. They were initially given time in class to discuss and plan their approach to this activity. They were particularly asked to pay attention to the context of the ritual or event: What was happening? Why was it happening? What was the significance? Who was taking part and why? They were instructed to be respectful at all times and were prompted in text chat on entering religious buildings on the UQ Religion Bazaar island to be respectful and wear appropriate clothing or to remove their shoes or to wash before entering. The nature of these requests depended on the religious building. For example, on entering the mosque, students are asked to remove their shoes and wear an appropriate head covering.

Students were familiar with the UQ Religion Bazaar island; they knew which buildings were available and they knew which outfits and religious implements they could use. Before they settled on a particular event or ritual, they had to notify me just so I could verify they had the outfits, avatars and implements they needed to successfully complete the role-play. Though the avatars were equipped with a wide range of outfits and implements covering all of the religious traditions represented on UQ Religion Bazaar, on a few occasions students wanted to recreate rituals and events for which the avatars were not suitably equipped. If only a few objects were required and they were readily available, we simply bought them for the avatars from the Linden Lab marketplace, XStreetSL. In some cases this was not possible and the groups were asked to select another ritual or event (Fig. 4).

When the students returned to the computer lab, they were able to work on their re-enactment of the ritual or event. They were given four hours over

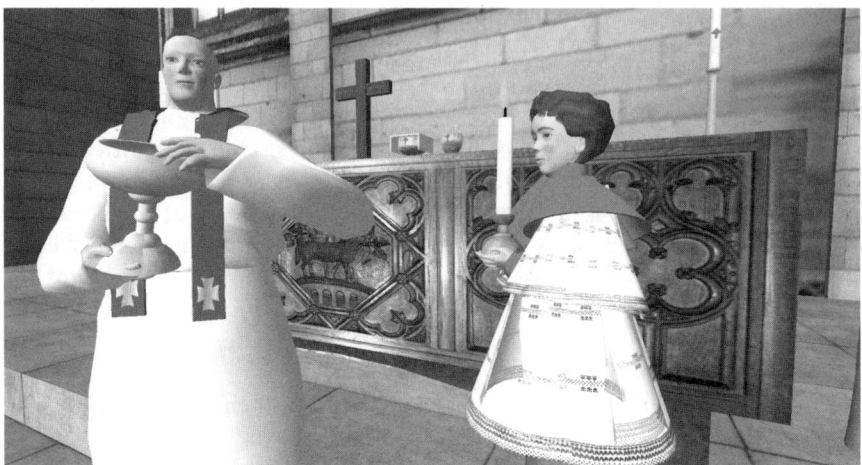

Fig. 4. Student Avatars Within the Church on UQ Religion Bazaar.

four weeks to do this. In the subsequent two weeks, two groups in each session would perform their role play while the rest of the tutorial class could watch. Students would explain the significance of what they were doing and saying as well as the symbolic significance of the implements they were using. Further, they would explain the relevance and importance of that event or ritual to the religion.

This exercise met with varying degrees of success but was mostly enjoyed by the students. Some expressed frustration with their lack of familiarity with the interface or with the limitations of the technology. For example, the range of movement, gestures and facial expressions is limited and are difficult to instigate when the student was busy doing something else such as typing, speaking or moving (Ediringha, Nie, Pluciennik, & Young, 2009, p. 471). No data was collected though instructors talked to students about their perceptions, difficulties with the environment and motivation towards the exercise. Students reported feeling a sense of identity with the avatar that they were using, even though they did not create that avatar. This was especially interesting when talking to students who adopted an avatar of the opposite gender to their own. In some cases, they reported identifying and feeling more like they expected that gender to feel. For example, one male student said he felt more submissive and less outspoken when he was operating through a female avatar.

Because they had to research and then re-enact the ritual or event in an immersive environment, students reported feeling they had a bigger investment in the activity. As they were a part of the ritual, the research informing the role play became more personal to them. Some students said that it gave them much greater insight into what it felt like to belong to a different religion (or to a religion). Many reported that they felt they understood people from different religions better than they did before the exercise and that they would probably be more tolerant of differences between religions in the future.

We originally intended that this exercise would form part of the assessment for the courses in which it ran. Our idea was to record the role play using some screen capture software such as Fraps and assess that recording according to prescribed criteria (Fraps: Real-time video capture and benchmarking, 2010). We also intended that group work would be assessed within the group, as it is in the other major assessment piece in the courses. There were a number of reasons why we decided against this in the initial iteration. Firstly, we were unsure how successful the activity would be. We had not used Second Life before for a class and were dubious about the stability of the platform. It would be unfair to assess students if the success of the activity was going to be hindered by technical difficulties. Secondly, we felt challenged enough learning how to use Second Life without having to master a screen capture programme as well. We were concerned that students too would be overwhelmed by this technical challenge. As it turns out, Fraps is very easy to use and probably would not have posed any significant problems. However, large file sizes could still be an issue. Thirdly, the school administration was sceptical about our use of Second Life. We had already encountered difficulties and prejudice because we used group assessment in the courses. We believed that a change of culture at that level would be needed before we could effectively assess learning using group work and Second Life.

As mentioned previously, we did not formally evaluate this learning design. It was intended to be exploratory with a view to refining the design for future iterations. A survey of the relevant literature seemed to suggest that this approach held significant promise in terms of improved student engagement through immersion, and that the environment would support learning through andragogical and constructivist approaches. In this iteration of the Second Life activities, we were broadly looking for major problems with either the design or with implementing Second Life that would preclude us from proceeding with future iterations of the activities. We talked informally with students during the courses, either during or after the class sessions. We used no pre- or post-questionnaires to assess learning outcomes, perceptions of the technology or changing attitudes to various religious issues. We did not use focus groups to ensure a range of opinions and acknowledge that students who spoke out to us may not be representative of the class groups or hold representative opinions.

Overall, we had the sense that students enjoyed the experience and felt they gained something from it, but adjustments to the design still needed to be made. For example, we felt that students did not have enough lab time to adequately prepare for the role play. In the future iterations of this design, more class time will be allocated to group work in preparation for the role-play; they will have five weeks instead of four. Also, students tended to be too ambitious in what they were trying to do. This was a product of both our inexperience as well as the enthusiasm of the students. As instructors, we now have a better idea of what can be achieved in the allocated time and can advise students accordingly. A modified version of this exercise will run in the summer semester of 2010 in the course *Introduction to World Religions*, using a design-based methodology (e.g. see Ma & Harmon, 2009).

GUIDELINES FOR THOSE CONSIDERING TEACHING AND LEARNING IN MUVEs

Though we felt our experience taking students into Second Life was largely positive, we did learn a number of lessons. After reviewing the literature, reflecting on our own experiences, talking with my teaching colleagues who had a familiarity with the courses and as a result of conversations with and observations of the students during the tutorial sessions, the following guidelines were formulated. The guidelines take into account several key issues including both the advantages and disadvantages of using MUVEs for education, the principles of constructivist learning and andragogy, the importance of maintaining engagement through immersion and presence, and the minimization of technical difficulties.

1. Don't recreate real life learning environments.

If students already have an on campus classroom experience, there is little point in trying to recreate that real life learning environment. Though it may reduce the cognitive dissonance associated with entering a novel environment (Addison & O'Hare, 2008, p. 8), it fails to leverage the unique characteristics of the MUVE in order to create genuinely engaging learning designs. Second Life affords the opportunity to create whatever can be imagined, enabling learners through their avatars to interact with the environment in a very personal and hands-on way, encouraging the acquisition of 'first-person knowledge' as described by Winn (1993, p. 4).

Though some 'traditional' learning environments are present on the UQ Religion Bazaar island, these facilitate inworld events rather than enable the recreation of real life learning contexts. For example, a religious leader may be invited to address a class but not be available on campus; that person could address a class inworld from the amphitheatre. Similarly, students could run a religious studies conference inworld, collaborating with peers enrolled in other universities anywhere in the world.

2. Make sure that the preparation is adequate.

As with all teaching contexts, good preparation is paramount to a successful learning session in Second Life. If technical problems become too intrusive, learners will disengage from the activity. Ensure that all computers meet the required specifications (Linden Lab, 2010d). Some schools and faculties are reluctant to have Second Life installed on their computers. In this case, the Second Life programme can be run from a flash drive (Bixler, 2007). In addition, for Second Life to run successfully the computers require an external IP address; without this learners will be unable to log on. As technical problems are reasonably common, it is advisable to have alternative activities planned. Because of the large amount of preparation that is often involved in preparing for a learning activity in Second Life, it can be tempting to persevere in trying to solve a technical problem so that the activity can ultimately proceed as planned. If problems do arise, even though it is necessary to make some attempt to resolve them, it is not desirable to take too long before refocusing on the alternatives. Learners soon become bored watching educators battle with their computers and this may prejudice them against future activities in the environment.

3. Introduce Second Life in a safe and reliable way.

With such a small number of Second Life users in Australia and New Zealand, it is unlikely that a significant proportion of religious studies undergraduate students will be familiar with the environment. It is possible that more will have had some experience of MMORPGs such as World of Warcraft so will be more used to moving around and the general layout of the interface.

An effective way to introduce Second Life to students is to show them a short video which highlights the features of this environment. There are a number made about various aspects of Second Life which can be downloaded from YouTube for no cost. Alternatively, it is not difficult to make a video (known as machinima in Second Life) to suit the specific aims of the class, using a programme such as Fraps (Fraps: Real-time video capture and benchmarking, 2010). Ideally, for Studies in Religion course, this machinima would show some of the religious spaces already established inworld such as the elaborate cathedral on Epiphany Island or the peaceful Buddhist Shrine of Varosha. Showing machinima or other video removes the potential for technical problems sometimes associated with a live demonstration of Second Life; it is not uncommon for some Second Life regions to be offline or experiencing significant lag. Because this will probably be the learners' first experience of Second Life, it is desirable that it goes smoothly so as to reduce any anxieties associated with entering the environment.

In addition, it will be helpful if learners are made aware of the online resources available specifically for newcomers: when difficulties do arise, they will be able to resolve most problems on their own. Among the most useful of these resources is a web page of video tutorials (available at http://sltutorials.net/) or through YouTube.

Second Life is a public space so learners must be made aware of the expectations of their behaviour. Linden Lab has created a set Community Standards. Contravention of these standards could result in an immediate suspension or expulsion from the environment (Linden Lab, 2010a). It is similarly important to give learners some guidelines in dealing with the bad behaviour of others in Second Life. If the space you are going to use is open to the public, this is a potential problem. Individuals that purposefully harass another or willingly damage buildings or objects are known as 'griefers' (Mansfield, 2008, pp. 26–28). The most convenient way for learners to deal with griefers is to immediately log out of Second Life.

4. Scaffold learners for their first few times inworld.

Meditation and Soul Journeys: Religious Practice East and West and *Introduction to World Religions* are both offered on campus during regular class times. Because of the high-end hardware and broadband requirements, it is not feasible to expect students to complete activities in Second Life in their own time or at home. Therefore it will be necessary to schedule times in a computer lab for Second Life activities. This does not preclude students from entering the environment in their own time if they want to explore further or practise their inworld skills. In fact, this should be encouraged by suggesting interesting inworld locations to visit or Second Life events to attend.

Another advantage in holding Second Life sessions during class time is that the educator is available to offer guidance and encouragement through those difficult first visits. An additional way to lessen the slope of the learning curve is to provide learners with avatars that have already been created. This has the added advantage of allowing learners to emerge inworld in a place of the educator's choosing rather than at one of the orientation points such as that provided by NMC. When learners are more comfortable with the environment, they can create their own account and avatar.

Collaboration within the classroom can be enhanced by asking two students to share a computer and an avatar. In this way they will work through the novel problems of Second Life together. Even the technical problems often associated with the environment, can be turned into an opportunity for interpersonal communication, problem solving and collaboration.

Once the learners are successfully inworld it can be helpful to use an inworld device known as a 'Pied Piper' or another called a 'Magic Carpet'. These devices enable students to sit behind the educator on a seat (in the case of the Pied Piper) or on a carpet (in the latter case), following the instructor as he or she moves around. This removes the immediate necessity of the learner being able to navigate with an avatar, which can be difficult proposition at first, while still demonstrating some of the advantages of the environment.

5. Keep the student at the centre of learning.

Both constructivist and andragogical methodologies advocate keeping the student at the centre of their own learning. Within certain broad guidelines, individuals should be responsible for both the direction and the nature of his or her own learning. To facilitate this, learners should be encouraged to manipulate the environment and take an active role in the process of learning (Dickey, 2003, p. 106). One way of embedding this would be to offer learners a number of choices as to which religion each would like to more fully investigate; letting the individual research a topic of keen interest to them rather than one of less interest but which is prescribed. This could take the form of researching and then role-playing the rituals of a particular religion in a particular context. It could also include students creating their own religious spaces in which to enact rituals or role plays or creating machinima of the activities of others.

It is vital that educators do not slip into more traditional role of a 'talk and chalk' teacher, acting as transmitters of knowledge rather than facilitators. PowerPoint presentations and whiteboards are readily reproducible in Second

Life. Overreliance on these media may lead to a more teacher-centred, didactic approach to learning.

6. Cater for all learning styles.

Fleming identified four types of learning styles: (a) visual; (b) auditory; (c) reading/writing; and (d) kinaesthetic, tactile or exploratory, all of which can be accommodated within Second Life (Fleming & Baume, 2006, p, 6; Bonk & Zhang, 2006, p. 250). Visual learners will respond to the detail of the builds and diagrams. They will respond positively to being able to move around builds, looking at how they are put together and watching historical re-enactments and ritual re-creations. Podcasts, music broadcasts and voice chat conversations will satisfy more the learning style of auditory listeners. They could listen to and comment on performances. Those learners that favour reading and writing will find information displayed on notecards, in inworld books and newspapers to be more suited to their style. Treasure hunts that require them to find information displayed on a build may also cater to their particular taste in learning. A Blog HUD (Heads Up Display) enables students to write and add pictures to their real world blog while they are still immersed in Second Life. Finally, kinaesthetic learners will respond best to designs that incorporate some aspect of movement or the application of knowledge within realistic simulations into the design. These examples are just a small proportion of the possibilities available in Second Life. There is no real barrier to incorporating several of these ideas into the one design in order to facilitate the learning of those utilizing different learning styles.

7. Encourage communication and collaboration.

Second Life affords many opportunities for communication via voice chat, text chat and instant messaging. Text chat may be particularly helpful for those students that lack confidence in the traditional classroom setting, such as shy students or those with English as a second language. Texting allows students to think about and proof read what they are going to say before they display it. In addition, it is relatively easy to form groups which have further options for communication via group notices and group calls.

Another of the features of Second Life is the ease with which it enables collaboration. Students can collaborate in the real life classroom to direct an avatar or deal with technical problems. Inworld, collaboration between peers, between learners and teachers and between groups of students from different universities is relatively common, facilitated by the multiple communication channels. But also through other web 2.0 technologies used in conjunction with Second Life such as Google Docs, Second Life Facebook applications and Twitter in Second Life. Further, the user interface of Second Life allows students to keep track of their 'friends' (designated as such by the participants), being alerted as they log on and log off. Collaboration can be encouraged through the careful design of group activities, peer assessment and establishing connections with peers in other institutions.

8. Evaluate the success of your learning designs.

This is an important part of implementing novel learning designs. It allows the educator to build on the work of previous semesters, refining the design, responding to changing circumstances or cohorts. It also helps educators justify new designs and provides data which can be used to document designs to be reported in the academic education literature. Evaluation can be done informally and formally. Informal evaluation would include such things as noticing how students interact with the environment and each other. Informal conversations with students about the design in class would also be considered an informal evaluation. Formal evaluation could take the form of structured focus groups, anonymous class surveys and so on.

Another useful tool to assess the level of engagement and presence in this educational design would be the Multi User Virtual Environment Education Evaluation Tool (MUVEET), developed by Ross McKerlich and Terry Anderson (see McKerlich & Anderson, 2007). This builds on the work of Garrison, Anderson, and Archer (1999) who developed the Community of Inquiry (COI) model to assess the context and learning process in computer-mediated, text-based environments. In this model, social presence, cognitive presence and teaching presence are thought to be the key elements forming an educational experience (McKerlich & Anderson, 2007, p. 36). McKerlich and Anderson supplemented the COI indicators to incorporate the additional features of MUVEs including real-time text, verbal and visual cues (McKerlich & Anderson, 2007, p. 48).

These guidelines provide a starting place for educators, when planning educational designs within the MUVE of Second Life. They take into account the principles of andragogy and constructivist learning, while leveraging the unique qualities of the virtual environment in order to generate presence and immersion.

CONCLUSION

Second Life is the most popular MUVE used by educators with hundreds of universities offering courses entirely or partially delivered in the environment.

The UQ is also represented there, with the Studies in Religion discipline purchasing a Second Life island in 2007 and constructing a number of religious builds for use in two large first-year courses.

The UQ Religion Bazaar island in Second Life provided the venue for group role-play activities designed around constructivist and andragogical principles for first-year Studies in Religion students at the UQ. Even though a formal evaluation of learning outcomes and perceptions was not undertaken, informal evaluation consisted of classroom observations and informal discussions throughout the courses under consideration. Though students were frequently frustrated by the technical aspects that interfered with the execution of the assigned tasks in the supervised Second Life sessions, some reported feeling a sense of identity with their avatar and an increased understanding of how it feels to belong to another religion (or to a religion). These students also reported that they believed that these activities would increase their understanding and tolerance of those of another religion.

Overall, we were largely satisfied both with the stability of the Second Life platform and the effectiveness of the task design. Even so, some modifications will be made for future iterations. For example, students will be given more time in class to prepare for the role plays. Also, the role plays will be recorded using screen capture software so that they may be displayed on the course Blackboard site to be viewable by all class members. It is hoped that we will be able to use this activity as an assessment piece in the future once issues within the institutional culture have been dealt with.

I have formulated guidelines for the creation and evaluation of educational designs in Second Life in light of our experiences in the first iteration of these activities. This set of nine suggestions was informed by lengthy consultation with colleagues and subsequent to our reflection on the informal evaluations of students during their time in Second Life. It is hoped that they will help smooth the way for educators considering teaching in this exciting and dynamic space.

REFERENCES

Addison, A., & O'Hare, L. (2008). How can massive multi-user virtual environments and virtual role play enhance traditional teaching practice? Paper presented at the Researching Learning in Virtual Environments International Conference. Milton Keynes, UK.

Annetta, L., Klesath, M., & Holmes, S. (2008). V-Learning: How gaming and avatars are engaging online students. *Innovate: Journal of Online Education*, 4(3). Available at http://www.innovateonline.info/index.php?view = article&id = 485.

Bixler, B. (2007). Second Life on a Flash Drive. Retrieved on 8 June, 2008, from http://ets.tlt.psu.edu/gaming/SLOnAFlashDrive

Bonk, C. J., & Zhang, K. (2006). Introducing the R2D2 Model: Online learning for the diverse learners of the world. *Distance Education, 27*(2), 249–264.

Bricken, M., & Byrne, C. M. (1992). *Summer students in virtual reality: A pilot study on educational applications of virtual reality technology* (Available at http://ftp.hitl.washington.edu/projects/education/psc/psc.html). Seattle: Pacific Science Center.

Brown, S. W., & King, F. B. (2000). Constructivist pedagogy and how we learn: Educational psychology meets international studies. *International Studies Perspectives, 1*(3), 245–254.

Calleja, G. (2008). Virtual worlds today: Gaming and online sociality. *Online-Heidelberg Journal of Religions on the Internet, 3*(1), 7–42.

Calongne, C., & Hiles, J. (2007). Blended realities: A virtual tour of education in second life. Paper presented at the Technology, Colleges and Community Online Conference, 2007. Available at www.edumuve.com

Castronova, E. (2001). Virtual Worlds: A First-Hand Account of Market and Society on the Cyberian Frontier. *The Gruter Institute Working Papers on Law, Economics, and Evolutionary Biology, 2*(1). Available at http://www.bepress.com/giwp/default/vol2/iss1/art1.

Clarke, J., & Dede, C. (2005). Making learning meaningful: An exploratory study of using multi-user environments (MUVEs) in middle school science. Paper presented at the American Educational Research Association. Available at http://muve.gse.harvard.edu/rivercityproject/documents/aera_2005_clarke_dede.pdf

Conklin, M. S. (2007). 101 Uses for second life in the college classroom. Paper presented at the Games, Learning, and Society Conference 2005. Retrieved from http://trumpy.cs.elon.edu/metaverse

Dickey, M. D. (2003). Teaching in 3D: Pedagogical affordances and constraints of 3D virtual worlds for synchronous distance learning. *Distance Education, 24*(1), 105–121.

Duridanov, L., & Simoff, S. (2007). 'Inner listening' as a basic principle for developing immersive virtual worlds. *Online-Heidelberg Journal of Religions on the Internet, 2*(3). Available at http://archiv.ub.uni-heidelberg.de/volltextserver/volltexte/2008/8299/.

Ediringha, P., Nie, M., Pluciennik, M., & Young, R. (2009). Socialisation for learning at a distance in a 3-D multi-user virtual environment. *British Journal of Educational Technology, 40*(3), 458–479.

Fleming, N., & Baume, D. (2006). Learning styles again: VARKing up the right tree. *Educational Developments, 7*(4), 4–7.

Fraps: Real-time video capture and benchmarking. (2010). Retrieved on 30 April 2010, from http://www.fraps.com/

Garrison, D. R., Anderson, T., & Archer, W. (1999). Critical inquiry in a text-based environment: Computer conferencing in higher education. *The Internet and Higher Education, 2*(2–3), 87–105.

Kafai, Y. B., Fields, D. A., & Cook, M. (2007). Your second selves: Avatar designs and identity play in a teen virtual world. Paper presented at the Digital Games Research Association Conference 2007, pp. 1–9.

Kay, J., & FitzGerald, S. (2008). Educational uses of second life. Retrieved on 8 June, 2008, from http://sleducation.wikispaces.com/educationaluses

Kennedy, G., Dalgarno, B., Bennett, S., Gray, K., Waycott, J., Judd, T., et al. (2009). Educating the net generation: A handbook of findings for practice and policy. Available at http://www.netgen.unimelb.edu.au/outcomes/handbook.html

Knowles, M. (1980). *The modern practice of adult education: From pedagogy to andragogy.* New York: Cambridge Books.

Knowles, M. (1984). A theory of adult learning: Andragogy. In: *The adult learner: A neglected species* (pp. 27–63). Houston, TX: Gulf Publishing.

Leeming, D. (Ed.) (2001). *A dictionary of Asian mythology* (Oxford Reference Online ed.). London: Oxford University Press.

Linden Lab. (2003, 2010). What is second life? Retrieved on 30 April 2010, from http://secondlife.com/whatis/

Linden Lab. (2010a). Community standards. Retrieved on 30 April 2010, from http://secondlife.com/corporate/cs.php

Linden Lab. (2010b). Economic statistics. Retrieved on 30 April 2010, from http://secondlife.com/whatis/economy_stats.php

Linden Lab. (2010c). Educators-SL Educators. Retrieved on 30 April 2010, from https://lists.secondlife.com/cgi-bin/mailman/listinfo/educators

Linden Lab. (2010d). System requirements. Retrieved on 30 April 2010, from http://secondlife.com/support/sysreqs.php

Ma, Y., & Harmon, S. (2009). A case study of design-based research for creating a vision prototype of a technology-based innovative learning environment. *Journal of Interactive Learning Research, 20*(1), 75–93.

Maher, M. L. (1999). Designing the virtual campus. *Design Studies, 20*(4), 319–342.

Mansfield, R. (2008). *How to do everything with second life.* New York: McGraw-Hill.

McInnis, C. (2001). Researching the first year experience: Where to from here?. *Higher Education Research & Development, 20*(2), 105–114.

McInnis, C., & James, R. (1996). *First year on campus: A report on Australian first year students.* Canberra: Committee for the Advancement of University Teaching.

McKerlich, R., & Anderson, T. (2007). Community of enquiry and learning in immersive environments. *Journal of Asynchronous Learning Networks, 11*(4), 35–52.

Merriam, S. B. (2001). Andragogy and self-directed learning: Pillars of adult learning theory. *New Directions for Adult and Continuing Education, 2001*(89), 3–14.

Morningstar, C., & Farmer, F. R. (1991). The lessons of Lucasfilm's habitat. In: M. Benedikt (Ed.), *Cyberspace: First steps.* Cambridge: MIT Press.

Nelson, B. C., & Ketelhut, D. J. (2007). Scientific inquiry in educational multi-user environments. *Educational Psychology Review, 19*(3), 265–283.

New Media Consortium. (2010). New Media Consortium: About Us. Retrieved on 30 April 2010, from http://www.nmc.org/about

Ondrejka, C. (2008). Education unleashed: Participatory culture, education, and innovation in *Second Life.* In: K. Salen (Ed.), *The ecology of games: Connecting youth, games, and learning* (pp. 229–252). Cambridge: The MIT Press.

Prensky, M. (2003). Digital game-based learning. *ACM Computers in Entertainment, 1*(1), 1–4.

Ritzema, T., & Harris, B. (2008). The use of second life for distance education. *Journal of Computing Sciences in Colleges, 23*(6), 110–116.

Robinson, P. (2008). UQ in Second Life. *UQ News,* June, p. 15.

Salmon, G. (2009). The future for (second) life and learning. *British Journal of Educational Technology, 40*(3), 526–538.

Sanders, R. L., & McKeown, L. (2008). Promoting reflection through action learning in a 3D virtual world. *International Journal of Social Sciences, 2*(1), 50–55.

Sardone, N. B., & Devlin-Scherer, R. (2008). Teacher candidates' views of a multi-user virtual environment (MUVE). *Technology, Pedagogy and Education, 17*(1), 41–51.

Schapiro, S. A. (2003). From andragogy to collaborative critical pedagogy: Learning for academic, personal, and social empowerment in a distance-learning Ph.D. program. *Journal of Transformative Education, 1*(2), 150–166.

Stevens, V. (2006). Second life in education and language learning. *TESL-EJ, 10*(3), 1–4.

Tashner, J. H., Bronack, S. C., & Riedl, R. E. (2005a). 3D web-based worlds for instruction. Paper presented at the Society for Information Technology and Teacher Education International Conference, Phoenix, AZ, pp. 1–5.

Tashner, J. H., Riedl, R. E., & Bronack, S. C. (2005b). Virtual worlds: Further development of web-based teaching. Paper presented at the Hawaii International Conference on Education, Honolulu, Hawaii, pp. 1–10.

Winn, W. (1993). *A conceptual basis for educational applications of virtual reality*. Washington: Human Interface Technology Laboratory.

Witmer, B. G., & Singer, M. J. (1998). Measuring presence in virtual environments: A presence questionnaire. *Presence, 7*(3), 225–240.

PART IV
BLOGGING AND
MICRO-BLOGGING IN A
NEW EPOCH OF TEACHING
ARTS AND SCIENCE

FOSTERING AN ECOLOGY OF OPENNESS: THE ROLE OF SOCIAL MEDIA IN PUBLIC ENGAGEMENT AT THE OPEN UNIVERSITY, UK

Linda Wilks and Nick Pearce

ABSTRACT

This chapter illustrates the ways in which The Open University (OU), one of the leading distance learning universities in the world, uses a range of social media to engage members of the public in learning. The OU has been an early adopter of innovative technologies which enabled public engagement right from its inception, forty years ago, contributing to fulfilling its ethos of social justice. It is this aim to remove barriers and provide learning materials to a wide audience, including those who may be excluded from other learning institutions, which has been a major strategic driver of recent changes. Today the OU harnesses a range of social media to continue to develop this strategic policy. The OU's ecology of openness includes a presence on externally developed social media such as YouTube, iTunesU, Facebook and Twitter, which are used as platforms to transfer knowledge and expertise to interested members of the public and encourage academic debate. Alongside these, the OU has also developed its own cutting edge social media platforms, which also allow public engagement. Key OU platforms include OpenLearn, a

Teaching Arts and Science with the New Social Media
Cutting-edge Technologies in Higher Education, Volume 3, 241–263
Copyright © 2011 by Emerald Group Publishing Limited
ISSN: 2044-9968/doi:10.1108/S2044-9968(2011)0000003015

website that gives free access to a vast range of OU course materials;
and Cloudworks, a site for finding, sharing and discussing learning
and teaching ideas, experiences and issues. This chapter explores
the achievements of the OU in using social media to engage with public
audiences, as well as highlights the challenges and issues encountered.

INTRODUCTION

This chapter explores how the latest developments in The Open University's
(OU) long-standing ecology of openness are engaging members of the public
in learning. It examines how the OU is using new technologies to further
open up access to its learning resources, as well as looking at the addition of
a new dimension to their use through the incorporation of social dialogue
media. It also looks at the ways in which exploration of the open access
educational resources has been encouraged, as well as the ways in which
web-based social dialogue has been facilitated.

A CONTINUUM OF OPENNESS AT THE
OPEN UNIVERSITY

The OU was established in the United Kingdom (UK) in 1969 to meet the
growing demand for university-level education amongst the general public.
It is now one of the leading distance learning institutions in the world.
Openness has always been a key feature of the OU's ethos: the university's
mission is still, in 2010, as it has been since the start, to be 'open to people,
places, methods and ideas' (The Open University, 2010a). The OU has over
180,000 enrolled HE students, with over 25,000 of those living outside the
UK. Nearly all these students are studying part-time and around 70% of its
undergraduate students are also working full-time. Although research
degree students are based on the campus at Milton Keynes, the majority of
students study at home using the OU's 'supported open learning' or
'distance learning' style of teaching.

From the start the OU has acknowledged the importance of using
technology to fulfil its mission by promoting 'the advancement and
dissemination of learning and knowledge...by a diversity of means such
as broadcasting and technological devices' (Open University, 2005 [1969]).
The OU relies on a particular mix of media to deliver supported open

learning to students, a mix which has been modified over the years. In its early years, the OU relied on off-peak transmissions of its lectures on the British Broadcasting Corporation's (BBC) television and radio. This was not only an effective asynchronous way of delivering lecture materials to students, but was also an opportunity to engage with the wider public, as the programmes were freely available to all television viewers and radio listeners. Synchronous learning took place through face to face group tutorials, tutor to student telephone calls, and at residential summer schools, where students would get intensive face to face learning support (Daniel, 1996).

Today the OU has added various media, including online social media, to deliver its content to its registered students. As residential summer schools have declined, more and more synchronous, as well as asynchronous, learning now takes place on the virtual learning environment (VLE), as well as by using venues such as the OU's islands on Second Life (Linden Research & Inc., 2010). Person to person synchronous contact is still available through telephone calls and tutorials. Recorded material on CDs and printed books and journal articles also deliver the asynchronous content. In 2010, television and radio programmes are not part of the registered student's specifically course-related resources, although the OU does collaborate with the BBC on popular programmes with an educational dimension, which are also loosely linked with specific OU courses.

An inclusive approach, where social justice is achieved through the provision of life-long learning that is open to all, continues to be a core value of the OU. A third of the OU's undergraduates have entry qualifications lower than those normally demanded by other universities in the UK and their ages range from 16 to 80 and beyond. Thus educational opportunity is made available to all who wish to realise their ambitions and fulfil their potential.

At the same time as being the first university in the UK to welcome students with no prior formal educational qualifications, thus opening higher education to those who had previously been denied access, the OU also promised at its inception 'to promote the educational well-being of the community generally' (op cit). This aim of enabling the educational well-being of the wider community is particularly relevant when considering the transfer and building of knowledge through the new social media.

Echoing the earlier open access of learning materials on BBC television and radio, web-based modes of delivery now encourage a new set of learners to draw on the OU's materials and teaching. These new learners are members of the public with various motivations, including career or personal development. They may also be learners who wish to build confidence before opting for formal registration for a course, or those who feel that they lack the time to commit to a complete course. Learners may also be located outside the global reach of the formal support given by the OU to its registered learners. All these new learners can benefit by accessing a selection of the OU's learning materials which have been made freely available online under Open Educational Resources (OER) licences (OER Commons, 2007). Furthermore, the inclusion of social dialogue facilities such as online discussion sites hosted by the OU and an OU presence on popular sites such as Facebook (Facebook, 2010) have added a new dimension to the learning experience for registered students and informal learners alike.

It should also be noted that the introduction of this combination of educational resources and education-related social dialogue opportunities that are open to all has begun to challenge the simplistic dichotomy of synchronous/ asynchronous learning connections. New multi-way synchronous and asynchronous connections may be enjoyed by learners online, alongside the asynchronous open educational content.

Today the OU harnesses a range of social media to continue to develop its strategic policy of public engagement, as well as to be in line with its core values of being inclusive, innovative and responsive (The Open University, 2010b). The term 'ecology of openness' highlights the diversity and interdependence between the different social media forms used by the OU, as well as their combination in the provision of an open platform for public engagement. The OU's social media may be divided into those that offer OER and those that offer social dialogue opportunities either alongside or instead of the OERs. The OU offers its open social media through various platforms. Some of these OER and social dialogue media are hosted by the OU itself, whilst others sit on the platforms of hosts such as YouTube and Facebook (Table 1).

Alongside the open access educational resources hosted on OpenLearn (Open University, 2010c), which are primarily designed for individual use, the OU has added social dimensions through the introduction of a 'Learning Space Forum' (Open University, 2010d). This space for online discussion aims to facilitate the interaction of learners around the open access course materials in a spirit of mutual support.

Highlighting the blurring of the interface between research and teaching within the field of OER, the OU provides open access to a large proportion of the peer-reviewed research publications written by its academic staff on its online research repository, Open Research Online (ORO) (Open

Table 1. The Ecosystem of OU Open Social Media.

	Open Educational Resources	Social Dialogue Media
OU hosted	OpenLearn topics Open Research Online HESTIA Reading Experience Database	OpenLearn forums Platform open2.net Cloudworks SocialLearn Vital Creative Climate
Externally hosted	BBC/OU TV programmes iTunesU	YouTube Facebook Twitter Second Life

University, 2010e). Open access is also provided to research outputs such as those of the Faculty of Arts' Reading Experience Database (RED) (Open University, 2010f) and that of the Project HESTIA (Herodotus Encoded Space-Text-Imaging Archive) (Open University, 2010g). This is part of the wider open access publishing movement that is gaining ground globally and at the OU (Willinsky, 2006). Each of these research-related open media also encourages social dialogue, as well as offering research outputs to learners. ORO has a blog attached where the manager of the resource highlights issues and answers questions; while RED encourages members of the public to submit material for potential inclusion in the resource; and HESTIA maximises public understanding by drawing on popular web-mapping tools, such as Google Earth and Google Maps.

Alongside its selection of formal course materials and research outputs available on open access, the OU has also developed a range of social media which are not directly linked to formal course materials. The aim is that the materials will excite and inspire members of the public, while providing bite-sized learning objects and the opportunity to interact online with fellow learners.

These new social dialogue-based learning materials sit on both OU-hosted and externally hosted platforms and vary in format and type of content. OU-hosted social dialogue media include Platform (Open University, 2010h), a mixture of news items, blog highlights and discussion; open2.net (Open University, 2010i), which, in conjunction with the BBC, provides in-depth articles and commentary on academic topics relating to BBC/OU TV and radio programmes; and Cloudworks (Open University, 2010j), a website for finding, sharing and discussing learning and teaching ideas, experiences and issues. The OU also has significant presence on externally-hosted new social media websites, including a range of podcasts on iTunesU (Open University, 2010k), participatory videos and related comments and discussion on YouTube (YouTube, 2010) and two islands on SecondLife (Linden Research & Inc., 2010) designed to act as settings for student seminars, as well as being open to the general public. The chance to interact online with other OU students and fans of the OU is offered through the OU's Facebook page (Facebook, 2010); while members of the public can also follow and discuss the OU's activities on Twitter (Twitter, 2010).

Finally, to complete the ecosystem of openness, the OU continues to work closely with the BBC to commission and produce a range of general interest programmes which air on the public broadcasting channels and are then available to view online through the BBC iPlayer (BBC, 2010) for a limited period. The OU's particular role in the partnership is to provide academic advice, input on content and a proportion of the funding. This partnership has resulted in some very popular programming content such as 'Coast', which examines the coast of the UK from a range of angles, including geology, ecology and history; and 'A History of Christianity'.

The following sections look more closely at this ecosystem of open social media, after first looking at the concept of social media.

DEFINING SOCIAL MEDIA

Social media can be contrasted with more traditional broadcast media in that broadcast media are unidirectional and one-to-many; social media have the capacity to be bidirectional or even multidirectional. In digital terms this may be characterised by a shift from a relatively static web page to more interactive user-created media such as wikis and blogs, commonly labelled Web 2.0 (Anderson, 2007; O'Reilly, 2005) but perhaps more usefully characterised as the read/write web, which emphasises openness and collaboration (Hemmi, Bayne, & Land, 2009) and is 'profoundly social' (Alexander, 2006, p. 33). Within the educational arena, social media may also encompass open access educational resources, as well as provide opportunities for education-related social dialogue.

It is worth noting at this point that, despite the possibilities for openness and interaction offered by social dialogue media, this offer is not always taken up, as seen for example by contrasting the number of readers of Wikipedia with the number of editors, or the numbers who view videos on YouTube (YouTube, 2010) with those who upload them. Thus the read/write web is often still the read web for many users.

This distinction is further refined in the idea of social technographics where users of social media are characterised on a ladder with six (and later seven) categories fro inactives to creators (Li & Bernoff, 2008). This breaks down the simple dichotomy between producers and consumers by introducing intermediate roles such as conversationalists and spectators, but nonetheless this further highlights that the full potentiality of the social media is not being realised.

Social media have the potential to impact upon all aspects of scholarship, including research, teaching and public engagement. In each case the new media have the potential to lead to more open forms of scholarship, although the adoption of new technologies does not make change inevitable as established closed practices can persist and even be reinforced (Pearce, Weller, Scanlon, & Kinsley, 2010). Social media may not, however, necessarily be welcomed within HE, as Hemmi et al. (2009) suggest, when highlighting the strangeness of the volatile and challenging nature of a social media environment for formal learning.

The impact of new technologies will be constrained by the extent to which established norms and values persist. In a learning context, one way of limiting the disruptive potential of new technologies is by appropriating them with limited functionality, as with the adoption of closed blogs under the control of the academy and only accessible to a student and their tutor, rather than to the community (Hemmi et al., 2009). Here the interactive and egalitarian aspects of blogging have been reduced to a more traditional tutor–student relationship.

Similarly, although it has been suggested that the internet may be used as an empowering tool at the personal, interpersonal, group and citizenship level (Amichai-Hamburgera, McKenna, & Tald, 2008), others such as Castells (2007) remind us that corporate media and mainstream politics have also invested in this new communication space and still retain a level of power, resulting in a convergence of mass media and horizontal communication networks.

Within this framework of debate about new social media, the OU has developed a range of tools. Many of these are still in the experimental phase and although evaluations and assessments are constantly being made, showing overwhelmingly positive outcomes, it is still too early to confirm, which will prove most useful in the long term.

OU-HOSTED OPEN AND SOCIAL MEDIA

OpenLearn

The OU's OpenLearn website provides free access to a selection of high-quality course materials to anyone who wants to pursue an interest or sample the OU's materials. In contrast to the first two, the OU is responsible for activity on OpenLearn so is able to moderate where necessary. The OpenLearn scheme is in line with the OER movement (OER Commons, 2007), which aims to increase the freedom to learn by providing resources that can be adapted and shared. The site also offered learners the chance to use scholarly networking to connect with and help each other using Web 2.0 tools. The rights-cleared resources offered on these platforms are designed to raise the profile of the OU with current students, alumni and potential students, as well as to give people across the world the chance to boost their learning.

Open2.net

The OU has been working closely with the BBC since 1998 to produce the open2.net website (Open University, 2010i), which was aimed at the general public. The site is jointly badged by the OU and the BBC and its content linked to BBC/OU television programmes and to related OU courses.

One aspect of the content of this site was a set of subject-organised blogs linking in to the BBC television programmes, with individual blog entries commissioned from staff or research students to relate to individual broadcasts. These are polished extended blogs with academic content, rather than the stream of consciousness style characteristic of many blogs. The blogs include links to OU course information to encourage readers to take the opportunity to follow up on the broadcast programme by also studying a course at the OU. The blogs are also used by the academics who write them to draw attention to their related published work and to provide links to other related websites. The academic authors aimed to present their blogs in a format which is accessible in style to the general public, whilst also giving the potential to attract attention from fellow academics or interested experts from across the world.

Platform

Platform (Open University, 2010h) is a news, comment, social and networking website hosted and maintained by the OU and open to the general public.

It also includes polls, such as 'have you decided yet who to vote for in the General Election?' and a feed from the OU's Twitter stream. The website stressed that the space is for the 'OU community', which is defined as including anyone interested in the OU and education more generally.

Platform includes links to blogs posted by OU staff and students, as well as by members of the public, audio clips which highlight issues relating to OU courses, and a 'time out' section of puzzles and activities. The Forum invited website visitors to discuss books or current affairs, for example. Visitors to the site could also log in to rate articles, tag them or send them to a friend.

Platform was launched in December 2008 and was built using the OpenSource software, Drupal. Usage in 2009 was estimated at 50:50 OU:non-OU users, although as 90% of the readers were browsers, rather than logged in, accurate numbers were difficult to assess.

SocialLearn

The SocialLearn website, in test mode early in 2010 and due to be launched to the general public later in the year, is designed around the principle of actively encouraging users of the OU's open learning materials, as well as other web-based materials with learning potential, to engage in focused conversations around their learning. The aim is that by getting users of the learning materials actively to define their question or problem, their learning would be grounded and enhanced (Buckingham Shum, 2010). Learners and educators are invited to make contributions to SocialLearn on several different levels: by asking or answering a question, for example or by posting a comment. They may also support or challenge any of these, or add resources to enhance contributions, then perhaps add reflections on the points raised in order to consolidate learning. Thus a learning path, which attempts to provide an answer to the original question, will be built by the collaborators.

SocialLearn aimed to provide the 'glue' to cohere other OU-related open and social learning sites, including OpenLearn, ORO, Cloudworks (The Open University, 2010j), the Research Skills website (The Open University, 2010l), and the OU's VLE. The user community activities in signposting these resources was seen as the key to drawing on these resources for informal collaborative learning.

Cloudworks

Cloudworks is a social networking site for sharing and discussing learning and teaching ideas and designs (Conole, 2010). Development of the site began in February 2007 with the latest version launched in September 2009. It is developed and hosted by the OU and emerged from a series of related projects to do with promoting open learning and teaching practices. There are four key ideas associated with the site:

Clouds
These can be anything to do with learning and teaching. A cloud might be a description of a specific element of teaching practice or a question to stimulate debate or ask for advice. For example a current cloud asks whether the use of Twitter has peaked. These clouds are social in the sense that users can comment on the cloud and co-create it. This builds on Engeström's notion of social objects, where he argues that successful social networks build around collective social objects (Engeström, 2005). Hence Cloudworks is an object-centred rather than ego-centred site (Dron & Anderson, 2007). Clouds can be cumulatively improved, anyone can add additional content to the core cloud or tags, links, references or embedded content. In addition, each cloud has an associated social space to foster debates and discussion. It is these elements of clouds that emphasise and encourage the social aspects of the site.

Cloudscapes
Clouds can be grouped into community spaces or 'clusters of interest'. So, for example, a cloudscape can be set up to support and enhance a particular event, such as a lecture or conference. Alternatively a cloudscape might consist of a collection of clouds relating to a specific course or resources and references around particular research topics. Clouds are mobile and can belong to more than one cloudscape; all the collective intelligence associated with the cloud travels with it.

Activity Streams
These are dynamic filters of new activity. There are four different types of activity streams. The first is the public activity stream, which is shown on the homepage of the site. This lists all recent activity on the site. The second type is the activity streams associated with cloudscapes, again these are tabbed and they show all the latest activities associated with a particular cloudscape. The third type is the activity stream associated with an individual and their latest activity on the site. These appear on a user's profile page. The final type is an individual personal activity stream, which shows any activities associated with things (cloudscapes and/or people) that a person has chosen to follow.

Follow and Be Followed

It is possible to 'follow' both people and cloudscapes, this has a dual function in terms of acting as a form of peer recognition in the site and also technically anything a user follows is added to their personal activity stream. This adds another social layer to the site, where users can help filter content as well as create it.

At the time of writing, in 2010, the Cloudworks site has over 1,800 registered users and a similar number of clouds. In addition to registered users the site has a large, but difficult to quantify, number of 'lurkers'. Unique page views stood at 40,000 as at April 2010.

Plans are also underway to further push the boundaries of opportunity for the use of Cloudworks. The OU's annual conference where practice and research around learning and technology teaching is shared, 'Openness in Education', was hosted entirely on the OU's Cloudworks site with synchronous elements taking place in Illuminate in 2010, and will be open to the general public as well as to members of academia.

iSpot

The iSpot website (The Open University, 2010m) is an OU-hosted website where members of the public can share their observations and images of nature. Users can share images they have taken and they can help in identifying the content of other users' images. They gain reputation points as they interact with the site, and as their identifications are corroborated by other users, with increasing 'reputation' awarded when large numbers of users agree, or where these users themselves have high levels of reputation. In addition to this there are 'expert users' who have been identified and validated and whose opinions and corroborations count for even more. Through encouraging the input of its users in this way, the iSpot site is attempting to maximise the value of social media, with the users co-creating the content.

The site works closely with external naturalist groups, for example, there are currently 'badges' available recognising input from 19 groups, including the British Dragonfly Society and the British Trust for Ornithology.

In addition to this the site is linked with a more traditionally delivered OU course 'S159 neighbourhood nature', and it is often integrated with BBC television and radio broadcasts. These interactions benefit the site through increased publicity, which drives usage and in turn improves the quality of the content. The integration with the formal course also works in two ways, it provides a more formal platform for providing and recognising the learning that takes place, and it also operates as a way for general public users of the site to become formal students of the OU, with the possibility of going on to complete formal qualifications.

Vital

The open-access Vital website (The Open University, 2010n) has been set up by the OU in partnership with eSkills UK consortium (eSkills UK, 2010), the government-sponsored Sector Skills Council for Information Technology, to help schools and colleges to raise the numbers of young people with strong ICT skills. Resources are offered through the open access website to teachers and a network of exam boards, learning centres, local authorities, independent training providers and other universities is encouraged to learn through engaging in discussion, and sharing expertise and good practice.

In a free sign-in area, community members can also make use of a range of tools such as blogs, online forums and community wikis. Fifteen minute CPD events will also be offered in which practitioners, enthusiasts and subject specialists can share experiences. Vital is encouraging both those teaching ICT as a specialist subject as well as those wanting to use ICT more effectively in teaching other subjects to get involved.

Creative Climate

Creative Climate (The Open University, 2010o) is an OU project that will run from 2010 to 2020 and is designed to document public and expert understanding and response to environmental change issues across the globe. It aims to encourage people to contribute online their understanding and responses to knowledge claims about environmental change.

It has also been developed in response to the recognition by environmental research and policy communities of the need for more innovative and interactive forms of knowledge exchange and public engagement. To this end it is a 10-year diary project that will make use of the web as a platform for participation, learning and multimedia communication. Diarists will be encouraged to write substantive thoughtful, personal entries, tagging their entries to promote access, and members of the public will be encouraged to browse and reflect on the entries. However, in the spirit of diary writing, which essentially stresses personal reflection, there will be no opportunity for readers to make direct comments on the diary entries. The project will also be promoted through primetime BBC programming, as well as on radio and iTunesU, with the Creative Climate website providing viewers with the chance to make web-based journeys off the programmes. It is hoped that mobile devices, including audio, will also be supported as a means of making contributions.

DEVELOPING AN OU PRESENCE ON EXTERNALLY HOSTED OPEN EDUCATIONAL RESOURCES AND SOCIAL DIALOGUE PLATFORMS

In early 2010, the OU is using several externally hosted platforms to develop new ways of engaging the general public. Freely accessible learning materials are available on YouTube and iTunesU, while comment and discussion was encouraged through YouTube, Twitter and Facebook.

The Open University on iTunesU

The OU has been offering audio and video-based materials on iTunesU since June 2008 and is one of the largest presences on this site, with 250 hours of content available in March 2010. Since the launch of the OU site on iTunesU there have been over 20 million downloads by nearly 2 million visitors, the first institution to reach this milestone (The Open University, 2010o). Approximately 89% of these visitors were from outside the UK. In March 2010 301 'albums' containing 2,600 tracks were on offer. Of these, 242 albums included audio visual material taken from 140 of the courses in presentation at the time at the OU, while the others included general interest material and material related to OU research projects. One in fourteen of the iTunes downloaders also went to visit the main OU website and many have been inspired to sign up for formal OU courses. To ensure accessibility to those with disabilities, transcripts were provided for 96% of the courses.

Podcasts presented on iTunesU have been found to be beneficial for their potential to convey ideas and enthusiasm effectively. Training has been offered across the OU to academics interested in using this medium to transfer their ideas to the general public, using the software package Audacity (Audacity Development Team, 2010). Academics have been encouraged to create podcasts as part of their usual working routine, for example, as they deliver a seminar presentation or create learning materials. They have also been encouraged to use animation tools, such as animoto (Animoto, 2010), which is useful when individuals are reluctant to see themselves on screen. Budgeting and resources for open access podcasts are being routinely included in plans for new courses at the OU and course teams are encouraged to consider transferring a selection of existing audio-visual course material to the iTunesU platform.

RSS feeds enable the users of these open iTunesU materials to integrate them into their own RSS readers and therefore to gain easy access to new material as it is released. Viewers were also encouraged to embed links to the content on their own website, profile or blog, thus distributing knowledge of the learning materials further still. Users are also encouraged to share comments on the materials by clicking on the links provided to networking sites such as Twitter and Facebook, thus entering social dialogue domains, or to add the link to their own social bookmarking listings within sites such as Delicious (Yahoo, 2010).

Open University Channels on YouTube

The OU has four official channels on YouTube in April 2010, providing open access to over 700 videos. The OUView channel allows people to find out more about the OU and its courses by inviting viewers to watch short information clips about course content. A second channel, OU Life provides an insight into life at the OU by staff and students and OU Research showcases some of the OU's world-leading research. By far the biggest site is OU Learn which is offering 492 videos of OU course materials and OU/BBC programmes in April 2010. By March 2010, the OU's channels had received 2.7m downloads since their launch, with the United States being the country with the biggest number of views of OU materials at 27% of total views, followed by users in the UK at 21% of total views. In March 2010 a total of 67,300 unique users were identified.

Like other YouTube channels, the OU's channels also invite comments from viewers. These may be comments about the videos they had just watched, opinions on the OU as an education or information provider, or requests for further information. By 2010, comments have come from 6,000 OU registered channel subscribers, who include members of the public as well as current or past students of the OU.

The Open University on Facebook

The OU has several institutional presences on Facebook in 2010. Facebook offers the chance for 'fans' of the OU to interact online with each other on this social networking site, as well as to interact with OU representatives.

The Open University's 'official' Facebook page had around 22,500 'fans' in March 2010, whilst the OU Library's page had around 5,500. Smaller, more specialised groups also exist on Facebook, such as the Art Historians Group aimed at linking OU students, tutors and Faculty who shared a passion for Art History. All of these groups' content is completely open so could be read, and the group joined, by any member of the public who is interested. There are also 'unofficial' Facebook groups, which have been set up and run by students: 'I ♥ the OU', for example had around 9,000 members in July 2010 and can be found by searching for 'Open University'; as well as several groups set up by students and aimed at people studying a particular course. Again, all content on the organisation pages was open, and social interaction was encouraged.

The aim of the OU's official presence on Facebook is to provide the distance learning students with a way of building community oriented around their studies, as well as to offer a showcase to prospective students. Discussion often centres around course content, making contact with people on the same course, or asking for advice on which course to study. As highlighted earlier, all these Facebook groups are open to all to read and differed from the OU's secure VLE: students were warned not to use the site to share assignments, for example, due to their open nature.

The official OU Facebook page includes announcements about OU activities and achievements, as well as videos about the Library. It should be noted that the OU had no control over what happens on its Facebook sites: it is up to Facebook to control users who are breaking the terms and conditions of Facebook. However, it is also possible for the OU staff who are looking after these 'organisation' pages to take down comments as an administrator. Indeed the need for OU representatives to 'listen' to the views of the students is seen as important and a key benefit of an OU presence on the site. One example of the 'listening' OU was when students complained that a supermarket voucher system, which allowed students to get money off courses, was withdrawn: the outcry through Facebook persuaded the OU to reinstate the scheme.

Following Facebook's decision to allow access to their full application programming interface (API) to third party developers, the OU created two add-ons. One was called Course Profiles, which encouraged students to record a list of courses they had studied, wanted to study or were studying at the time: this allowed them to link to others studying that course with the possibility of finding a 'study buddy'. It also allowed them to leave a rating, or to link to the OU course page or to free materials on OpenLearn, and so on. The other add-on tool was 'My OU Story' which allowed users to microblog about the courses they were studying and to make the contents available to everyone if they choose to do so.

The Open University on Twitter

Alongside Facebook, the OU also has an institutional presence on the Twitter microblogging social networking website, @OpenUniversity with just over 6,000 followers in July 2010. In line with OU policy on social media, the 'tweets' it sends out include OU official press releases, from which it takes an RSS feed; and automatic tweets about OU/BBC co-production television broadcasts, such as Coast and Bang Goes the Theory, just before they begin. A recent initiative on the OU's Twitter stream, designed to tie in with the General Election campaign, which was underway at the time, was the inclusion of tweets relaying 'news' which relate to the 1992 General Election in the UK. Links to a related blog on open2.net were provided as well as to archive photographs. The OU library also manages a separate Twitter stream, @OU_Library using it in a similar way to the OU's Twitter stream. Library news of new resources and library seminars were included, as were links to the Library's Facebook page, highlighting videos which demonstrate how to use the new Web2.0 tools. The OU also uses both its main Twitter presence and that of the OU Library to engage with its community, by 'following' some of its followers and also by including a selection of course-related 'lists' on its homepage. Responses to followers by the OU were limited to items that related to the OU rather than the personal issues of its followers. It is also possible for Twitter users to set up the Library's Twitter stream as an RSS feed to their aggregator, such as My Yahoo or Google Reader so that the news comes to them directly rather than having to log in to Twitter.

THE BENEFITS OF USING THE NEW SOCIAL MEDIA FOR PUBLIC ENGAGEMENT

A range of benefits may be identified as arising from the OU's use of the new social media. The benefits vary according to the characteristics of the stakeholders.

For learners, for example, the OU supports the concept of lifelong learning by enabling members of the general public to dip into the freely available open learning materials and social discourse opportunities whenever they like. There are no charges, and learners can read or view as little or as much as they like, whether experimenting with new subject areas, or refreshing a previous interest. The OU encourages learners to set up mutually supportive study groups to enhance and co-construct their learning experience, although many learners choose to study alone. The OU's reputation ensures that learners feel confident in engaging with the materials, and the OU's openness in encouraging user ratings also helps to raise trust. Experimentation with the open learning materials may also lead to informal learners gaining the confidence to sign up to formal courses with the OU.

The OU itself also benefits from using the new social media for public engagement. Feedback is gained on the open learning materials and a wider audience is gained for materials which have taken many hours to put together. The learning materials may also be reinforced and extended 'in the cloud' by informal contributions from new perspectives by learners who have made use of the formal materials. Younger learners, an expanding section of formal OU learners, are encouraged to draw on the materials due to their easy availability on the internet, a medium with which most young people are familiar. Presentation of research outputs in an open arena may stimulate dialogue with academic peers from other institutions, resulting in mutual benefits, as well as benefits to the general public, knowledge is further developed. The newly developed knowledge may then feed back into further development of the teaching materials and thus be ultimately pushed back to the general public.

Society as a whole may also be identified as a beneficiary of the ecology of openness at the OU. Social inclusion is promoted by the removal of barriers to access to learning materials. Barriers between formal and informal learners are also broken down, promoting equality. Cross-fertilisation of ideas and feedback from worldwide users of the open materials and social discourse provides benefits to all in promoting consideration of a range of perspectives. The transfer of knowledge back to the taxpayer may also be said to be a repayment of public investment in this higher education institution, enabling business development as well as skills enhancement.

THE CHALLENGES OF USING THE NEW SOCIAL MEDIA FOR PUBLIC ENGAGEMENT

Although the benefits of using the new social media for public engagement are many, as explained above, several issues have been considered by the OU (Wilks, 2009). The OU is still grappling with many of these, with situations continually changing and new challenges emerging.

Open Educational Resources versus Social Dialogue

As highlighted earlier, the new social media varies in the level of dialogue expected or facilitated from its users. OER, which are basically about institutions providing one-way online access to their formal learning materials, may be distinguished from social dialogue media, which promote learning through multi-way social interaction and may not include any formal learning material element. The examples above demonstrate that the OU uses both approaches, although the boundaries between the two are blurred: OER media may include links to social dialogue opportunities; whilst social dialogue media may include links to formal materials. Institutions need to align the approach to the learning experience envisaged, as well as to consider the desired learning outcomes. An appropriate mix of approaches can help to stimulate learning by enabling users to select the media which is most appropriate to their learning style.

Intellectual Property Issues

Several intellectual property issues need to be considered within an ecology of openness. OU materials which are on open offer have been licensed under the Creative Commons system (Creative Commons, 2010), which allows members of the public to copy, distribute and display the work, as well as to make derivative works. However, this allowance is under the condition that the works must not be used for commercial gain, that the original author must be given credit, and that any subsequent work which builds on the materials offered must also be distributed under the same licence.

Similarly, social dialogue deposited on OU websites is also subject to the same open licensing system. This also means that members of the public who participate are not giving up their rights to their own contributions. However, the OU has no control over the licensing approaches on most externally-hosted sites. Facebook, for example, is criticised for its approach to intellectual property, particularly its license terms which enable it to use any IP content that members post on or in connection with Facebook. The OU therefore chooses carefully where it puts its learning materials and advises its fans not to deposit their own learning materials, such as essays or assignments on externally hosted sites.

Consideration of where the boundaries lie in connection with blogs or tweets written and published by OU staff is an interesting issue. These could be considered as OER as well as providing the chance for contributions in the form of social dialogue. Some of the more popular authors of these media attract followers in their thousands. OU staff are advised to be aware that they are blogging in a public arena and that their comments will reflect on the OU if they identify themselves as staff. Media training, such as the new podstars course run by the OU's Institute of Educational Technology to encourage OU staff to engage with the public in innovative ways is also offered to highlight the use of the new media by OU staff for public engagement purposes.

Usability, Accessibility and Learner Support

Usability and accessibility are key features for users of new social media. The OU ensures that its own websites are designed and tested to enable all learners to find what they are looking for or make interesting discoveries. The OU also ensures that users with disabilities have access to the sites, making sure that the platforms on which it builds support accessibility tools and are continually tested.

Users of the open learning materials are not directly supported by the OU's technical team, but are directed towards an open user forum which encourages users to offer support to each other. Members of the public who are engaging with the OU on externally hosted sites, such as Twitter or Facebook, would be directed towards the technical support offered by these platforms, although informal advice from the OU moderators keeping an eye on the activity on external sites may occasionally be offered.

Worries are sometimes raised that the introduction of new forms of technology can exclude some potential learners from the learning experience. However, as Martin Bean, the OU's Vice Chancellor commented (Bean, 2010), the OU sees it as important to support learners by introducing (and retiring) appropriate technology at the appropriate time.

Getting the Public to Engage in the New Social Media

In 2010, getting members of the public to access the OU's new social media is not really a problem for the established sites and presences. The OU's OpenLearn website had 10 million individual users by April 2010, after three years of operation, with 60% of these being located outside the UK, and the OU's Facebook page had 22,600 fans in April 2010, for example. However, this engagement does not often go beyond short factual comments or ratings. As mentioned above, some social media initiatives, such as the OU's Creative Climate, however, have made a conscious decision that they will not facilitate multi-way discussion between website visitors. However, the Creative Climate site will be promoting input from the general public, aiming to encourage this input to be at length and of high quality, and making those contributions open to all. The creators of the site are awaiting the results of this aim.

Barriers to participation have purposely been kept low, with OU OER websites not requiring sign-up. The OU has also gone to where the public is, by setting up a presence on Facebook and Twitter, for example.

Sustainability

Sustainability of the social media offerings needs to be considered at the outset. Issues of staffing, technical support and troubleshooting are all issues which the OU has had to explore. Externally hosted sites, such as Facebook, have the technical issues taken care of; in-house supported sites need programmers and technical support staff to maintain them. However, it has been important at the OU in groundbreaking initiatives using social media, to experiment with new ideas, pushing technical and creative boundaries in the conception and testing of the initiatives.

Platforms

As illustrated earlier, the OU's ecology of openness is spread over various platforms. This has been a conscious policy of risk spreading and also aims to avoid loss of control over the content. Partnerships, such as that with the BBC, have also been a major feature of the policy. In 2010, plans are about to come to fruition which bring together some of the disparate open social media which the OU offers onto one platform. This will combine the resources on open2.net with those from OpenLearn into a new portal, which will simply fall under the OpenLearn brand. This aims to make access easier for the general public, as well as to gain better technical and IP control over many of the open access web resources. As well as OER in the form of OU course materials, material connected with OU television programmes made with the BBC, will be available. Feeds from Twitter will also feature. Members of the public will be encouraged to provide input in the form of social dialogue relating to the learning materials.

CONCLUSION

As this chapter has highlighted, OER and associated social dialogue opportunities are a major feature for the OU in 2010. Although the OU has made substantial progress in the resources and opportunities it offers, it is also a time of transition and experimentation and continuous improvement as new technological possibilities become available. It is too early to reach firm conclusions on the efficacy of the various resources reported above, although early evaluations suggest that they are certainly valued as learning tools by members of the public. Future research will be consolidating and extending evaluation of the tools, however.

The move towards developing the ecology of openness at the OU has provided the institution with many benefits, as well as being in line with the OU's ethos of social justice. Members of the public from across the globe who have engaged in the social media-enabled learning opportunities have gained many opportunities for personal development.

It is also hoped that other institutions will be able to learn from the OU's experience and that this sharing of achievements and approaches will prove useful to colleagues who are embarking on similar missions.

REFERENCES

Alexander, B. (2006). Web 2.0: A new wave of innovation for teaching and learning? *EDUCAUSE Review*, *41*(2), 32–44. Available at http://www.educause.edu/EDUCAUSE + Review/ EDUCAUSEReviewMagazineVolume41/Web20ANewWaveofInnovationforTe/158042. Retrieved on 2 July 2010.

Amichai-Hamburgera, Y., McKenna, K. Y. A., & Tald, S.-A. (2008). E-empowerment: empowerment by the internet [20010-07]. *Computers in Human Behavior*, *24*(5), 1776–1789.

Anderson, P. (2007). *What is Web2.0? Ideas, technologies and implications for education*. JISC Technology and Standards Watch. Available at http://www.jisc.ac.uk/media/documents/ techwatch/tsw0701b.pdf. Retrieved on 27 April 2010.

Animoto. (2010). *Animoto: The end of slideshows*. Available at http://animoto.com/. Retrieved on 29 April 2010.

Audacity Development Team. (2010). *The free cross-platform sound editor*. Available at http:// audacity.sourceforge.net/. Retrieved on 29 April 2010.

BBC. (2010). *BBC iPlayer*. Available at http://www.bbc.co.uk/iplayer/. Retrieved on 28 April 2010.

Bean, M. (2010). *Opening keynote. The learning journey: from informal to formal* [Video; Powerpoint slides]. Available at http://www.jisc.ac.uk/events/2010/04/jisc10.aspx. Retrieved on 30 April 2010.

Buckingham Shum, S. (2010, March 18). SocialLearn update. Available at http://www. open.ac.uk/blogs/sociallearn/

Castells, M. (2007). Communication, power and counter-power in the network society. *International Journal of Communication, 1*, 238–266.

Conole, G. (2010). Facilitating new forms of discourse for learning and teaching: Harnessing the power of Web 2.0 practices. *Open Learning*, *25*(2), 141–151.

Creative Commons. (2010). *Attribution-Non-Commercial-Share Alike 2.0: England and Wales.* Available at http://creativecommons.org/licenses/by-nc-sa/2.0/uk/. Retrieved on 26 April 2010.

Daniel, J. S. (1996). *Mega-universities and knowledge media: Technology strategies for higher.* London, UK: Kogan Page.

Dron, J., & Anderson, T. (2007). Collectives, networks and groups in social software for E-learning. In: T. Bastiaens & S. Carliner (Eds), *Proceedings of world conference on e-learning in corporate, government, healthcare, and higher education 2007* (pp. 2460–2467). Chesapeake, VA: AACE.

Engeström, J. (2005, April 13). Why some social network services work and others don't – Or: the case for object-centered sociality. Available at http://www.zengestrom.com/blog/2005/04/why_some_social.html

eSkills UK. (2010). *eSkills UK.* Available at http://www.e-skills.com/. Retrieved on 30 April 2010.

Facebook. (2010). *The Open University on Facebook.* Available at http://www.facebook.com/theopenuniversity. Retrieved on 16 April 2010.

Hemmi, A., Bayne, S., & Land, R. (2009). The appropriation and repurposing of social technologies in higher education. *Journal of Computer Assisted Learning, 25*(1), 19–30.

Li, C., & Bernoff, G. (2008). *Groundswell: Winning in a world transformed by social media.* Boston: Harvard Business Press.

Linden Research, Inc. (2010). *Second Life.* Available at http://secondlife.com. Retrieved on 16 April 2010.

OER Commons. (2007). *Open educational resources.* Available at http://www.oercommons.org/. Retrieved on 29 April 2010.

O'Reilly, T. (2005). *What is Web 2.0: Design patterns and business models for the next generation of software.* Available at http://oreilly.com/web2/archive/what-is-web-20.html. Retrieved on 29 April 2010.

Pearce, N., Weller, M., Scanlon, E., & Kinsley, S. (2010). Digital scholarship considered: How new technologies could transform academic work. *Education, 16*(1). Available at http://www.ineducation.ca/. Retrieved on 2 July 2010.

The Open University. (2005 [1969]). *Statutes and charter 23rd April 1969 [Amended by the Privy Council to December 2005],* p. iv. Available at http://www7.open.ac.uk/gsh/Charter/Charter.pdf. Retrieved on 6 April 2010.

The Open University. (2010a). *About the OU: Our mission.* Available at http://www.open.ac.uk/about/ou/p2.shtml. Retrieved on 8 April 2010.

The Open University. (2010b). *The Open University's strategic priorities 2009-10: OU Futures – values and capabilities.* Available at http://www.open.ac.uk/ou-futures/values-core.shtm. Retrieved on 27 April 2010.

The Open University. (2010c). *OpenLearn Learningspace.* Available at http://openlearn.open.ac.uk/. Retrieved on 16 April 2010.

The Open University. (2010d). *OpenLearn LearningSpace Forum.* Available at http://openlearn.open.ac.uk/mod/forum/view.php?f=64. Retrieved on 16 April 2010.

The Open University. (2010e). *Open research online.* Available at http://oro.open.ac.uk/. Retrieved on 16 April 2010.

The Open University. (2010f). *The reading experience database (RED), 1450-1945.* Available at http://www.open.ac.uk/Arts/RED/. Retrieved on 30 April 2010.

The Open University. (2010g). *Project HESTIA*. Available at http://www.open.ac.uk/Arts/hestia/. Retrieved on 30 April 2010.

The Open University. (2010h). *Platform: The OU community online*. Available at http://www.open.ac.uk/platform/. Retrieved on 16 April 2010.

The Open University. (2010i). *open2.net*. Available at http://open2.net. Retrieved on 21 April 2010.

The Open University. (2010j). *Cloudworks*. Available at http://cloudworks.ac.uk/. Retrieved on 16 April 2010.

The Open University. (2010k). *The Open University on iTunesU*. Available at http://www.open.ac.uk/itunes/. Retrieved on 16 April 2010.

The Open University. (2010l). *Research School: Research Degree Skills*. Available at http://phdskills.open.ac.uk/. Retrieved on 27 April 2010.

The Open University. (2010m). *iSpot: Your place to share nature*. Available at http://ispot.org.uk/. Retrieved on 30 April 2010.

The Open University. (2010n). *Vital: Transforming lessons, inspiring learning*. Available at http://www.vital.ac.uk/. Retrieved on 30 April 2010.

The Open University. (2010o). *iTunes U downloads pass 20 million mark*. Available at http://www3.open.ac.uk/media/fullstory.aspx?id = 19109. Retrieved on 1 July 2010.

Twitter. (2010). *The Open University on Twitter*. Available at http://twitter.com/openuniversity. Retrieved on 16 April 2010.

Wilks, L. (2009). *It's like a permanent corridor conversation': An exploration of technology-enabled scholarly networking at the Open University*. Available at https://wiki.projectbamboo.org/display/BPUB/Open + University + Scholarly + Networking + Report. Retrieved on 11 June 2009.

Willinsky, J. (2006). *The access principle: The case for open access to research and scholarship*. Cambridge, MA: MIT Press.

Yahoo. (2010). *Delicious*, Available at http://delicious.com. Retrieved on 27 April 2010.

YouTube. (2010). *OUView: The Open University's Channel*. Available at http://www.youtube.com/theopenuniversity. Retrieved on 16 April 2010.

DISCONNECT36: A SOCIAL EXPERIMENT TO TEACH STUDENTS TO SHUT DOWN, TURN OFF, AND UNDERSTAND CONNECTIVITY

Monica Flippin-Wynn and Natalie T. J. Tindall

ABSTRACT

American teens are using online social networks more than ever before. According to a 2010 Pew Internet Project study, close to 75% of teens use social media sites and wireless connections (cell phones, game consoles, and portable gaming devices) to access the Internet (Lenhart, Purcell, Smith, & Zickuhr, 2010). These constant connections and ties to the Internet are fascinating to some scholars who see a tremendous value to the communities found and made online. Yet, this ability to be in constant connection is troubling to other scholars who believe that this constant ability to contact and connect is changing society for worst, not the better [Putnam, R. (2000). Bowling alone: The collapse and revival of American community. *New York: Simon & Schuster; Bugeja, 2005]. This chapter outlines a social media class experiment undertaken by the lead author to provide students with an opportunity to understand their reliance on the new media or media in general and add to the scholastic*

Teaching Arts and Science with the New Social Media
Cutting-edge Technologies in Higher Education, Volume 3, 265–281
Copyright © 2011 by Emerald Group Publishing Limited
ISSN: 2044-9968/doi:10.1108/S2044-9968(2011)0000003016

literature on teaching and technology in the classroom. In the Spring 2010 classes, the majority of students agreed to disconnect from all communication technology and social media for 36 hours. The assignment was worth 65 points. As they started to withdraw from the media, the class assignment provided students with insights into their constant connectivity and how they manage information through various mediated channels. After the assignments students were required to complete an 800-word blog or paper. To receive full credit for the assignment, students needed to complete the written component. All the students who participated completed the written requirement. The majority of the students completed their assignments on their blogs but about half of the students both turned in a written paper and posted the assignment on their blogs. The students that provided written permissions were selected for inclusion in this chapter. We were careful to make sure that the students in this chapter were representative of the entire population, including male and female, students who were bothered by the disconnect and those who were intrigued by the possibility of being disconnected, traditional and non-traditional students, and students who worked, had no outside employment, and students with other non-academic obligations. Our insight into students' issues of connectivity was drawn from these stories. This chapter further offers ideas on how to integrate such an experiment in other settings and provides pedagogical rationales for this type of assignment. The names of the students in this experiment were changed to safeguard student anonymity and personal privacy.

Like the Industrial Revolution before it, the Digital Revolution is changing the shape of our world. (Doyle, 2004, p. 371)

Technological developments in the last decade have resulted in a staggering number of new devices that can mediate information in ways that would have been deemed science fiction not 20 years ago. (Yzer & Southwell, 2008, p. 8)

In March 2010, for the first time, Facebook supplanted Google as the most visited website in the United States (Dougherty, 2010). Two recent studies from the Pew Research Center's Internet and American Life Project also confirmed what academics, pundits, and social researchers already knew: social media networking sites have become omnipresent in the lives of teens and college students. The Pew report, titled "Social Media & Mobile Internet Use among Teens and Young Adults," found that more and more teens and young adults are microblogging or using various platforms and mechanisms to constantly update and inform others about what is taking place in their lives. In fact, according to the report, 72% of young adults 18–29 years use social networking sites. Over 80% of young adults between 18–29 years are wireless Internet users, and 93% of young adults own a mobile phone (Lenhart, Purcell, Smith, & Zickuhr, 2010). Conversely, the authors asserted that the Internet is a "central and indispensable element in the lives of American teens and young adults" (p. 5). Further validation comes from a 2010 recent University of Maryland study that asked students to give up all forms of media for 24 hours, and the results declared that students were addicted to social media. The report stated, "Students are swimming in an ocean of media, but are often oblivious to their use of the technology all around them they so take it for granted" (Moeller, Chong, Golitsinski, & Guo, 2010).

In the classroom, the ubiquitousness of media is ever present. Students take notes on laptops and smart phones, multitasking between Facebook and Twitter and the document on which they are typing notes. Cell phones buzz or chirp during lectures and students (either blatantly or surreptitiously) text whenever they can. Turkle (2010) however believes all these connections and the media availability is doing a disservice to the educational environment, writing that (students) "have done themselves a disservice by drinking the Kool-Aid and believing that a multitasking learning environment will serve their best purposes. There are just some things that are not amenable to being thought about in conjunction with 15 other things" (para. 5). This raises the question for us: What is the price we are paying for this constant contact and connectivity?

This question became the primary focus of Disconnect36, a social experiment that is the focus of Flippin-Wynn's New Media Technology course during Spring 2010. As the semester progressed, the students had little sense or understanding of the critical aspects and the consequences of the media that they used. Thus, following a class discussion on new media and student management, the idea for Disconnect36 was born.

This chapter outlines the assignment and the student responses while presenting the literature that validates the concern seen with constant connectivity and the necessity of this assignment. This study and the analysis of the data are grounded theoretically in an understanding of social information processing and electronic propinquity, which adds value to any discussion on media usage. Following the results, a discussion of pedagogical relevance and implications of this assignment follows concerning both teachers and students.

REVIEW OF RELEVANT LITERATURE

McLuhan (1964) was very clear about his ideas on technology and change. It was not so much the content that concerned him, but it was the actual medium or the conduit that was more of pressing concern. Today, it is not so much the messages that are being communicated that are causing widespread discussion and interest; it is the new media that are being utilized to channel the messages.

From telephones, phonographs, and television to computers, video games, and social networking, new media is a constant component of our everyday landscape and experience. As a nation, we are supersaturated with technology and are truly wired at home, at work, and on the go. Although the technologies may differ, our environment has always had its share of technology saturation. If we look back to the start of the 20th century, we can find numerous researchers concerned with the proliferation of technology and its possible effects (Meszaros, 2004). The changes we see in technology are more than just new ways of access and communication. These new media are becoming a way of life, especially for the younger generation. Watkins (2009) states that these new tools allow young adults to manage their professional, educational, social, and intrapersonal lives and connect to the outside environments. In addition, the Internet and media technology are instrumental in allowing individuals to stay connected with families, friends, and other interpersonal communities (McMillan & Morrison, 2006).

Electronic Propinquity

The original idea central to electronic propinquity is "nearness" (Korzenny, 1978, p. 7), specifically focusing on face-to-face communication and the technologies of audioconferencing and videoconferencing. According to Korzenny, this organizational communication theory could be used and applied to not only "electronically mediated communication, but all symbolic human interaction conducted over a wide range of channels" (p. 4). As Walther and Bazarova (2008) noted, "The theory predated the Internet, yet its conceptual definitions and level of abstraction offer a broad and potentially powerful approach to understanding the effects of electronic media it did not originally consider" (p. 623). In fact, Walther and Bazarova expanded the idea to the following: "the feeling of nearness that communication experience using different communication channels" (p. 624). The basic conceptualization for propinquity through electronic, mediated ways is that a person's perceptions determine whether the person feels functionally close to someone.

To increase electronic propinquity, multiple factors such as bandwidth, mutual directionality, and individual communication skills must be present in the communication. In contrast, the decrease in electronic propinquity is related distinctly to information complexity, perceived communication rules, and perceived number of choices in communication channels. Walther and Bazarova (2008) summarized the factors succinctly: "Essentially, the theory argues that one experiences greater propinquity when there is greater bandwidth ... or when there is less information complexity, or greater mutual directionality ... or when there are greater individual communication skills, or fewer rules, or a smaller number of perceived choices among communication channels" (p. 624).

The value of electronic propinquity to this project lies in the fact that basic attributes that define the theory are ever-present in the social mediated and networked living environments of students.

Social Information Processing Theory

The social information processing theory (SIPT) suggests that it is possible for individuals to build community and reach their goals of interpersonal communication and connection through online channels (Walther, 2008). Westerman and Skalski (2009) posited that social information processing assumes that communicators make attempts to achieve communication goals in online settings as much as in offline settings. When the lack of cues available in an online setting presents obstacles to accomplishing their goals, users adapt their behaviors to the cues that are available. Thus, given enough time, people can utilize these circumventions to accomplish goals online just as well as they do face-to-face. SIPT focuses on the way people overcome limitation of a technology instead of only focusing on the limitations of the technology (as SPT would do) (pp. 74–75).

The importance and value of SIPT to this research is that online computer acts can imitate offline associations (McMillan & Morrison, 2006). The outcomes of online interpersonal communication are similar to face-to-face interactions and opportunities. Although nonverbal cues are missing from online interactions and communications, individuals still use online media to achieve their interpersonal goals.

This fact is important when considering the young adult media usage and when considering the number of hours that young adults spend online and conducting online relationship maintenance as opposed to the number of hours spent offline.

Media Usage of Young Adults

Most of the scholarly literature on being connected and individual use of media has been conducted on either young children or teenagers (Valentine & Holloway, 2002; Livingstone & Bovill, 2001).

As indicated earlier individuals, specifically young adults, are living their lives online, utilizing new media and technology tools for interpersonal communication and social structure. Research indicates that communities can be built up through the various forms of media that exist on the Internet (McMillan & Morrison, 2006).

According to Donath and Boyd (2004),

> [M]any of the most commonly used Internet applications, such as e-mail, IM, and social networking sites, are designed to foster interaction between two people. If through the use of these technologies, people have become accustomed to thinking of this world as a social place and have developed the medium-specific skills ... necessary to use these channels, it is possible that they have also become more accustomed to activating mental models when using these technologies. (p. 76)

Holloway and Valentine (2000) explained that especially children are very good at constructing their own individual environments within their own personal spheres and groups when using the Internet. McMillan and Morrison (2006), in a study looking at how young adults feel about the Internet, found that young adults construct their social relationships online. Because they may be using some features and language tools that are associated with oral communication, the communication may result in the quick acceptance of individuals as friends (Wilkins, 1991).

The technology that young adults and children today have grown up with and incorporate into their daily lives is not all a bad thing, but the key is gaining an understanding of what is enough, what is too much and where does the technology best fit? As instructors, we are teaching a millennial generation that has grown up connected to media at all times. Turkle (2010) explained that,

> One of the things I've found with continual connectivity is there's an anxiety of disconnection; that these teens have a kind of panic. They say things like: "I lost my iPhone; it felt like somebody died, as though I'd lost my mind. If I don't have my iPhone with me, I continue to feel it vibrating. I think about it in my locker". (para. 7)

It is important to teach our young adults how to manage their technology; how to disconnect and how to be able to utilize other basic skills to accomplish communication and everyday tasks. Disconnect36 was an opportunity to allow a group of students an opportunity to take inventory of their everyday media and technology use.

DISCONNECT36 ASSIGNMENT BASICS AND STUDENT PREPARATIONS

Demographics

Gender	Female: 48 students (66.7%)
	Male: 24 students (33.3%)
Race	2 biracial (Caucasian and Black) (2.78%)
	70 African American/Black (97.22%)
Classification	Freshman (9)/Sophomore (1)/Transfer (3): 13 (18.06%)
	Junior: 32 (44.44%)
	Senior: 19 (26.39%)
	Continuing: 8 (11.11%)
	Transfer: 3
Major	Mass communication: 67 (93.056%)
	Other majors: 5 (6.94%)

The Disconnect36 assignment was scheduled to begin on Monday, February 22 at 8:00 a.m. and end on Tuesday, February 23 at 8:00 p.m. The assignment was worth 65 points, yet some students believed that the assignment should have been worth more points based on all that they had to give up. Students were given information two weeks before the actual disconnect so they could make the necessary arrangements to disconnect (i.e., inform teachers, parents, friends, and employers of their online and mobile phone inaccessibility for those two February days). A total of 75 students participated, and 70 students completed the entire process. The ages of the participants ranged from 19 to 50 years old. The majority of the students were communication majors; however, several of the students majored in political science, psychology, and science.

The Dissconnect36 parameters were as follows:

• Students had to totally disconnect from all media, including television, movie theatres, video games, MP3 players, all computer use, social networking sites, cell phones, texting, any electric or battery-operated devices, landline phones, or other electronic media devices that would connect them to the Internet.

- In addition, they could not use other people's phones or devices to call or receive messages.
- Students were allowed 30 minutes in 10-minute increments a day to talk to parents, employers, or daycare providers on a mobile phone or landline, but they could not make contact through texting.
- All student-identified family, work, or caregiver numbers were verified in advance, so if students answered any other numbers and those calls appeared on their phone logs, they were disqualified for the points.
- The only media options students were allowed were terrestrial radio and print newspapers, not online or satellite versions.

One week before the actual disconnect, students in general got antsy and nervous and really started thinking about how they would fare without the media they had come to rely so heavily on. Some of their main concerns included how they would contact their teachers if they had questions, how to reach the police or security in case something happened, and how to do their online homework assignments. The night before the Disconnect36 project began, most students updated their Facebook statuses alluding to their participation in the project and the imminent 36-hour disconnect. They informed family and friends and some students even stayed online most of the night to somehow load up on their online activities to last them for the next few days. Chase, a mass communication senior, broadcast to all his Twitter and Facebook associates that he was not going to be able to communicate with them for 36 hours because of a class assignment. For the most part, he wanted to inform them, but he also believed that by letting his friends know about his involvement in the project, he would be less likely to cheat and not complete the assignment. Kristen, a mass communication junior, stayed up all night watching her fellow classmates tweet and update their status on Twitter (it was even a trending topic on Twitter that day). Alex, a mass communication junior, informed his friends on Facebook that he would be unavailable for a couple of days, and when he explained why, his friends and family placed a side bet that he would not be able to complete the assignment. Some other students elected to have a "pre-disconnect" party, where they played board games, talked, and did their homework together.

After the two-day disconnect, students were required to complete an 800-word blog or paper to discuss their experiences. The following section discusses the student responses. Each of the student papers or blogs was thoroughly assessed for their ideas, thoughts, concerns on new media influence and connectivity.

"All of sudden, our lives shut down": Student Responses to Disconnect 36

Initially, some students did not feel they would have any problems with disconnecting from all things media; however, as the day got closer, most of the students shared that they experienced some sort of panic and frustration. Students' names were changed to protect their privacy and involvement in this experiment and in this class.

Anxiety Related to Disconnecting

Mimi, a political science sophomore with a minor in mass communication, explained that without her phone she lost all perception of time. Since she did not own a watch she had several scheduling issues during the two days of Disconnect36. While waiting over 30 minutes for a friend, Mimi asked a complete stranger for the time. She noticed that being connected to her mobile phone and other media meant that she really did not have to talk to anyone she did not want to because the information needed could be found by searching using a predetermined set of tools. Mimi said,

> I was looking lost for about 15 to 20 minutes, so I stopped a complete stranger, which I never do, and asked what time it was. The person explained to me that it was only 12:35 and I laughed and walked off then sat down because I had lost my sense of time. I thought it was much later. But, I was so uncomfortable asking a complete stranger for the time.

For some students, being disconnected posed a safety issue. This was especially felt among those from out of state: Janice, a public relations junior, indicated that at first her parents were reluctant about Disconnect36 specifically because not having access to her mobile phone put her at a safety disadvantage. Janice made sure that she had contact with friends and classmates not involved in the same class, in case she needed help or access to a mobile phone during the assignment. She wanted to be ready just in case.

Emptiness without Social Media

Cameron, a public relations senior, said she felt an "unexplainable emptiness" and felt that being disconnected separated her from her world.

In addition, she said she very quickly saw how dependent she was on technology and saw herself grow frustrated trying to figure out how she was going to function without her media tools. She had never realized how much of her existence she owed to cell phones, texting, and online activities.

Courtney, a marketing and advertising junior, realized during Disconnect36 that she had been providing too much information about herself on some of the social sites. So during this assignment, instead of tweeting through her phone or computer, she started a journal and tweeted what she would normally put out on Twitter. Going through the information she tweeted in her journal, she found that being connected to all these social sites left her exposed and vulnerable by "allow[ing] audiences a real passageway into her private life."

Tisha, a mass communication sophomore, thought she would lose her mind. She could not remember a time that she had not been connected to some form of media. She felt depressed, isolated, and alone without the media connections because the tools provided her the only means in which she stays connected with family, friends and the news. She decided that she needed Twitter, YouTube, Facebook, and JamGlue too much to ever go through this type of withdrawal again, no matter how many points she'd earn. She is over-connected and finds no problem with it.

Using Social Media Time for Social Time

Another student, Alexis a graphic and production major, prepared for being disconnected by setting up several appointments to meet up with friends at the campus plaza. She realized she did all her socializing through social media and saw how different it was to sit and talk face to face with friends without the interruptions of texts, tweets, or phone calls. In addition, this student wanted to spend some time in her drawing room. At first, she explained, she was tensed because it was so quiet. Then she became aware of a persistent click sound. She went on to investigate and found that noise came from a clock in the room ticking. The room, she explained, was usually so full of all kinds of sounds: music, television, cell phones, and activities – she had not noticed that there was a clock ticking in the room before. This assignment made Alexis realize that there is too much that showed her that there is too much technology going on in her life.

Randi, a broadcast journalism junior, explained that the assignment cautioned her about the dependency her generation of millennials has with technology and social media. She felt she did not experience "social media withdrawal" as severe as some of her classmates because she only utilized Twitter and Facebook occasionally, but she did feel "technology abandonment" because of her addiction to television and her mobile phone. In fact, Randi said that this assignment made her realize that her time issues with finishing homework stemmed from media and technology distraction rather than not having enough time. Disconnect36 provided her the opportunity to "really buckle down and study without all the distractions."

Initially, according to Linda, an advertising and production major, she felt a real sense of anxiety even before the assignment began because being connected to technology was a huge factor in her life. After the assignment, Linda wrote that she realized how technology controlled and dictated her choices and relationships and how removed she was without other types of interpersonal communication: "Individuals are disconnected from interpersonal relationships because most of my emotions, thoughts or expressions are shared through email or text messages and rarely [occur in] face to face interaction. Society has no real idea of the effect technology and social media are [having] on the relationships and mentalities of individuals."

During this assignment, one student named David, a broadcast journalism and speech major, felt he was living in the 1920s instead of 2010 because he was cut off from all current events, news, information, and from his family. According to David, during these two days, any news or information he received was through word of mouth or found in regular newspapers, which he found to be a truly foreign concept. On a typical day, he would head to his room after classes. This held an abundance of technology, and he would listen to his iPod while searching his computer. During Disconnect36, he met up with his friends at the cafeteria and stayed until it was time to go to bed because he did not trust himself to remain disconnected.

Cassie, a mass communication sophomore, felt that disconnecting from her phone and other media left her with nothing to do. She said that her "social life, education, email, all work through her phone, so without it, I was lost." At the conclusion of the project, Cassie wrote in her essay that she realized that if you wanted to be connected today there was no way around utilizing technology and connectivity, because they are now constants in our daily lives.

Benefits of Disconnecting

Some students thought Disconnect36 was an excellent experience and suggested it should be expanded to 48 instead of just 36 hours. They utilized the opportunity of being disconnected to catch up with homework and spend some quality time with themselves and others. For instance Shari, an advertising senior, said Disconnect36 showed her how technology was taking over her life. So, Disconnect36 allowed her to focus on her school assignments and gave her time to focus on ways to relax like prayer and meditation.

James, a mass communication senior, felt that being online all the time and utilizing some of the social networking sites was sometimes more about other people needing him to be available than his individual desire or need to be continuously connected. Hence, with Disconnect36, he was grateful for the opportunity to really focus on completing some quality homework, more so than what he usually turned in for his assignments. He also was able to catch up on some reading and meditating in preparation for his various athletic responsibilities. James went on to say that although this assignment seemed extremely difficult, he believed it would benefit them in the long run; he thought their grades and scores would eventually be higher if they spent less time being connected to so many networks all the time. On the basis of this experience, he said disconnecting, "forces us to occupy ourselves and our time with constructive things which can ultimately improve a person's [life] and character."

Finally, Stephanie, a mass communication and English junior, found that her use of technology had made her distant and impersonal, so she utilized the extra time to make new friends among people who like the same things as she does. She indicated she may never have met these new acquaintances if she had not had her head up looking for different opportunities instead of down toward her mobile and texting device.

Some students missed rides to work, were late for class and other appointments, because they normally utilize their mobile phones as their time and calendar connection. To deal with the uncertainty and boredom, students played charades, engaged in multi-player card games, and had coffee dates with other students and friends. Many of the students stated that they had rarely communicated with people out of their personal circles, but Disconnect36 necessitated that they be open and communicate in different ways.

Cheating the System

Some students started off the assignment looking for ways in which they could "cheat the system." From the very first mention of the assignment, some students were thinking of ways in which they could score the points and still stay connected to their media and social networking sites. However, most students decided that the assignment was worth the effort and it would be a total loss of points if they were caught, but more importantly they were intrigued by the challenge and wanted to prove that they were not dependent on any media or social networking site.

For example, Sarah, a production junior, said this was one of the hardest things she has ever done, and initially she considered purchasing another communication device (mobile phone) that she could set up with an alternate e-mail address and Twitter account. However, after some consideration, her conscience would not let her, and she knew her grandmother and mother would be really proud if she completed the assignment. Maurice, a mass communication senior, had planned to secretly use his iPod as he walked to classes and off campus, but he said "I wanted to display some self control and besides my roommates knew about the assignment and they were not going to let me slide by, so I decided not to cheat and instead stay disconnected."

Ease of Disconnecting for Non-Traditional Students

Several non-traditional students participated in Disconnect36. Melody, a non-traditional advertising major, indicated that although she utilizes some of the social networking sites, she felt that they do not have an impact on her daily functions and instead saw this assignment as a self-improving opportunity and a chance to rid her daily schedule of people and things that had very little significance. Alice, another non-traditional media production student, agreed that she had fewer issues than her younger classmates with the Internet and social media. She saw this assignment as an opportunity to get some extra work done and finding some time to mediate and de-stress.

PEDAGOGICAL CONCERNS AND IMPLICATIONS

The goal of this assignment was to get students to recognize their dependence on the Internet and social networking sites. They realized that they are connected in several different circles, feel isolated when they are not, and have a hard time communicating through more traditional methods. As educators, it is important to be knowledgeable in some of the social networking and Internet tools, so they can be utilized as learning tools in the classroom. However, it is also important to continue the more traditional pedagogical outlets of teaching, like public speaking, group communication and written essays, and even blogs. Students should have the complete experience to fully function in today's society, and it may just be their only opportunity in our classrooms. Students should understand that while being adept with these new technological tools, it is the actual learning and content that is important (McMillan & Morrison, 2006).

Implications of Student Participation

Students are pulled in so many different media directions. They are updating their status, adding new friends on Facebook, shopping online, uploading pictures, reading the latest news and entertainment articles and trying to complete their homework. As Turkle (2010) stated, there is so much going on but some tasks cannot be completed in tandem with 10 other choices. By doing this assignment, students were able to scrutinize how they manage technology, communicate and connect over the Internet and social media. Students had the opportunity to confront their lack of certain basic skills and experience possible addictions to some of these sites. By recognizing their reliance, they can now begin to review how much time to spend on the Internet and on social networking sites, how to more efficiently manage how they connect, what they do when they are connected, and what information they share. Giving students the latest facts and figures on how young adults are over mediated will not always get them to pay attention or take the information seriously. But when students were able to experience going without technology, they could make the connections themselves. They felt the isolation, they feel the withdrawal symptoms. They struggled to get through some basic tasks, like finding a phone book to get an address or asking people face to face for information, because they are not as familiar or comfortable in accomplishing some of these tasks without some online or media assistance.

Achieving Balance

As we teach and instruct in this time of the millennial multi-tasker, how can we (as instructors) achieve a balance of injecting the right amount of technology into the classroom while also trying to reduce or minimize student's dependence on the Internet and social media? Many professors are utilizing various social media and technology devices in their classrooms. There are courses that focus on providing students with the necessary tools to work and live in a mediated society. According to a survey conducted by Babson, New Marketing Techniques and Pearson (2010, May 4), more than four of five professors use some type of social media and over half of the professors utilize, other media like blogs, traditional videos, and wikis as instructional tools in their classes. Many in academia believe that to keep this generation of students interested, teachers must learn to provide the course material and information in platforms in which students understand and feel comfortable with. There are many research papers and books written on this topic (Bracken & Skalski, 2009; Lister, Dovey, Giddings, Grant, & Kelly, 2009). College faculty have embraced social media and a majority have integrated some form of these tools into their teaching," said Jeff Seaman, Ph.D., co-director of the Babson Survey Research Group. "While some faculty remain skeptical, the overall opinion is quite positive, with faculty reporting that social media has value for teaching by over a four to one margin." Watkins (2009) states that there really is no choice for professors anymore, students are coming into the classroom armed with definite technological skills. Teachers cannot just plan regular lectures utilizing the learning portals such as Blackboard or ULearn and presentation technologies such as PowerPoint; they must come ready to incorporate an entire group of skills and technologies to teach this generation. Teachers will need to change how they organize their classes. More interactive communication, incorporation of video and social networking sites alleviate student boredom, lack of participation, and miscommunication. Students today are active learners not passive (Watkins, 2009), and the courses and platforms must adhere to that type of learning environment. If teachers work at incorporating bits of new technology and the new learning styles, students will be more prone to pay attention, and stay focused in class, and spend less time updating their status while in class. The new technologies will be a permanent element of the culture and academia must find a way to incorporate them into this critical process of teaching.

CONCLUSION

Disconnect36 was an important and enlightening experience for the students who participated. The students learned how they are linked, connected, and in some cases addicted to social media, and they also learned how they manage their time and information through social networking. Some students realized that they were addicted to social media and noted that during this process they experienced the symptoms of withdrawal, including headaches, and nervousness, isolation, frustration, and loneliness because they were disconnected from the Internet and other types of media. In addition, some had not realized the full force of media in their lives until they disconnected. The assignment, if only 36 hours, gave students an opportunity to disconnect from the bells, chimes, and tweets. It gave them an excuse to be still. According to Bugeja (2005),

> When we temper use of technology and consumption of media, the first thing that we gain is time for mainly, friends, neighbors, colleagues and others. We do not waste time online or on couches. We streamline our lives, communicating clearly at the right moment through the proper medium or none at all and experience fewer misunderstanding at home and at work. (p. 187).

If we want to make changes, then we have to determinedly examine the role that media and technology play in our lives.

REFERENCES

Bracken, C. C., & Skalski, P. D. (2009). Immersed in media: Telepresence in everyday life. New York: Routledge.

Bugeja, M. (2005). *Interpersonal divide: The search for community in a technological age.* New York: Oxford University Press.

Donath, J., & Boyd, D. (2004). Public displays of connection. *BT Technology Journal, 22,* 71–82.

Dougherty, H. (2010, March 15). Facebook reaches top ranking in US. Available at http:// weblogs.hitwise.com/heather-dougherty/2010/03/Facebook_reaches_toptop_ranking_i.html

Doyle, K. O. (2004). Introduction: Psychology and the new media. *American Behavioral Scientist, 48,* 371–376.

Holloway, S. L., & Valentine, G. (2000). Spatiality and the new social studies of childhood. *Sociology, 34,* 763–779.

Korzenny, F. (1978). A theory of electronic propinquity: Mediated communications in organizations. *Communication Research, 5,* 3–24.

Lenhart, A., Purcell, K., Smith, A., & Zickuhr, K. (2010, February 3). Social media & mobile internet use among teens and young adults. Available at http://pewresearch.org/pubs/ 1484/social-media-mobile-internet-use-teens-millennials-fewer-blog

Lister, M., Dovey, J., Giddings, S., Grant, I., & Kelly, K. (2009). *New media: A critical introduction.* New York: Routledge.

Livingstone, S., & Bovill, M. (2001). *Children and their changing media environment: A European comparative study.* Mahwah, NJ: Lawrence Erlbaum Associates.

McLuhan, M. (1964). *Understanding media: The Extensions of Man.* New York: McGraw Hill.

McMillan, S. J., & Morrison, M. (2006). Coming of age in the e-generation: A qualitative exploration of how the Internet has become an integral part of young people's lives. *New Media and Society, 8,* 73–95.

Meszaros, P. S. (2004). The wired family: Living digitally in the post information age. *American Behavioral Scientist, 48,* 377–390.

Moeller, S., Chong, E., Golitsinski, S., & Guo, J. (2010). A day without media. Available at http://withoutmedia.wordpress.com/study-conclusions/dependence/. Retrieved on April 28, 2010.

Turkle, S. (2010). Digital natives: Introduction. Available at http://www.pbs.org/wgbh/pages/frontline/digitalnation/etc/synopsis.html

Valentine, G., & Holloway, S. L. (2002). Cyberkids? Exploring children's identities and social networks in online and off-line worlds. *Annals of the Association of American Geographers, 92,* 302–319.

Walther, J., & Bazarova, N. N. (2008). Validation and application of electronic propinquity theory to computer-mediated communication in groups. *Communication Research, 35,* 622–645.

Walther, J. B. (2008). Social information processing theory: Impressions and relationship development online. In: L. A. Baxter & D. O. Braithwaite (Eds), *Engaging theories in interpersonal communication: Multiple perspectives* (pp. 391–405). Thousand Oaks, CA: Sage.

Watkins, S. C. (2009). *The young and the digital: What the migration to social network sites, games, and anytime, anywhere media means for our future.* Boston: Beacon Press.

Westerman, D., & Skalski, P. D. (2009). Computers and telepresence: A ghost in the machine. In: C. C. Bracken & P. D. Skalski (Eds), *Immersed in media: Telepresence in everyday life* (pp. 63–87). New York: Routledge.

Wilkins, H. (1991). Computer talk: Long-distance conversations by computer. *Written Communication, 8,* 56–78.

Yzer, M. C., & Southwell, B. G. (2008). New communication technologies, old questions. *American Behavioral Scientist, 52,* 8–20.

IS PODCASTING AN EFFECTIVE RESOURCE FOR ENHANCING STUDENT LEARNING?

Sheila Scutter

ABSTRACT

Podcasting can be an effective resource for enhancing student learning, if its pedagogical use aligns with best practices. Podcasting is easy, requiring only cheap and simple technologies that educators can learn to use quickly. Student feedback is very positive, and this has become one of the major drivers for providing podcasts of teaching material. This chapter discusses the way students use podcasts and the possible impacts on learning. Despite concerns about students reducing attendance at lectures, most studies have shown that lecture attendance is not diminished by the provision of podcasts. Students do not tend to use MP3 players to listen to podcasts "on the go"; most students listen to podcasts directly from home computers, often while replaying PowerPoint slides. The academic staff perspective of podcasting is discussed in relation to advantages and concerns about their use.

The ease with which podcast materials can be uploaded into learning managements systems has made podcasting a simple technique for providing audio learning resources to students. Podcasting has been very

Teaching Arts and Science with the New Social Media
Cutting-edge Technologies in Higher Education, Volume 3, 283–295
ISSN: 2044-9968/doi:10.1108/S2044-9968(2011)0000003017

widely taken up across the tertiary sector; however, although it has been extensively reported, the impacts on student learning are largely under-researched (Savin-Baden, 2010). As with all teaching and learning resources, its use needs to be thoughtfully integrated into the curriculum, with an understanding of the way students use it and the impact on learning. Just as providing an exemplary lecture, interactive webpage or simulations in virtual worlds cannot ensure student engagement with the process of learning and assimilation of graduate attributes, providing podcasts of materials does not necessarily result in student learning (Henry & Meadows, 2009).

There has been some discussion in the literature about the differences between podcasting and the uploading of audio recordings of lectures (Cebeci & Tekdal, 2006). Although technologies are now being used that automate the process of both audio and video recordings of lectures, this chapter will focus only on audio recording. For this purpose, simple digital audio recorders achieve the same outcome as sophisticated automated equipment, except perhaps requiring more intervention by academic staff in the process. The use of podcasting in education has taken many forms, and this list is expanding as its potential is realised and new approaches incorporated into teaching and learning. Podcasts may consist of entire lectures shortly after a lecture has been presented. These recordings may also be edited and indexed (Lee & Chan, 2007), although this requires additional academic work and can delay the time of uploading. Podcasts can also consist of explanations of key concepts and themes, condensed down into short segments, including pre-recorded points accessible in advance of class time, reviews of assessment or other 'housekeeping' tasks (Guertin, 2010).

There are many 'how to' resources for podcasting (Doe, 2007; Stephens, 2007; Raymond et al., 2010) that contain technical information about hardware and software and discussion about copyright and intellectual property issues (Gordon-Murnane, 2005). Ractham and Zhang (2006) discuss the potential for podcasts to increase communication and social networking of students and staff, while Forbes and Hickey (2008) include a useful checklist before recording a podcast, including checking the battery in the recording device and repeating student questions before answering them. These are useful hints about the mechanics of podcasting; however, this paper focuses more on what has been gleaned about the ways in which students use podcasts and the impact of podcasts on their learning.

STUDENT FEEDBACK ABOUT PODCASTING

Numerous studies have reported positive feedback from students concerning podcasting of lectures. Convenience, mobility and control over the time and place of learning have been reported as benefits of podcasting (Maag, 2006; Ractham & Zhang, 2006). Students suggest that podcasting of all lectures should be 'routine', so that they are available to them if they need them. Once available, podcasts seem to rapidly become an expectation from students (Jowitt, 2008).

The positive feedback from students about the use of podcasts has been a major driver in their introduction. As they are easy to provide, assist in achieving positive feedback from students, increasingly required for academic performance management, promotion, awards and indeed institutional status, then it seems as if the choice to provide podcasts is an obvious one.

However, the difficulty with such a rapid introduction is that there is no time to reflect on how podcasting interfaces with other learning resources. Podcasting has been introduced in many settings without understanding its impact on student learning (Cook, 2009; Henry & Meadows, 2009). As discussed above, there are a large number of resources available that provide advice about how to podcast, how to plan a podcasting session and how to engage students. However, many of these resources are not based on sound evidence of how students use podcasts, how this may vary between students in different disciplines and with different learning needs. This evidence is needed to allow academics to make an informed decision about their pedagogical place in curriculum (Boulos, Maramba, & Wheeler, 2006).

How do Students Listen to Podcasts?

What do we know about how students listen to podcasts? Do we assume that because students appear to be listening to MP3 players when jogging, walking, on the bus or in lectures, that they will listen to podcasts of teaching material 'on the go'? Can students really *listen* to a lecture podcast while driving the car? Or would we hope that they are paying more attention to the traffic around them? Podcasting may be attractive to students because it enables the students to increase the number of hours of studying without removing something else from their daily activities. However, is this congruent with the actual evidence about listening and downloading of podcast materials? (Bell, Cockburn, Wingkvist, & Green, 2007).

We undertook an online survey of students in our faculty (Scutter, Stupans, Sawyer, & King, 2010). Programmes in the faculty include a range of health sciences programmes, including physiotherapy, occupational therapy, nursing and pharmacy. We received responses from about 25% of the 6,750 students. We were surprised to find that only about a third of students normally download their podcasts to MP3 players, with the remaining two-thirds listening to podcasts through their computers (Scutter et al., 2010). As students regularly arrive in class, work in the library or laboratory or walk around campus listening to MP3 players, we had expected that they would listen to podcasts through MP3 players. However, other researchers have also reported students showing a strong preference for listening to podcasts via computers rather than mobile devices (Lane, 2006; Copley, 2007; Deal, 2007; Guertin, 2010). Our study found this to be an even stronger preference among EAL (English as an additional language) students who preferred to listen to podcasts via a computer at home (Scutter et al., 2010; O'Flaherty, Scutter, & Albrecht, 2010).

Copley (2007) monitored the download of podcasts by students and showed that each podcast was downloaded several times, with the highest level of download activity being directly after the podcast was provided, with continuing low-level download peaking again before exam time. This is consistent with our findings (Scutter et al., 2010) that students either listened to podcasts within a few days of the lecture, often in conjunction with the PowerPoints provided with the lecture, or listened to podcasts as revision for exams.

One of the aims of podcasting is to make studying more 'flexible' for students, so that they can learn when and where is appropriate for their lifestyle. But, does this mean that students can learn by listening to podcasts at the same time as cleaning the house or other activities? Our research showed that students in different programmes listened to podcasts at different times. Most striking was the difference between the nursing and pharmacy students. Nursing students listened to podcasts while doing housework, whereas pharmacy students usually listened in front of the computer while viewing PowerPoint slides. This difference could be due to a range of factors including but not limited to socio-economic status, age, family responsibilities, and deserves further interrogation.

Laing and Wootton point out that the 'learner will probably be listening to your podcast whilst carrying out another task such as walking, sitting on a bus or exercising ... podcasts should avoid dense complex material which is better covered in a lecture' (Laing & Wootton, 2007, p. 8). However, it is usually lectures that are podcast, making this a somewhat contrary argument.

Our studies show that students in later years of a programme listened to more podcasts compared to students in their first year. Students' podcast use also varied according to the programme in which they were enrolled. Across the health sciences faculty, students in the pharmacy programme used podcasts considerably more than other students. Whether this was due to the high percentage of EAL students in our pharmacy programme or the nature of the material provided in lectures (heavy content and complex concepts) warrants further investigation.

Some (Lightbody, McCullagh, & Hutchinson, 2006; Malan, 2007; Grabe & Christopherson, 2008), but not all (Bongey, Cizadlo, & Kalnbach, 2006; Shannon, 2006; Fietze, 2009), suggest that the provision of podcasts could lead to poor attendance at face-to-face classes. Our studies show that although students use podcasts to catch up on missed lectures, there was little evidence that students actually plan to miss lectures if they know that the podcasts was going to be available.

We found that students in a pharmacy programme missed less face-to-face classes as they progressed through their studies despite the availability of audio reproductions of the lectures (O'Flaherty et al., 2010). One might ask why students attend a lecture despite the PowerPoint slides and podcasts being available. Students clearly feel that they gain something extra from attending a lecture. Some students responding to our survey indicated that they attended lectures as they liked to see the lecturer speaking, and that gestures including pointing at PowerPoint slides were not available with podcasts. As stated by Copley (2007), a good lecture is about more than the transmission of information and provides social interaction, discussion and debate providing a '... multi faceted experience which cannot be provided by a podcast alone' (Copley, 2007, p. 389).

When providing podcasts, an assumption is made that students have access to an MP3 player and a computer with Internet access. However, Brabazon (2006) found that 23% of students had never used the web to download MP3 files and 14% relied on dialup Internet access at home, inadequate for downloading MP3 files. The consensus seems to be that around 30% of students in Australian Universities don't have a dedicated MP3 player (this may have changed in recent years) (Kennedy, Judd, Churchward, Gray, & Krause, 2008). Without an MP3 player, podcasts can only be accessed via a computer, compounding the issue of access for students with dialup or slow Internet.

Much interest is focussed on how students learn, with a preference for active, deep learning over passive and superficial learning, but the impact of podcasting on the approach to learning has not been clearly determined.

Although students report that podcasts support their learning, there are mixed reports on outcomes. Abt and Barry (2007) demonstrated that undergraduate exercise physiology students gained direct benefit from being provided with supplementary materials, in the form of either printed text or podcasts. The use of podcasts provided little additional benefit over the printed text (Oliver & Goerke, 2007; Kennedy et al., 2008; Dyson, Litchfield, Lawrence, Raban, & Leijdekkers, 2009) and compared learning by students who attended a lecture to those who received podcasts synchronised with PowerPoint slides used in the lecture. Not surprisingly, students who used the podcasts took notes while listening, and who listened to them again to clarify difficult issues performed better than students who just attended the lecture.

Podcasting has been argued to lead to passive learning with students focussing on the audio facility rather than actively engaging with the lecture content (Palmer & Devitt, 2007). Although listening to podcasts may be considered to be a passive activity, recent work on learning describes listening as an active, creative and demanding process of selecting and interpreting information from auditory clues (Kavaliauskiene, 2008). McKinney, Dyck, and Luber (2009, p. 619) describe the 'explaining voice'

> ... doesn't just convey information; it shapes, out of a shared atmosphere, an intimate drama of cognitive action in time. The explaining voice conveys microcues of hesitation, pacing, and inflection that demonstrate both cognition and metacognition. When we hear someone read with understanding, we participate in that understanding, almost as if the voice is enacting our own comprehension ...

This demonstrates the importance of the way in which the podcast is recorded and the voice that is used. Similarly, Cameron and Van Heekeren (2008) discuss the use of an 'explaining voice' in presenting podcasts. When investigating whether a more 'radio-like' approach to podcasting would impact on the acceptance and effectiveness of podcasting as a teaching tool, they argued that using an explaining voice in podcasting would improve learning, due to the 'nuances, hesitations and emphases' that can be included in a voice comparable to a radio announcer. Although this is an interesting concept, there is no evidence yet to support the use of a different 'voice' for recording podcasts.

Listening to a podcast may also enable a student to take more effective notes. These are active approaches to learning, especially when students pause podcasts in order to refer to other resources. However, students in our studies also commented that listening to podcasts allows them to memorise the lecturers' words, an activity that most lecturers would not encourage.

We asked students whether knowing that a lecture would be podcast changed their behaviours during the lecture (Scutter et al., 2010). The major change identified was that some students took the opportunity to listen to and watch the lecture, rather than focus on taking extensive notes. As noted by one student, 'It is much easier to concentrate on what the lecturer is SHOWING at a certain time, because I know I can go back over it and listen more carefully to what the lecturer was SAYING at that point in time'.

Furthermore, as many as 25% of students choose not to access podcast because they believe their own note taking during lectures was sufficient to grasp the key concepts conveyed during lectures and choose not to access podcasts (Campbell, 2005).

One group of students for whom podcasts are potentially a very useful resource are EAL students (Dyson et al., 2009). There are currently few published studies which have investigated the use of podcasting by these students (Read, 2005; O'Bryan & Helgelheimer, 2007). Listening has been shown to be an effective tool in the education of all students including EAL students, not just because the spoken word adds lucidity and meaning, but it also is an effective means of motivating students (O'Bryan & Helgelheimer, 2007; Nataatmadja & Dyson, 2008).

Although not reported in the literature, prior to the introduction of podcasting EAL students, the author frequently observed students recording lectures on small tape recorders. Providing podcasts allows all students access to such recordings and also allows the lecturer to discuss the issue of intellectual property with students before podcasts are uploaded.

EAL students showed a strong preference for listening to the whole podcast from a computer in conjunction with the PowerPoint presentation of the lecture allowing these students to work at their own pace and take more effective notes (Durbridge, 1984). We found that EAL students very rarely listen to podcasts when performing other activities. The usefulness of m-Learning is based on the premise that students can deal with the cognitive load of multitasking, that is listening to the audio podcasts while performing other activities (Forbes & Hickey, 2008). It seems that most students don't use podcasts for m-learning, a trend that is even stronger with EAL students.

Podcasts of entire lectures are regularly provided for students studying in external mode, along with copies of PowerPoint presentations. Lee and Chan (2007) have reported a decrease in the sense of social isolation and lack of connection to university life experienced by external students. Hence, podcasts appear to have benefits beyond mere access to teaching materials. Kasper (2000) used short informal podcasts of supplementary learning material and evaluated their use in distance students. Although these short podcasts received very positive feedback, students still used the podcasts as formal learning opportunities, rather than listening to them while undertaking other activities.

STAFF PERSPECTIVES

An online survey was sent via email to all contract and continuing teaching staff within the faculty (Elliot, King, & Scutter, 2009). Academic staff were also invited to participate in an in-depth interview to gain further insight into their motivations to podcast, or not podcast their teaching sessions. Purposeful sampling was used in order to gain a range of views about podcasting.

A total of 92 out of 167 (55%) teaching staff responded to the questionnaire. Of these, 65% used podcasts in some or all of their teaching and 35% chose not to use podcasts. For those staff that chose to podcast, the majority had made the decision based on requests from other people (staff or students) or as part of a team decision. Only one staff member had initiated podcasting because he/she wanted to try something new to see if it assisted learning. Most of the academics who chose to podcast were lecturing to large classes (>150 students) and delivering content-rich material in an environment where student interaction with the lecturer is often minimal due to class size. Not surprisingly, those who were not podcasting were more likely to have smaller classes where there was a lot of student–staff interaction or 'hands-on' activities which could not be adequately conveyed in an audio recording.

Other reasons for podcasting included 'students were doing this themselves, often without permission, so at least all could have access if I did it too' to 'I felt that it would affect my student evaluation of teaching (SET) data badly if I didn't'.

When asked about the benefits of podcasting, staff indicated that they provided a useful resource for students who missed lectures due to sickness or other commitments and allowed students to revise lecture content later. They also indicated that podcasts provided an opportunity for students to learn the correct pronunciation of new or difficult words, particularly in subjects like anatomy and physiology. Staff also indicated that they could use the podcasts themselves when preparing for subsequent lectures, especially in team teaching situations (Elliot et al., 2009).

Many staff expressed concerns about the potential of podcasts to inhibit student learning by facilitating passive learning. The important skill of taking effective notes was also mentioned. Staff felt that students needed to develop this skill for their future studies and careers. There was also a concern that providing podcasts would encourage students to rely on the podcast as the sole source of information on the topic (Elliot et al., 2009).

There is also the possibility that instructional material will be uploaded from year to year, without being updated and thus becoming stale. This is a potential problem in situations where a lot of time and effort has gone into the development of sophisticated podcasts, rather than direct recordings of lectures (Lee & Chan, 2007).

Two-thirds of the academic staff surveyed chose to podcast their lectures and almost one-third of those surveyed did so in response to expectations of students (Elliot et al., 2009). Student expectations can be a strong driver of academic practices and in an era where student evaluations of teaching are used for staff promotion many academics feel pressured to provide students with as many extra resources as practicable. In the words of one academic '... there is a tension between giving students what they want and providing an educational experience' (Gibson, 2008, p. 8).

Others reported that their teaching style was more inhibited when they were podcasting and that they were less likely to use anecdotes, particularly about friends or patient case scenarios. In the words of one staff member '... I feel like a different person when I am Podcasting lectures. I am less relaxed and less animated when I am Podcasting'.

Chan, Lee, and McLoughlin (2006) have provided a set of recommendations for best practice when podcasting and these include short lively podcasts rather than lengthy lecture monologues; not podcasting just for the sake of it, rather considering the purpose and refraining from duplicating content that is available elsewhere. Nevertheless, while these recommendations are sound, if podcasts were to be used in this way, they would add to an academic's workload and probably few staff would be willing to provide these additional resources. Podcasting an entire lecture is easy and not onerous on staff time. Sharma and Kitchens (2004), Corbeil and Valdes-Corbeil (2007) and Duncan-Howell and Lee (2007) surveyed academic staff about their concerns of podcasting. Concerns were raised about privacy and about potential misuse of the recordings by students. Staff were concerned that elements of the lectures could be taken out of context, and that their material could even end up posted on YouTube or other public platforms. In addition, staff were concerned about use of podcasts as a management tool to evaluate teaching performance and content delivery. The issue of ownership of images or recordings of one's voice do not seem to have been addressed in universities where the use of podcasts is mandated.

Although we are gaining an understanding of how students use podcasts, when they listen to them and that lecture attendance is not greatly affected by podcasting, there still remain questions about the impact of podcasts on learning. How do we design podcasts so that they encourage active learning by students, rather than memorising? Should academic staff have the choice about whether lectures should be podcast, or should this be routine? How do we control the distribution of podcasts beyond our students, therefore maintaining intellectual property? There are still many questions to be answered about podcasts.

SUMMARY

Students listening to podcasts of lectures provided by academics tended to use the podcast in conjunction with replaying PowerPoint presentations that are also provided. The concept of podcasts as providing m-learning does not seem to have played out practically. Student feedback about podcasts is positive and this can influence the choice of academics to podcast. Although some students use podcasts with PowerPoints to replace attendance at lectures, this has not been found to be a general trend. Although students report that the provision of podcasts improves their learning, there is mixed evidence for changes in learning outcomes. Podcast resources may also increase students' reliance on the lecturer's interpretation of particular topic materials, rather than encouraging students to seek additional resources.

REFERENCES

Abt, G., & Barry, T. (2007). The quantitative effect of students using podcasts in a first year undergraduate exercise physiology module. *Bioscience Education e-Journal, 10*. Available at http://www.bioscience.heacademy.ac.uk/journal/vol10/beej-10-8.aspx

Bell, T., Cockburn, A., Wingkvist, A., & Green, R. (2007). Podcasts as a supplement in tertiary education: An experiment with two computer science courses. University of Canterbury Conference collections. Available at http://hdl.handle.net/10092/482

Bongey, S., Cizadlo, G., & Kalnbach, L. (2006). Explorations in course-casting: Podcasts in higher education. *Campus-Wide Information Systems, 23*(5), 350–367.

Boulos, M., Maramba, I., & Wheeler, S. (2006). Wikis, blogs and podcasts: A new generation of Web-based tools for virtual collaborative clinical practice and education. *BMC Medical Education, 6*(1), 41.

Brabazon, T. (2006). Socrates in earpods? The ipodification of education. *Fast Capitalism, 2*(1), 23–29.

Cameron, D., & Van Heekeren, B. (2008). Hello, and welcome to the show: Applying radio's 'explaining voice' to educational podcasting. Hello! Where are you in the landscape of educational technology? *Proceedings Ascilite Melbourne 2008*, Melbourne, Australia.

Campbell, G. (2005). There's something in the air: Podcasting in education. *Educause Review*, *40*(6), 33–44.

Cebeci, Z., & Tekdal, A. (2006). Using podcasts as audio learning objects. *Interdisciplinary Journal of Knowledge and Learning Objects*, *2*, 47–57.

Chan, A., Lee, M., & McLoughlin, C. (2006). Everyone's learning with podcasting: A Charles Sturt University experience [Electronic Version]. Available at http://www.ascilite.org.au/conferences/sydney06/proceeding/pdf_papers/p171.pdf

Cook, D. A. (2009). The failure of e-learning research to inform educational practice, and what we can do about it. *Medical Teacher*, *31*(2), 158–162.

Copley, J. (2007). Audio and video podcasts of lectures for campus-based students: Production and evaluation of student use. *Innovations in Education and Teaching International*, *44*(4), 387–399.

Corbeil, J., & Valdes-Corbeil, M. E. (2007). Are you ready for mobile learning? Frequent use of mobile devices does not mean that students or instructors are ready for mobile learning and teaching. *Educause Quarterly*, *30*(2), 1–3.

Deal, A. (2007). *Podcasting: A teaching with technology White Paper*. Carnegie Mellon University.

Doe, C. (2007). The podcasting phenomenon. *MultiMedia & Internet@Schools*, *14*, 27–31.

Duncan-Howell, J., & Lee, K. (2007). M-learning: Finding a place for mobile technologies within tertiary educational settings. Ascilite conference, Singapore.

Durbridge, N. (1984). *The role of technology in distance education, part 2: Media in course design*. London: Croom Helm.

Dyson, L. E., Litchfield, A., Lawrence, E., Raban, R., & Leijdekkers, P. (2009). Advancing the m-learning research agenda for active, experiential learning: Four case studies. *Australasian Journal of Educational Technology*, *25*(2), 250–267.

Elliot, E., King, S., & Scutter, S. (2009). Motivating science undergraduates: Ideas and interventions. Uniserve conference, Sydney.

Fietze, S. (2009). Podcast in higher education: Students usage behaviour. In same places, different spaces. *Proceedings Ascilite Auckland 2009*, Auckland, New Zealand.

Forbes, M., & Hickey, M. (2008). Podcasting: Implementation and evaluation in an undergraduate nursing program. *Nurse Educator (September/October)*, *33*(5), 224–227.

Gibson, J. W. (2008). A comparison of student outcomes and student satisfaction in three MBA Human Resource Management classes based on traditional vs online learning. *Journal of College Teaching and Learning*, *5*(8), 1–9.

Gordon-Murnane, L. (2005). Saying "I Do" to podcasting. *Searcher*, *13*(6), 44–51.

Grabe, M., & Christopherson, K. (2008). Optional student use of online lecture resources: Resource preferences, performance and lecture attendance. *Journal of Computer Assisted Learning*, *24*, 1–10.

Guertin, L. A. (2010). Creating and using podcasts across the disciplines. *Currents in Teaching and Learning*, *2*(2), 4–12.

Henry, J., & Meadows, J. (2009). An absolutely riveting online course: Nine principles for excellence in web-based teaching. *Canadian Journal of Learning and Technology*, *34*(1), 1–3.

Jowitt, A. L. (2008). Creating communities with podcasting. *Computers in Libraries*, *28*(4), 54–56.

Kasper, L. (2000). New technologies, new literacies: Focus on discipline researchand ESL learning communities. *Language Learning and Technologies*, *4*(2), 105–128.

Kavaliauskiene, G. (2008). Podcasting: A tool for improving listening skills. *Teaching English with Technology: A Journal for Teachers of English*. Available at http://www.iatefl.org.pl/call/j_techie33.htm

Kennedy, G., Judd, T., Churchward, A., Gray, K., & Krause, K.-L. (2008). First year experiences with technology: Are they really digital natives? *Australasian Journal of Educational Technology*, *24*(1), 108–122.

Laing, C., & Wootton, A. (2007). Using podcasts in higher education. *He@lth Information on the Internet*, *60*(1), 7–9.

Lane, C. (2006). *Podcasting at the UW: An evaluation of current use*. The Office of Learning Technologies, University of Washington, Washington, DC.

Lee, M. J. W., & Chan, A. (2007). Pervasive, lifestyle-integrated mobile learning for distance learners: An analysis and unexpected results from a podcasting study. *Open Learning: The Journal of Open and Distance Learning*, *22*(3), 201–218.

Lightbody, L., McCullagh, P., & Hutchinson, M. (2006). The supporting role of emerging multimedia technologies in higher education. *Higher Education Academy*, 54–59. Available at http://www.ics.heacademy.ac.uk/Events/HEADublin2006_V2/papers/Gaye%20Lightbody%209.pdf

Maag, M. (2006). Podcasting and MP3 players: Emerging education technologies. *Computers Informatics Nursing*, *24*(1), 9–13.

Malan, D. J. (2007). Podcasting computer science. *38th SIGCSE Technical Symposium on Computer Science Education*, Kentucky.

McKinney, D., Dyck, J., & Luber, E. (2009). iTunes university and the classrom: Can podcasts replace professors? *Computers and Education*, *52*(13), 617–623.

Nataatmadja, I., & Dyson, L. E. (2008). The role of podcasts in students' learning. *International Journal of Interactive Mobile Technologies*, *2*(3), 17–21.

O'Bryan, A., & Helgelheimer, V. (2007). Integrating CALL into the classroom: The role of podcasting in an ESL listening strategies course. *European Association for Computer Assisted Language Learning*, *19*(2), 162–180.

O'Flaherty, J., Scutter S., & Albrecht, T. (2010). Informing academic practice about how podcasts are used by diverse groups of students. In: *HERDSA, Reshaping Higher Education*, Melbourne, Australia, 6–9 July.

Oliver, B., & Goerke, V. (2007). Australian undergraduates' use and ownership of emerging technologies: Implications and opportunities for creating engaging learning experiences for the Net Generation. *Australasian Journal of Educational Technology*, *23*(2), 171–186.

Palmer, E., & Devitt, P. (2007). A method for creating interactive content for the iPod, and its potential use as a learning tool: Technical advances. *BMC Medical Education*, *7*(1), 32.

Ractham, P., & Zhang, X. (2006). Podcasting in academia: A new knowledge management paradigm within academic settings. *Proceedings of the 2006 ACM SIGMIS CPR Conference on Computer Personnel Research: Forty four years of computer personnel research: Achievements, challenges and the future*, Claremont, CA (pp. 314–317).

Raymond, Y. K. L., Rachael Kwai Fun, I., Chan, M. T., Ron Chi-Wai, K., et al. (2010). Podcasting: An internet-based social technology for blended learning. *IEEE Internet Computing*, *14*, 33–41.

Read, B. (2005). Lectures on the go. *The Chronicle of Higher Education*. Available at http://chronicle.com/weekly/v52/i10/10a03901.htm

Savin-Baden, M. (2010). The sound of feedback in higher education. *Learning, Media and Technology*, *35*(1), 53–64.

Scutter, S., Stupans, I., Sawyer, T., & King, S. (2010). How do students use podcasts to support learning?. *Australasian Journal of Educational Technology*, *26*(2), 180–191.

Shannon, S. J. (2006). Why don't students attend lectures and what can be done about it through using iPod nanos? *Proceedings of the 23rd annual ASCILITE conference: Who's learning? Whose technology?*, Sydney (pp. 753–756).

Sharma, S. K., & Kitchens, F. L. (2004). Web services architecture for M-learning. *Electronic Journal on e-Learning*, *2*(1), 203–216.

Stephens, M. (2007). All about podcasting. *Library Media Connection*, *25*(5), 54–57.

INTRODUCING STUDENTS TO MICRO-BLOGGING THROUGH COLLABORATIVE WORK: USING TWITTER TO PROMOTE CROSS-UNIVERSITY RELATIONSHIPS AND DISCUSSIONS

Tricia M. Farwell and Richard D. Waters

ABSTRACT

The job market for communication majors increasingly expects those graduating in these specializations to not only know how to create strategic plans for using social media in both one-way and two-way communication environments, but also maintain proper social media etiquette and virtual culture norms for their clients. To better prepare students for this expectation, two faculty members at separate universities designed and implemented a course assignment intended to promote cross-university collaboration, foster discussion, and bring students to use microblogging via Twitter. This assignment was designed so that it would not only have the students construct the meaning and best practices in a social setting using social media, but also encourage them to experience Twitter from a user perspective while building relationships in a manner

Teaching Arts and Science with the New Social Media
Cutting-edge Technologies in Higher Education, Volume 3, 297–320
Copyright © 2011 by Emerald Group Publishing Limited
All rights of reproduction in any form reserved
ISSN: 2044-9968/doi:10.1108/S2044-9968(2011)0000003018

that their future employers may have to work with their publics or customers. Overall, the educators involved in this project did feel that it was a beneficial assignment for students in both classes. While the students may not appreciate the assignment while it is being conducted, many of them have expressed the value in it now that the assignment is completed.

INTRODUCTION

Social media has become a primary way for individuals to stay connected and communicate with others throughout society, including friends, family members, communities, and organizations. It has become so prevalent in our society that one study found that faculty members who gave up Facebook for Lent ended up feeling disconnected from their family and friends by not being able to read their updates (Kist, 2008). While it may seem simple for the average person to sign up and participate in social media, the situation is more complex for businesses and nonprofit organizations. Many organizations believe they should be on social media but are unsure how to strategically incorporate it into their public relations and advertising campaigns (Reid, 2009). Part of the confusion stems not only from the fact that the organization is unsure how to monitor social media, but also from the uncertainty regarding the roles of employees online. A recent poll found that 26% of the companies polled reported that they fired someone for what was seen as misuse of the Internet (Grensing-Pophal, 2010).

While organizations are struggling to find their place in the social media world, organizational communication disciplines (e.g., advertising, marketing, and public relations) increasingly are expecting new hires to not only be aware of social media technologies but also be savvy in their strategic usage (Solis & Breakenridge, 2009). The job market for these majors increasingly wants and expects students to not only know how to create strategic plans for using social media in both one-way and two-way communication environments but also to maintain proper social media etiquette and virtual culture norms for their clients.

As the Pew Internet Project has demonstrated, the vast majority of today's college students are savvy with social media usage (Pew Research Center, 2010). Research has shown that these students warmly embrace blogging, social networking sites (e.g., Facebook and MySpace), and video and photo-sharing sites (Pew Research Center, 2010). However, industry analysts indicate that the micro-blogging service Twitter is the social media application that strategic communicators most often use in their organizational campaigns (Stelzner, 2009). Research on the demographics of Twitter users indicates that the millennial generation have not embraced Twitter as much as other social media applications despite its adoption by organizations (Nielsen Online, 2009). This generation, which includes those born from the 1980s to the early 2000s, has been characterized by an increased use and familiarity with communication, media, and digital technologies. However, this group not fully embraced Twitter although their predecessors have.

Given the growing impact that Twitter is having on organization's strategic communication efforts and the lack of reception for the service by today's college students, the purpose of this chapter is to introduce a collaborative Twitter assignment between advertising and public relations students at two separate universities and to discuss student and faculty experiences in promoting collaboration, fostering discussion and encouraging students to explore micro-blogging through Twitter.

LITERATURE REVIEW

Millennial Students and Social Media

It seems that most students approach courses with an understanding of basic Internet and computer skills. While some may assume that the students hold prior knowledge of applications such as email, blogging, video blogging, MySpace, Facebook, it still can be a challenging experience for those who have never experienced the tool in question (Väljataga & Fiedler, 2009). However, once being introduced to the new application, studies have found that many students are willing to use the applications after the course has ended (Väljataga & Fiedler, 2009).

Despite their interest and attention in some social networking sites, students (and others) may not be as aware of or concerned with the issues surrounding social networking. Numerous stories have been told about the person who called in sick to work only to be caught being completely healthy and usually on the beach somewhere through Facebook posts. Misusing sick days for personal reasons and then carelessly revealing this online is just one misstep people have taken online. In 2010, an episode of *Kell on Earth* revealed that an intern was turned down for the position in part because of a series of Twitter posts that documented the interview and the interviewee's belief that she was definitely going to be offered the position. These missteps reveal that many millennials fail to grasp the impact of Twitter on their lives. While each social network site will have its unique elements, some advice can apply across all media. Null (2009) points out that people on social networking sites need to remember that their posts are available for all to see. Additionally, people need to be thoughtful of sharing too much information with an unintended audience (Null, 2009).

Even with all the potential complications from using social media, 99% of people surveyed had an active social networking profile, and 52% of adults had multiple profiles (Participatory Marketing Network, 2009; Jones, 2009; Vankin, 2009). Facebook is still the most popular social networking site among adults as 73% of those polled maintained a Facebook profile compared to only 48% on MySpace (Lenhart, Purcell, Smith, & Zickuhr, 2010). Age demographics regarding several of the social networking sites has been shifting. Although the median age of people participating on LinkedIn and MySpace has dropped since 2008 (down from 40 to 39 and 27 to 26, respectively), Facebook's median age has risen (from 26 to 33), and Twitter's median age has stayed at 31 (Fox, Zickuhr, & Smith, 2009).

Researchers have found that students do know how to use social media for specific practices. For example, Kidwai (2010) found that students were good at using social media to promote their projects and could be used to be advocates for education. This finding is in line with Greenhow and Reifman's (2009) observation that students knew how to use social media to promote projects, inspire others and share information they found relevant to their lives. Along with these practices, many in the 18–28 age range have used instant messaging, written a blog or responded to a post, played online games, downloaded music, downloaded an application for their profile page, and posted photographs (Kidwai, 2010; Participatory Marketing Network, 2009). Interestingly, one study found that blogging may be dropping in popularity among teens and millennial students. Since 2006, blogging has dropped 14% among teens and 9% among those aged 18–29 while responding to blogs has dropped 26% for teens (Lenhart et al., 2010). Despite the decline in blogging among this age group, they are not migrating to micro-blogging sites such as Twitter as only 8% of those aged 12–17 are on it and one-third of those aged 18–29 reported using it (Fox et al., 2009; Lenhart et al., 2010).

Micro-blogging Using Twitter

According to the About Twitter section of the web site, "Twitter is a real-time information network powered by people all around the world that lets you share and discover "what's happening now" (About Twitter, 2010). Tweeple or Tweeps, people who use Twitter, are asked to answer the question "What's happening?" in a maximum of 140 characters. This ability to communicate in real-time short note style has garnered attention from organizations and additional groups. In May 2009, the site had 17 million unique visitors (Fox et al., 2009). However, among millennials, Twitter usage is reported to be around 22% (Jones, 2009; Participatory Marketing Network, 2009; 78% Gen Y, 2009; Vankin, 2009). Those who fall in that group tend to use Twitter more to follow friends and celebrities than to establish themselves as a voice or expert on a certain topic of interest (78% Gen Y, 2009).

It has been posited that the millennial generation is not interested in Twitter as a social media tool because the group is more interested in developing self-branding than community discussions (Vankin, 2009). Twitter's structure does not easily allow for the quick development of a personal fan base, and the ability to learn what others are doing can be limited if one follows a large number of people (Vankin, 2009). Additionally, Twitter may function more as background conversation for the average user (Crawford, 2009). While it does allow for following and posting updates, the ability to listen is more important than focusing on the self-brand. In essence, this may put the user in a predominantly "lurker" mode online instead of the center of attention. For businesses and nonprofit organizations this role can be beneficial as they listen in to what the community is saying and can use the information gathered to build relationships (Crawford, 2009).

Thus, being put in the role of lurker may not appeal to the millennial generation that leads to resistance in adopting the tool. This resistance to use Twitter makes it a prime opening for educators to step in and incorporate Twitter into a classroom to show students the benefits, successes and missteps when using it. Another benefit of incorporating Twitter into learning settings is that there is a low learning curve and the lack of needing to create a high-end, multi-tool virtual world to use it. Additionally, the site does allow for some personalization by allowing users to upload pictures and customize backgrounds. However, even with this customization, the pages and functions are still easy to navigate.

Collaboration and Learning Online

Researchers have found that Twitter is one of the easiest social media applications to learn and use for collaborative efforts (Honeycutt & Herring, 2009). The service's conversational format lends itself to individuals engaging with one another and sharing information. Through the knowledge transfer created by sharing the URL to a noteworthy blog posting from industry leaders to answering another's questions, conversations and collaborations on Twitter come naturally (Cheong & Lee, 2010).

Whether the collaborative efforts occur on Twitter or in another virtual environment, online learning has been shown to make it easier for students to engage in collaborative learning and share knowledge (Chou & Min, 2009; Andreas, Tsiatsos, Terzidou & Pomportsis, 2010; Petrakou, 2010; Solimeno, Mebane, Tomai, & Francescato, 2008; Cadima, Ferreira, Monguet, Ojdea, & Fernandez, 2010; Calvani, Fini, Molino, & Ranieri, 2010). Learning, in essence, is social (Brooks, 2009; So & Brush, 2008; Petrakou, 2010; Pozzi, 2010; Tsai, 2010; Cadima et al., 2010; Ioannou, 2010). As Tapscott and Williams (2008) point out, one of the key benefits to being online is that the professor–student power structure is flattened online and replaced with a more collaborative model. This is especially true in online learning where student perception of belonging to a (learning) community increases their success in the class (De Lucia, Francese, Passero, & Tortora, 2009). However, some studies have found that while sharing may increase with a combination of online and face-to-face meetings among class members, there can be less depth of information shared (Chou & Min, 2009).

Collaborative learning, however, has been shown to encourage students to reach greater levels of success and promote higher levels of thinking (So & Brush, 2008; Solimeno et al., 2008; Tsai 2010; Ioannou, 2010). In many cases, students are open to the ideal of collaboration in the classroom. One study found that students felt better about testing and learning after collaborative assessments (Ioannou, 2010). The results of online discussion forums are mixed. While threaded discussion forums may not be the most effective (Calvani et al., 2010), online discussion forums are beneficial if the instructor presence is noted and discussion leaders are assigned (Lam 2004).

Online Education

Increased collaborative efforts and building a greater community of shared knowledge are just two reasons colleges and universities are increasingly turning to online course delivery methods. It appears that online courses have been increasing as evidenced by the fact that more than 4.6 million people took at least one online course in Fall 2008 (Allen & Seaman, 2010). While the majority of students enrolled are undergraduates, the online offerings do work for graduate level courses (Solimeno et al., 2008). Thus, it may be safe to conclude that students are comfortable with and accept the realities of online education. In fact, some have posited that students will be expecting to use technology more and more in classroom environments (Brooks, 2009). As universities are becoming more comfortable with the idea of online education, faculty members are reaching out to discover the best online tools to promote learning.

One of the online tools gaining attention with academics is Second Life, an online 3D environment where members create avatars, navigate the environment and can use voice to communicate (Chou & Min, 2009; Petrakou, 2010). While Second Life provides various opportunities for interaction, many complain that there is a high learning curve and that it does not provide all the necessary tools to facilitate a beneficial classroom experience (De Lucia et al., 2009; Andreas et al., 2010). However, more and more educators need to be careful to not just showcase technology in the classroom. Instead, they must provide students not only with the information on how to use technology, but also why the technological innovation is important and possible strategic applications (Brooks, 2009). For example, millenials are very familiar with Facebook as a method for staying in touch with family and friends, but many are unaware of how to use this application from an organizational point of view (Mulhern, 2009).

With all of this in mind, two university professors set out to incorporate Twitter into their classroom experience to encourage students to explore the microblogging service. They sought to determine the answers to the following questions:

RQ1: Can Twitter be used to enhance online learning in a collaborative environment?
RQ2: How effective is Twitter in online learning collaborative experiences?

METHODOLOGY

To find the answers to the research questions, two educators set out to devise an assignment for their classes that involved using Twitter to communicate with groups containing members in different courses, states and academic levels. This assignment, essentially, provided a collaborative learning environment using asynchronous communication across two universities and two strategic communication disciplines.

The courses involved in this assignment were an undergraduate senior-level advertising course that dealt specifically with the use and impact of social media in advertising and a cross listed graduate and undergraduate-level public relations course that dealt with the applications and uses of social media in various public relations specializations. It was the first time the advertising and social media course was taught at the university. Owing to the way the course was made available to students, there was various communication specializations, ranging from advertising to electronic media communication. It was also the first time the public relations course had been taught at the partner university. The course was available to communication specializations ranging from public relations to technical communication.

Each class was provided with a similar instruction sheet regarding the assignment, with the only changes being made to course specific names and information. Twelve advertising students from University A and 42 public relations students from University B were paired and introduced to their group over Twitter through introductions from the professors. There were 12 groups total consisting of 1 University A advertising student and either 3 or 4 University B public relations students. The challenge was to get the students involved and thinking about how to use Twitter in their various professions. Students were tasked with discussing and agreeing on a 140 character definition of an assigned term and 12 useful strategies or best practices relating to the term. For example, if a group was assigned the term "stakeholder engagement," they would create a 140-character definition of the concept based on their conversations and the information they shared with each other on Twitter. Then, after the definition was created, the group was responsible for using Twitter to create a list of top 10 tips involving their term in advertising and public relations. The terms ranged from tactical elements of strategic communication, such as Web 2.0 and virtual communities, to managerial items, such as return on investment and Web metrics.

To complete the assignment, each student created their own Twitter account, or in some cases used their already established account and personalized it to their liking. After creating the accounts, students were asked to send their usernames by introducing themselves to and following each educator. Once the usernames were gathered, each student was assigned a term and a group. For the purpose of this assignment, the terms used were segmentation, evaluation and measurement, engagement, return on investment (ROI), Web 2.0, Web metrics, virtual communities, relationships, research, influencers, communication strategy, and goal development. Group members were introduced through Twitter using their Twitter ID's. The term that each group was researching was also included in the introductory tweet. Additionally, each group was assigned a specific hashtag, a way of marking conversational threads on Twitter by a # and a short phrase that they were to include in each tweet. To allow the students with the most characters possible to work out their definition, hashtags were kept short, but easy to remember and relevant. For example, #AP01 for advertising-public relations group 1. The assignment was scheduled to last for two weeks to facilitate discussion. Students were asked to turn in their assignment through Twitter and submit a typed document sent by one member of the group.

Ideally, this assignment would not only have the students construct the meaning and best practices in a social setting using social media, but also to encourage them to experience the tool from a user perspective while building relationships in a manner that their future employers may have to work with their publics or customers.

To answer the research question a total of six focus groups were held with the students; two of these focus groups occurred in a classroom setting, and four occurred online using Eluminate web conferencing software. Of the 54 students who participated in the assignment, 43 students participated in the optional focus groups upon completing the project. Participation was purely optional as students were not given extra credit for participation; instead, they were asked to participate in the focus groups to discuss their experiences and feelings about the project. The professors led the conversations and covered a few specific topics, but largely kept the conversations open and free flowing to let the students discuss their experiences.

Following the focus groups, the professors transcribed audio recordings to ensure that notes taken during the focus groups accurately reflected the students' thoughts. The transcriptions were then analyzed thematically, which involves reading the transcriptions and comparing each one with the others while looking for similarities, which are grouped together by category (Lindlof, 1995). The researchers conducted a validity check by asking students to make sure that their words and experiences were accurately transcribed (Lincoln & Guba, 1985).

RESULTS

To answer the first research question, the professors reviewed the consensus comments made by the students during the focus group, and they deemed the Twitter collaboration assignment as a success based on student feedback and the final results of the assignment despite a slow start. At first, the majority of students were hesitant to post to Twitter; one female undergraduate student said, "I didn't really know how to get started even though I knew my group's term, hashtag, and who my group members were. I said hello to everyone and asked what the best plan to complete the assignment was but I never really knew if my tweet was heard." As might be expected, this initial reaction to the Twitter assignment was not uncommon given the low Twitter usage rates among college-aged students when compared to other social networking sites. Many of the students' initial tweets were often introductory and sought advice for the next step of the project. For example, one student might tweet "#AP01 Hi group! Nice to meet you. I'm new to Twitter but ready to get this project started. Any ideas?" Students new to Twitter often mentioned it in their initial tweet, such as "Hi all. I'm still learning this. Is everyone getting this message? #AP10."

Many of the first tweets involved students double-checking to make sure their comments were being seen and read by through group members. Experienced users of the service were quick to welcome those new to Twitter and encouraged them to get started on the discussion involving the concepts they were to define. "I've used Twitter for about two years so it was easy for me, but I noticed some of my group members really struggled to jump into the conversation. But once we let them know we did see their posts, things went well," said one male graduate student. This was a common sentiment among all of the experienced users.

After introductions were made, real discussions were created around the questions and answers students started posting about their assigned topics. The teams took different approaches in composing their term definitions. Reflecting the collaboration aspects of the assignment, some students carried on legitimate conversations in their Tweets. For example, after mentioning their ideas regarding the term, other members in the group would either express their agreement tweeting something similar to "Engagement is not just one-way communication...it is all about involvement I like that idea #AP05." Other teams, however, took a more democratic approach to defining their terms. For example, one student asked the group to each post their own definition, and then they would vote on which definition was best.

Most of the groups used a combined approach of these two strategies. The groups began posting ideas about their definition and shared articles and hyperlinks with one another that they thought might apply to creating their definition and worked out the definitions. For example, one student said that a way to segment audiences was to meet them where they wanted to be met and provided a link to an industry-leading blog on the topic.

Students in two groups were often frustrated with coming up with a consensus definition and tips for their assigned topics. One group provided 5 different definitions – one from each group member – since they were unable to settle on one per assignment instructions. Another group did not seem to form any sort of bond, which led to them all mentioning definitions without interaction and discussion. While they understood that conversation was a key part of the negotiation of constructing the definition, they expressed a desire to have points wrapped up and then move on to the next part of the task. In both of these cases, the end goal seemed to be to simply move through the assignment. Instead of trying to get their group to agree and move on through leadership and discussion, these groups simply ended up submitting the most recent definitions and tips that the group had worked on, rather than reflecting on items shared by group members.

Daily, the professors of the two courses would search Twitter using the groups' hashtags to see what work had been made on the assignment that day. Daily searching also allowed the professors to stay in contact with students so that they did not flounder in the assignment and lose faith that it could be completed. Several noticeable trends emerged from the daily scanning of the Twitter groups' activity, including the strong performance of established Twitter users, the displeasure with not receiving instant feedback from other group members due to their irregular participation, and individuals not understanding how to use the hashtag. These observed trends and comments helped the professors understand how Twitter could be used to enhance the students' educational experiences.

The first research question sought to determine whether Twitter could be used to enhance online learning in a collaborative environment. In short, the answer is yes. But, instructors seeking to use Twitter collaboratively must be prepared to handle various challenges. One of the first obstacles that the students and professors encountered centered on limitations of the Twitter service. Given its exponential growth rate, Twitter users' status updates often overload the system when significant global events occur, and users rush to comment on the events of the day. One female student commented, "I got very frustrated with Twitter overloads. Even though they try to make light of the situation by showing the *Fail Whale*, having to visit the site multiple times to post a comment is an annoyance that would keep me from using the service." (For those unaware of Twitter's Fail Whale, it is an illustrated error message of multiple birds hoisting a whale out of the ocean with a net accompanied by a message such as "Too many tweets! Please wait a moment and try again.") During the course of the Fall 2009 semester when the assignment was conducted, several events occurred that garnered significant interest from Twitter users. Among the situations that occurred during the time of the assignment included the publicity-seeking launch of the now infamous "Balloon Boy" over Colorado and pre-release controversies surrounding former Alaska Governor Sarah Palin's book, *Going Rogue*. On the basis of the experiences with Twitter overloads caused by these events, one female graduate student noted:

> I understand people's frustration with the overload problem. I found one way around it by using an outside client to manage my Twitter account. With ÜberTwitter, I was always able to get my group's tweets. I encouraged others to use it as well, but none did.

Another common challenge that was encountered focused on the 24-hour nature of the service. A male undergraduate student said, "There's so much going on and so many constant updates that if you don't get on the Web site almost every day, you get behind." This is particularly true for individuals who follow many people on Twitter. Even though students had the ability to use TwitterSearch to find comments that were made using the hashtag, the search function often failed to produce results. "I wish the search function worked more than it did. There were so many times that I searched using our hashtag, and nothing came up in the results even though I know comments were made," said one female graduate student.

Another limitation of the search function when it did work properly centered on its limited search results. "I was disappointed that the search function only really provided comments that were made over the past few days," said one male student. "Since it didn't pull up tweets made in the previous week, a lot of relevant items were lost because we couldn't access them without scrolling back through the entire Twitter feed." This insight was particularly helpful for the professors as it demonstrated that while Twitter provided an excellent online source for conversations and collaborations, it may not provide a strong platform for storing and archiving useful information. A female graduate student noted:

> Twitter lets you label tweets that you want to have easy access to with a gold star. In and of itself, that's easy to do. The problem with this archiving approach, especially with this assignment, is that you don't necessarily know what will be useful as the definitions and 10 tips evolve over time.

Students who had been using Twitter for personal reasons expressed a different challenge that the professors had not anticipated. A female undergraduate said, "I use Twitter for my personal life, and many people were like, 'What the hell are all these weird Tweets you keep posting?'" The student admitted that she could have created a separate account for the assignment, but that managing multiple Twitter accounts could have been problematic with tracking multiple usernames, passwords, and keeping conversations straight. Another experienced Twitter user noted that "I was worried that my followers would get annoyed at my group tweets. The opposite was the case. They really enjoyed watching my thoughts develop. That grew into other conversations about public relations, social media, this class, and technology."

While these additional conversations emerged and were helpful for a few groups, there were others that struggled to begin conversations after the initial introductions were made. One female graduate student commented that "it seemed like a message board that nobody knew when to check. Responses to direct posts and questions could take a very long time to be received." Another female undergraduate student noted:

> I would post my tips and definition but no one would really respond. So I just put it out there and asked if anyone receiving my posts. That's when I finally got a response. Even after the response I still noticed that conversations didn't really happen.

Despite these challenges, there were several successes that reiterated the main goal of the assignment, educating students to understand the potential benefits of Twitter. Many students were initially perplexed over how the service could be used; but as a group, they understood its benefits by the end of the assignment. One female student admitted, "Honestly, I enjoyed the project. School assignments are never really fun. But this one was different and encouraged me to really explore a technology that I knew little about." Another commented that "I liked the idea of the project when I read it because I enjoy when professors branch out away from the normal assignments and tests." The Twitter assignment forced students to reach out and use a technology that many had resisted despite the media hype about the service. "Before the assignment, I had a negative outlook about Twitter," said one male undergraduate. "But, now I'm no longer confused about what it is and how it can be used. I'm actually having fun with my Twitter account."

One female graduate student said, "I never would have used Twitter if it weren't for this course, but I'm glad it was assigned because it made me realize that Twitter is more user friendly than other social networking sites." A male graduate student felt that "Twitter seems like its more appropriate for organizations than Facebook or MySpace. Using the service made me realize that lumping everything under the umbrella of social media hides some of the distinguishing features of the applications."

The biggest benefit of Twitter for these students, who will be entering the field of strategic communications after graduating, is the realization that clear conversations can be had over social media. One female undergraduate student said, "On Facebook, it's easy to get lost with the extra applications and all the different places you can post comments (e.g., notes, "The Wall," status updates, comments on videos and pictures)." Twitter, however, facilitates an easily followed conversation between users with its system of replies. "I liked how Twitter can connect people through brief statements. This assignment illustrated how it can be used for group communication and helps facilitate brief but quick communication, much like a real back and forth conversation."

The major challenges of the Twitter assignment really focused on lack of experience and understanding of how the service is used. One female graduate student said, "I was very hesitant at first. But once I got the hang of things, it almost seemed easier than e-mailing members back and forth. And, there was no way a group member could say 'I didn't get it' because everything was publicly available."

Overall, the students seemed not only to enjoy the Twitter assignment but gain valuable insights into how the service could be used from a strategic communication perspective. Even though there were challenges that the students encountered at first that involved a learning-curve on how to use Twitter, the successes outweighed the struggles the students first experienced using Twitter. One student added, "During the first couple of days, I hated this assignment. I didn't see what the point was. But now, it's probably one of the most beneficial projects I've done in college. This assignment will help me in my career."

While the first research question found that Twitter could enhance online learning, the second research question focused on its effectiveness. The question remains as to whether Twitter is any better than the many other online collaboration tactics available to professors. In assessing the effectiveness of Twitter, distinct areas emerged from the in-depth interviews and focus groups with students. These areas not only help make Twitter an effective online learning tool for students, but they also benefit the professor.

One of the distinguishing features of Twitter, the 140-character limit, was one of the most commented upon element of effectiveness. By having students complete an assignment using only 140-characters at a time, it forced them to think strategically about word choice. One male graduate student said, "While the character limitation was frustrating, I later realized that it made me communicate more creatively." These thoughts were echoed by a female undergraduate, "We're forced to be more strategic in our messages and cut out the unnecessary words. That's something we'll have to do in the real world everyday."

The shortened space to discuss the advertising and public relations topics ultimately resulted in students taking more time to review their work. One female undergraduate admitted that "I don't proofread my work for assignments or papers, but because I could only use 140 characters, I had to go back and edit my work." The assignment required students to follow traditional grammar and spelling rules so that popular internet abbreviations and slang terminology were not incorporated into the definitions and 10 tips. One male graduate student said, "I think the forced proofreading serves us well in the end. It's not something that I would normally do, but it shows how important professional communication is on the Web."

The brevity of the tweets made for the assignment helped to reinforce editing and grammar skills, which was not one of the main goals of the project though certainly it was an added benefit. The reminder of the importance of good communication skills easily could have been lost in a group wiki, a blog, or a discussion forum where users are free to ramble with their thoughts. With only 140 characters, run-on sentences and distracted thoughts rarely surfaced because there was no room for them. Additionally, the public nature of Twitter leads more students to proofreading their tweets before posting them. "I didn't want my group thinking I couldn't spell or write. I went back and checked my tweets for misspelled words and improper punctuation before I posted what I was thinking," said one female undergraduate.

Another advantage of using Twitter for online collaborative projects focuses on the productivity and output of the workgroups. As a whole, students dislike group projects even though it is something commonplace in the workforce. Twitter provided a new channel for doing group work, and many felt that it enhanced the overall performance of the groups. "Overall, I liked this method of group work more than what I've done in other classes because I didn't have to try to set up meeting times with people who had very different schedules," said one female undergraduate student. A non-traditional, female undergraduate student said, "I'm not the normal undergrad. I work full-time and have a family. This assignment allowed me to participate when I could rather than having to be available at a certain time that the majority of group members decided upon."

With the continued expansion of Web technologies into the classroom and course assignments, students are routinely being challenged to create podcasts, viral videos, Web sites, and other tactics within the domain of social media. When done in a group setting, individual contributions to these assignments are difficult to determine. Students appreciated that the

Twitter assignment provided an element of accountability since every tweet during the assignment was made available to the professors. One student summed it up very simply saying:

> Let's face it. There's always group members that don't do as much as others or chime in at the last minute. I loved this assignment because everything we did is documented. I did miss out on some discussion by not signing in for a few days, but I was able to contribute elsewhere to make up. This project held us accountable for the work that we did. I couldn't say that I did more work than I actually did because it was archived.

Even though the assignment was effective in encouraging students to be selective with their words and actually participate in the project, it did have limitations. As previously discussed several students were initially hesitant to become involved with Twitter and were unsure how to initiate conversations with one another. This reluctance to participate is a major obstacle that professors must overcome for Twitter to be an effective learning strategy. One student suggested that one way around the initial hesitation is to have a warm-up activity similar to a "getting to know you" activity done on the first day of class. Ideas that were mentioned for this activity ranged from personal introductions and discussions of hobbies and current events to having the professor throw out industry-related opinion questions and have the students discuss those topics to facilitate learning how to use Twitter before diving into the assignment.

Conducting a class assignment on the Internet presents unique challenges for students and professors regarding privacy. Twitter's nature made it problematical to shield students from potentially negative comments from other users not enrolled in the class. However, one key benefit of Twitter is the open discussions to which anyone can contribute. There were numerous times when industry leaders also became involved in the groups' discussion and used their hashtags to participate and share with everyone. To prepare the students for the possible privacy issues, throughout the semester both professors explained to the students that there was going to be an increased level of exposure regarding student work and the Twitter assignment. Examples of possible negative outcomes were reviewed and the students were told that if something happened that they felt uncomfortable with, they were to bring it to the professors' attention immediately.

For both courses, the vast majority students did not appear to be overly concerned regarding revealing their work product online. They were not worried about the transparency of their contributions; instead, they were looking forward to exploring the service and seeing who they could engage in the conversation beyond the group. However, there were a handful of students who were uncomfortable communicating openly on the Internet. One male graduate student said, "I didn't like knowing that non-students could see my tweets. I worried that my questions and comments might be perceived as naïve by the PR practitioners I followed and that it could be held against me in the job search."

Another student was more concerned with general privacy issues and did not feel comfortable participating in the assignment at all through virtual collaboration. The student's fears of communicating online were acknowledged and respected. The professors created an environment to allow the student to still participate while being able to retain strong feelings of privacy on Twitter. This student was asked to create an account, which was assigned to a group and hashtag, and follow the professor on Twitter; this enabled the student to send private messages to the professor directly. The student would then do a daily search for the hashtag and relay relevant comments for the group to the professor through private message. The professor would then repost the message without the student's identifying information so the group had it to complete the assignment. While this extra step made the project slightly more complicated for this one group, the concerned student said, "This was one of the only computer-based group projects I've ever been able to complete and not feel exposed and open to people I don't know."

While it was impossible to predict what kinds of reactions passing Internet users may have to the students' posts, no negative comments were seen by the professors or brought to the professors' attention. Instead, outsiders praised the students at times for engaging in intellectually-stimulating conversation on Twitter, and they thanked them for sharing the information they learned. The students were excited to see that respected industry leaders acknowledged them. One student even tweeted, "I hope my prof (sic) saw that @PR_Leader was impressed with our definition. That's worth an A, right?" (Note: The industry leader's Twitter account was left out to preserve their anonymity.)

Reflecting the broad range of Twitter experience among the students, others wanted the assignment to be more challenging and explore more aspects of Twitter. One male graduate student said, "I would have gone beyond the definition and 10 tips and asked us to go out into the Twitterverse to find current trends for the topics, reach out to professionals in the area and engage them in conversation." Another student suggested that doing interviews with working advertising and public relations practitioners over Twitter would provide additional insights into the site's usage. One female undergraduate suggested that "what if the assignment challenged us to go find five people who have no idea about the topics and get them involved in conversation. Their outside perspective might be very informative for us."

Perhaps one of the greatest measures of effectiveness for this assignment focuses on the final results of the workgroups. Only one of the professors that participated in the assignment had completed a similar project before using Twitter to have students collaborate on definitions and tips. The previous time, the assignment was done in a classroom setting with students taking an entire 75-min class session to use classroom resources to define strategic communication concepts. Qualitatively, the results produced by the students in the two classes were very similar in terms of the completeness and accuracy of the definitions and tips. Quantitatively, the students in the Twitter class earned slightly higher grades on the assignment. The boost in grades, in part, was due to the groups' usage of more industry-specific resources that they used through Twitter conversations with practitioners and the incorporation of this material into their submissions.

While the instructors did not use quantitative data to evaluate the students' views of the Twitter assignment, the in-depth interviews and focus groups resulted in very favorable comments for the assignment. One male graduate student said, "I would encourage you to use this assignment again in the future. It's not an easy assignment, but the struggle to learn a new technology and discover its relevance to your chosen career is a real reward." The end-of-term course reviews for both classes were overwhelmingly positive. Quantitatively, the advertising students rated both the course assignments and value of the class at a 4.8 on a 5-point Likert scale, and the public relations students evaluated the assignments as a 4.9 and the class as a 4.8 on a similar scale. In the open-ended questions, students said that even though the assignments were difficult and challenged them to think in ways they never had, they valued the lessons that the assignments taught. In particular, one student said:

> If you would have told me before the class that I would love Twitter by the end of the course, I would have laughed. But, the Twitter assignment opened my eyes to its potential. That assignment in particular stands out in my mind because it is so different than everything else I've ever done in school. It was great, and I'm really glad to have had this experience.

CONCLUSION

Overall, the educators involved in this project did feel that it was a beneficial assignment for students in both classes. While the students may not appreciate the assignment in its early stages, many of them recognize its value in retrospect. They understand that social media is not something to be entered into casually and that users must be committed to frequent participation. In this regard, they are probably further along than many businesses or organizations that see social media as a quick and cheap way to send messages to their customers or publics. In that sense, the assignment did reach one of the goals of having students participate as users and understand how relationships could be built.

One factor that may have to come into consideration regarding incorporating social media in courses is the perception that social media is something that can be done simply, quickly and then left alone. As demonstrated by the students who did not join the discussion until late in the project, it is increasingly difficult to join in the conversation once it has started without all participants. This information could be essential for businesses and organizations who are attempting to use social media. They cannot just jump in the discussion at the last minute, provide their perspective and hope for the best. Instead, relationships and conversations need to be cultivated over time.

Sadly, it appears that roughly 40% of the students involved in both courses have abandoned Twitter – or at least the account they used for the class – once the courses ended. The majority of the students are still using Twitter in varying amounts for different purposes. The Twitter assignment, indeed, made an impact on the vast majority of students in a positive manner. Six students enrolled in the two courses used their experiences on Twitter to help secure social media internships, and three students have gone on to be employed as social media managers for Fortune 500 companies.

While time allowed student and instructor frustration to resolve the challenges that emerged during the Twitter assignment, it is important to touch upon the a few tips to help other educators with the incorporation of a Twitter assignment into their own courses. Reviewing these suggestions while planning a similar assignment would help ensure that future iterations of this assignment run more smoothly for students and help eliminate some of the frustrations.

Preparation

Consistent with findings by Väljataga and Fiedler (2009), students felt the desire to have more preparation before the assignment launch. Even though Twitter was seen as an easy entry point by the professors, several of the students expressed the need for a more detailed preparation for using the service prior to the assignment. It may be necessary, given the level of expertise of the students, to actually have the students create Twitter accounts and "practice" sending Tweets to other members in the individual class before setting them to discuss the assignment with members in a different university or class. Additionally, it may be beneficial to have students act as lurkers for a while, observing how conversations work before starting the assignment. This may increase their understanding of building relationships through conversation.

Class Sizes and Matching

Ideally, it may be best to try this assignment with classes that are roughly the equivalent in size so that no one university group will appear to be more prevalent than the other. Additionally, pairing undergraduate students with graduate students may seem intimidating for some involved. Thus, it may be necessary to pair undergraduates with other undergraduates or to make sure the undergraduates feel secure in collaborating with those of a higher academic level.

Twitter Account

Consider creating a Twitter account that can be accessed by the faculty members involved in the assignment. This way, the students do not feel that one professor is more involved than another and it may cut down on paperwork and account monitoring. If the decision is made to use two separate accounts, it is best to have introductions made by all educators involved to show that the project is truly a collaborative effort instead of being owned by one participant.

Leaders

As Lam (2004) points out, it might be wise to consider assigning group leaders to the task. While this may impose somewhat of a bit of a hierarchy to the assignment, it will also help overcome student self-consciousness of being the first one to post or tweet. However, it is recommended that although a leader may be chosen to get the conversation started, that conversation is encouraged among all and that no one person is responsible for the project on his or her own.

Mutually Exclusive Hashtags

While this may seem to be relatively straightforward, Twitter is an open forum for anyone to use and create any hashtag. Make sure that the ones used for the assignment are specific enough that when educators search for the postings online, they retrieve the class related postings and do not have to navigate through ones that do not apply.

Grading

While it is possible to grade the assignment on a purely quantitative level, it is not recommended. Counting the number of tweets a student makes can show the amount of interaction, but not the quality of the posts. Ideally, one should look at whether each student contributed significant intelligent comments to the discussion or was just a "cheerleader" for the group. While both roles have their place in collaboration, it will be up to the instructors to determine if just posting a "good job" comment is significant enough for a passing grade.

The collaborative Twitter assignment proved to be a valuable learning tool for the advertising and public relations students at the two universities. The students created 140-character definitions for 12 contemporary tactical and managerial terms in strategic communication and generated active discussions on how to best incorporate the terms into the practices of the two industries. Through focus groups, the professors found that students' thoughts about Twitter shifted from one of being unaware of the service's potential to recognizing its benefit of information sharing and to spreading knowledge through communication and collaboration.

Reflecting research indicating that online collaborations can be more fulfilling because of their social nature (e.g., Brooks, 2009; So & Brush, 2008; Petrakou, 2010), the students enjoyed the conversations they were able to create with one another even though they were separated by 450 miles. Although there were several different dichotomies that could be used to classify the students (e.g., advertising/public relations, undergraduate/ graduate student), the students came together to produce solid submissions on the assignment's due date. Similar to Tapscott and Williams (2008) suggestions, the online nature of collaborations flattened out any potential power hierarchies as everyone worked together cooperatively.

Although the professors deemed this assignment a success, there are a few limitations that should be acknowledged. First, these results are qualitatively based and not grounded in statistical analysis. Negative comments were made and explored in the focus groups; however, students generally remained positive about their experiences. Perhaps the 11 students who did not participate in the post-assignment focus groups had different experiences that would have provided different insights. Additionally, this assignment was designed around an application that is highly used in the industries in which the students hope to work. It may be more difficult to get students in a broader liberal education course to jump into Twitter and see its relevance to the course material. Finally, the Twitter assignment was assessed after its incorporation in only two classes. After incorporating it into additional classes, more data needs to be collected to measure its true potential and impact.

The success of this assignment also triggers possibilities for future research and other course-related activities. Given the dominance of research indicating that millennials distaste for Twitter, an attitude assessment of the site before and after the assignment would provide insights into how students perceive Twitter's utility. Additionally, content analysis of the groups' tweets could provide valuable metrics for creating grading rubrics for others looking to use a version of this assignment. While professors may not be ready to incorporate a full Twitter-based assignment into their courses, they may be more comfortable with having students participate in Twitter conversations about lectures and course materials using a dedicated course hashtag, such as #SOC101. Using the hashtag may help introduce students to the service and its potential and help ease students into a future Twitter assignment.

REFERENCES

About Twitter. (2009). Available at http://twitter.com/about. Retrieved on April 14, 2010.

Allen, E. I., & Seaman, J. (2010). *Learning on demand: Online education in the United States, 2009.* Available at http://www.sloanconsortium.org/publications/survey/pdf/learningondemand.pdf. Retrieved on April 3, 2010.

Andreas, K., Tsiatsos, T., Terzidou, T., & Pomportsis, A. (2010). Fostering collaborative learning in Second Life: Metaphors and affordances. *Computers and Education*, Advance online publication. doi: 10.1016/j.compedu.2010.02.021

Brooks, L. (2009). Social learning by design: The role of social media. *Knowledge Quest*, *37*(5), 58–60.

Cadima, R., Ferreira, C., Monguet, J., Ojdea, J., & Fernandez, J. (2010). Promoting social network awareness: A social network monitoring system. *Computers and Education, 54*, 1233–1240.

Calvani, A., Fini, A., Molino, M., & Ranieri, M. (2010). Visualizing and monitoring effective interactions in online collaborative groups. *British Journal of Educational Technology, 41*(2), 213–226.

Cheong, M., & Lee, V. (2010). Twittering for earth: A study on the impact of microblogging activism on Earth Hour 2009 in Australia. *Computer Science, 5991*, 114–123.

Chou, S. W., & Min, H. T. (2009). The impact of media on collaborative learning in virtual settings: The perspective of social construction. *Computers and Education, 52*, 417–431.

Crawford, K. (2009). Following you: Disciplines of listening in social media. *Continuum: Journal of Media and Cultural Studies, 23*(4), 525–535.

De Lucia, A., Francese, R., Passero, I., & Tortora, G. (2009). Development and evaluation of a virtual campus on Second Life: The case of SecondDMI. *Computers and Education, 52*, 220–233.

Fox, S., Zickuhr, K., & Smith, A. (2009, October). Twitter and status updating, fall 2009. *Pew Internet and American Life Project.* Available at http://www.pewinternet.org/Reports/2009/17-Twitter-and-Status-Updates-Fall-2009.aspx. Retrieved on April 12, 2010.

78% of Gen Y is 'meh' about Twitter; they also retweet less. (2009). Available at http://www.marketingvox.com/78-of-gen-y-is-meh-about-twitter-they-also-retweet-less-044246/. Retrieved on April 12, 2010.

Greenhow, C., & Reifman, J. (2009). Engaging youth in social media: Is Facebook the new media frontier? *Neiman Reports, 63*(3), 53–55.

Grensing-Pophal, L. (2010). The new social media guidelines. *Information Today, 27*(3), 1, 46–47.

Honeycutt, C., & Herring, S. (2009). Beyond microblogging: Conversation and collaboration via Twitter. *Proceedings of the Forty-Second Hawai'i International Conference on System Sciences (HICSS-42).* Los Alamitos, CA: IEEE Press.

Ioannou, A. (2010). Learn more, stress less: Exploring the benefits of collaborative assessment. *College Student Journal, 44*(1), 189–199.

Jones, K. C. (2009, June 4). Gen Y not into Twitter. *InformationWeek.* Available at http://www.informationweek.com/story/showArticle.jhtml?articleID = 217701840. Retrieved on April 12, 2010.

Kidwai, S. (2010, March). How to mobilize students using social media. *Techniques: Connecting Education and Careers, 85*(3), 8–9.

Kist, W. (2008). I gave up MySpace for lent: New teachers and social networking sites. *Journal of Adolescent and Adult Literacy, 52*(3), 245–247.

Lam, W. (2004). Teaching tip: Encouraging online participation. *Journal of Information Systems Education, 15*(4), 345–348.

Lenhart, A., Purcell, K., Smith, A., & Zickuhr, K. (2010). Social media and mobile internet use among teens and young adults. *Pew Internet and American Life Project.* Available at http://pewinternet.org/Reports/2010/Social-Media-and-Young-Adults.aspx. Retrieved on April 11, 2010.

Lincoln, Y. S., & Guba, E. G. (1985). *Naturalistic inquiry.* Beverly Hills, CA: Sage.

Lindlof, T. R. (1995). *Qualitative communication research methods.* Thousand Oaks, CA: Sage.

Mulhern, F. (2009). Integrated marketing communications: From media channels to digital connectivity. *Journal of Marketing Communications, 15*(2/3), 85–101.

Nielsen Online. (2009). *Twitter's sweet smell of success*. Available at http://blog.nielsen.com/nielsenwire/online_mobile/twitters-tweet-smell-of-success/. Retrieved on February 26, 2010.

Null, C. (2009). How to avoid Facebook & Twitter disasters. *PC World, 27*(8), 97–103.

Participatory Marketing Network. (2009, 1 June). *Participatory marketing network study: Gen Y's are not yet taking flight on Twitter*. Available at http://thepmn.org/pressreleases/060109. Retrieved on April 12, 2010.

Petrakou, A. (2010). Interacting though avatars: Virtual worlds as a context for online education. *Computers and Education, 54*, 1020–1027.

Pew Research Center. (2010). *Millennials: Confident. Connected. Open to change*. Available at http://pewsocialtrends.org/assets/pdf/millennials-confident-connected-open-to-change.pdf. Retrieved on February 26, 2010.

Pozzi, F. (2010). Using jigsaw and case study for supporting online collaborative learning. *Computers and Education, 55*, 67–75.

Reid, C. (2009). Should business embrace social networking? *EContent, 32*(5), 34–39.

So, H. J., & Brush, T. A. (2008). Student perceptions of collaborative learning, social presence and satisfaction in a blended learning environment: Relationships and critical factors. *Computers and Education, 51*, 318–336.

Solimeno, A., Mebane, M. E., Tomai, M., & Francescato, D. (2008). The influence of students and teachers characteristics on the efficacy of face-to-face and computer supported collaborative learning. *Computers and Education, 51*, 109–128.

Solis, B., & Breakenridge, D. (2009). *Putting the public back in public relations: How social media is reinventing the aging business of PR*. Upper Saddle River, NJ: FT Press.

Stelzner, M. A. (2009). *Social media marketing industry report*. Available at http://www.socialmediasummit09.com/. Retrieved on February 26, 2010.

Tapscott, D., & Williams, A. D. (2008). *Wikinomics: How mass collaboration changes everything*. New York: Portfolio.

Tsai, C. W. (2010). Do students need teacher's initiation in online collaborative learning. *Computers and Education, 54*, 1137–1144.

Väljataga, T., & Fiedler, S. (2009). Supporting student to self-direct intentional learning projects with social media. *Educational Technology and Society, 12*(3), 58–69.

Vankin, S. (2009, June 23). Generation Y: We're just not that into Twitter. Available at http://news.cnet.com/8301-17939_109-10265060-2.html. Retrieved on April 12, 2010.

SOCIAL MEDIA ASSEMBLAGES IN DIGITAL HUMANITIES: FROM BACKCHANNEL TO BUZZ

Alexander Reid

ABSTRACT

While the term "humanities" is not in itself a particularly contentious one among academics, the addition of the term "digital" creates all sorts of problems, even the superficially illogical contention that digital humanities are not humanities at all. The fundamental rupture between digital and print humanities lies in the turning of a materialist, object-oriented analysis upon the practices of humanistic scholarship. That is, in their newness, the digital humanities are unsurprisingly self-reflective about the materiality of their scholarly practices. This self-reflection has been largely absent from traditional humanities where we had all but naturalized the material composition of dissertations, journal articles, monographs, and so on. As a result, even as we continue to pursue traditional scholarly methods, it becomes increasingly difficult to do so without a self-reflective awareness of the historical-material contingency of these practices. In short, they are no longer the same. To explore this issue, this chapter takes up assemblage theory, and actor-network theory to investigate the intersection of mobile technologies and social media in the digital humanities including conference backchannels and networked research communities mediated through Twitter, Google Buzz, and

Teaching Arts and Science with the New Social Media
Cutting-edge Technologies in Higher Education, Volume 3, 321–338
Copyright © 2011 by Emerald Group Publishing Limited
ISSN: 2044-9968/doi:10.1108/S2044-9968(2011)0000003019

similar applications. The chapter considers how, even for those who continue to publish in traditional genres on traditional subjects, the development of these digital assemblages are transforming compositional practices.

In 2010, the digital humanities remains a still nascent and ambiguous descriptor of academic activity. On the one hand, there is the continuing development and proliferation of digital humanities centers at universities from Los Angeles to Michigan to Maryland, as well as consortia, such as the Humanities, Arts, Sciences and Technology Advanced Collaboratory (HASTAC) and the New Media Consortium (NMC), and the emergence of digital-specific grants from the National Endowment for the Humanities' (NEH) Office of Digital Humanities to the MacArthur Foundation's Digital Media Learning Competition. Conferences, journals, and book series focused on various aspects of the digital humanities, many published in digital formats continue to appear, such as the University of Michigan Press' series in digital humanities and recent announcement that the press would publish all of its "books" primarily in digital formats (with print versions available on request). While there may be few specifically "digital humanist" positions to apply for, an increasing amount of facility with digital media as both a teacher and researcher have become desirable attributes for academic job seekers across the humanities. On the other hand, specific examples of digital work aside, the humanities remain largely attached to their traditional print modes of scholarly publication (perhaps extending to PDFs of essentially print articles) and, more notably, humanists generally remain attached to the scholarly methods developed in the print era, methods typified by working independently to produce interpretive works based largely on close readings of a small body of texts. The single-author monograph remains the hallmark of tenurable humanistic scholarship. Furthermore, there remains skepticism, caution, and antipathy within the humanities toward the digital, particularly in regard to emerging social media. One does not have to look far into the *Chronicle of Higher Education* or *Inside Higher Ed* to find complaints and cautionary tales about the potential professional dangers of social media use. That said, a growing number of humanists have taken up social media in support of their research and teaching from Facebook groups, blogs, and wikis to microblogging applications such as Twitter, scholars employ technologies to establish and maintain relationships with colleagues, share developments and interests in their research and teaching, and to collaborate on new intellectual ventures.

This chapter examines these activities, particularly the role of microblogging, as they impact upon traditional scholarly practices.

Social media and the digital humanities might be understood in the deeper context of humanities computing, a field which traces its history back to 1949 and Father Robert Busa's project to produce a concordance of the works of Aquinas and related medieval authors (Hockey, 2004). Logically, one might imagine that the use of social media within the humanities would be a form of digital humanities, which might, in turn, be viewed as one part of, or as an evolution of, humanities computing. Given all that, one might then imagine that humanities computing would be one part of the humanities in general. However this is hardly the case. These technological and humanistic activities do not fit neatly into one another. Instead, it might be more accurate to view each of these as a rupture or departure from the other. I do not wish to over-dramatize the discontinuities among these practices, but at the same time, the scholarly practices emerging with new technologies cannot be conceived simply as a continuation of the humanities by new means. As I will discuss, in its earliest forms, the digital humanities did operate largely as an extension of humanities computing, if not the broader humanities. One saw the use of computers to aid in the analysis of literary texts, including the large-scale digitization projects undertaken by librarians and others. The other common first-wave use of digital media was for the electronic distribution of humanities scholarship, generally in the form of PDF versions of otherwise print-published articles. While these both remain significant practices within digital humanities, this chapter focuses on new trends in digital humanities, where social media practices promise to impact humanistic research in more dramatic ways. As this chapter will describe, contemporary humanistic research practices formed in the context of a particular technological mediascape in the late nineteenth and twentieth centuries; a mediascape that is now clearly shifting. Social media offer new means of collaborating among scholars, analyzing texts, producing scholarship, and distributing/publishing that scholarship (to say nothing of the entire new arena of humanistic investigation into the aesthetic, rhetorical, ethical, cultural, and political concerns of digital and social media itself).

Despite the discontinuities among the humanities, digital humanities, and social media, to understand the response of the humanities to social media, it is useful to consider the longer history of humanities computing as it enters into the social media era. Clearly computers did not play a significant role in humanistic scholarship or teaching during the 1950s, 1960s, or 1970s. In the 1980s, with the development of the PC and word processing, one began to see the appearance of computer labs in the humanities, at least for first-year writing courses, and at least some professors and graduate students began using computers to compose their scholarly texts. During this time, humanities computing operated largely in obscurity, competing for access to university mainframes with scientists and engineers. It is in this context that the development of dedicated technology resources for humanities research and teaching become an important goal for humanities computing. As Alan Liu (2009) observed of his institution's decision to invest in humanities computing in the mid-nineties, "there is a vast difference between requesting permission for experiments on a campus server and improvising projects on a departmental server with its smaller trust community of users and higher tolerance for failure" (p. 22). The development of digital humanities centers at universities across the United States suggests that Liu's experience was not singular. In her survey of digital humanities centers in the United States, Diane M. Zorich (2008) found that the median and mode founding date for the 32 organizations in her survey was 1999. Given the rise of the Internet during this time, it is fairly evident that the emergence of the digital humanities was as much, if not more, of a response to these changing external conditions than a response to any internal disciplinary developments within humanities computing. In this context, digital humanities centers are related to broader adoptions of the web across the campus, such as course management systems, online registration, student account management, and the web marketing of institutions. In short, digital humanities has a bifurcated history, with the longer, though marginal, history of humanities computing on the one side and the more recent, though less disciplinary-specific, cultural and institutional developments of the Internet on the other.

As such, while the rise of the Internet has opened many new opportunities for, indeed arguably birthed, the digital humanities, humanities scholars have also viewed digital media as an extra-disciplinary influence and thus have looked to define carefully the role of the humanities in relation to technology. James O'Donnell (2009) echoes a common concern in the humanities in observing that "Everyone recognizes that waiting for technologists to provide tools and, worse, tell us what to do with them is no solution, for the questions of scholarship must come from scholars" (p. 102). However Liu suggests this division between technologists who provide tools and humanist scholars who ask questions may not hold in a digital culture that

> requires a full team of researchers with diverse skills in programming, database design, visualization, text-analysis and -encoding, statistics, discourse analysis, Web-site design,

ethics (including complex "human subjects" research rules), etc., to pursue ambitious digital projects at a grant-competitive level premised on making a difference in today's world. Humanists working on collaborative teams with engineers and social scientists will thus need to contribute perceived value. (2009, p. 27)

It is in this context that the digital humanities presents a future that is potentially quite different from humanistic traditions. O'Donnell writes, "Someday we will no longer speak, I am sure, of the 'digital humanities'; but for now the phrase is needed to distinguish the new object, techniques, and contexts of study" from prior ones (p. 99). However one might wonder if the future will not be one where digital technologies are digested into scholarly practices that remain fundamentally recognizable as the humanities. Instead, as Liu suggests, the digital might lead to "an encounter with other disciplines that far exceeds the now domesticated familiarity of 'interdisciplinary studies' to become a monstrous exodisciplinarity" (p. 31). Between O'Donnell and Liu one can observe the range of responses to technologies in the humanities: from mechanisms for continuing humanistic practices to forces for radical transformations. Zorich (2008) observes a similar, though perhaps more tame, ambivalence in the mission statements of the digital humanities centers she surveyed where she noted a common emphasis placed on the "enduring value of the humanities" but also commitments to interdisciplinarity, openness, and "questioning sacred cows" (p. 11).

While none of these perspectives are necessarily oppositional, they are indicative, I will argue, of a brewing reformation of humanistic endeavor that has only intensified with the development of social media. Liu points to a constellation of contemporary theories of evolution, emergence, and complexity to investigate this condition (p. 30). Here, this investigation will take up this task through the work of Manuel DeLanda (2006) with social assemblage theory and Bruno Latour (2005) with actor-network theory. While interesting differences certainly exist between DeLanda and Latour, in this chapter the emphasis will be on the linkages between their approaches and the insights they offer together into the role of social media in the digital humanities. Social media itself points to a broad range of applications and practices from the relatively more established blogs, wikis, and social networks to increasingly mobile applications such as Twitter and Google Buzz. As such, it is not possible to offer precise analysis of the entire scope of social media. This chapter focuses on the latter, more recent modes of mobile social media. At this point, it would appear that the hype surrounding Twitter and microblogging has crested. The excitement and consternation over backchannel tweets at academic conferences has already come and gone. However, mobile microblog networks remain a largely untapped resource for developing research communities that offer a new temporality to scholarship as well as a new discursive mode.

SOCIAL ASSEMBLAGE NETWORKS

DeLanda's social assemblage theory and Latour's actor-network theory share a common philosophical context in the work of Deleuze and Guattari (1983, 1987, 1994), particularly in their exploration of relations of exteriority (as opposed to relations of interiority) as key to understanding the emergence of objects, identities, thoughts, and actions. That is, conventionally, the identity of an object or individual is imagined to be a product of essential characteristics that are internal. However, as DeLanda (2006) explains

> the reason why the properties of a whole cannot be reduced to those of its parts is that they are the result not of an aggregation of the components' own properties but of the actual exercise of their capacities. These capacities do depend on a component's properties but cannot be reduced to them since they involve reference to the properties of other interacting entities. Relations of exteriority guarantee that assemblages may be taken apart while at the same time allowing that the interactions between parts may result in a true synthesis. (p. 11)

Put simply, in the context of social media and the humanities, one would recognize that particular practices of humanism do not result solely from some internal disciplinary set of properties but rather from the interaction of the humanities with other entities. As such, for example, there is nothing inherently "humanistic" about writing scholarly journal articles and monographs but rather these practices represent a synthesis resulting from the interaction of the humanities with other entities, ranging from media technologies to tenure committees. In some respects, this realization is familiar to the humanities in that it reflects some commonly held warrants that humanists apply to their objects of study, if not necessarily to their own practices. That is, within the interdisciplinary practices of cultural studies, one regularly investigates texts as the product of historical and ideological contexts. However, as Latour argues, his approach differs in important ways from the cultural studies practice he identifies with "critical sociology," particularly in the way power/ideology is investigated: "To the *studied* and *modifiable* skein of means to achieve powers, sociology, and especially critical sociology, has too often substituted an invisible, unmovable, and homogeneous world of power for itself" (2005, p. 86). That is, where critical sociology and humanistic practices of cultural studies might focus upon a broad, generic cultural force of ideological power, which impinges upon the interiorized relations on which subjectivity is produced, social assemblage-network theory (if I might jam together Latour's and DeLanda's theories) focuses upon the operation of networks of real objects/actors and the forces they mediate.

This is, importantly, not a distinction between local objects and global, ideological explanations. Assemblages and networks are never simply local but rather map relations temporally and spatially stemming from a specific event or practice. Instead the distinction between critical sociology/cultural studies and social assemblage-network theory lies in their understanding of the ontological status of the objects being studied. As DeLanda puts it, what is required is "an ontology in which the existence of institutional organization, interpersonal networks and many other social entities is treated as conception-independent. This realist solution is diametrically opposed to the idealist one espoused by phenomenologically influenced sociologists, the so-called 'social constructivists'" (2006, pp. 2–3): this is a position Latour echoes, though he comes to it from a different angle.

> To go from metaphysics to ontology is to raise again the question of what the real world is really like. As long as we remain in metaphysics, there is always the danger that deployment of the actors' worlds will remain too easy because they could be taken as so many representations of what the world, in the singular, is like. In which case we would not have moved an inch and would be back at square one of social explanation – namely back to Kant's idealism. (Latour, 2005, p. 117)

Either way, the central issue is to eschew idealist conceptions of social relations in favor of an approach where the social entities we study are neither treated as "matters of fact" nor as social constructions that are merely ideological. A theory of social assemblage-networks allows the study of social media to investigate how the exposure of objects/technologies and people (where both the object and people are actors) to each other results in the emergence of affects, thoughts, and social relations. The traditional objects within the humanities – library stacks, offices, seminar rooms, lecture halls, conference presentations, journal articles, monographs, etc. – have been thoroughly territorialized for the purpose of maintaining disciplinary structures, but there is nothing inherently humanistic about these traditional networks. The humanities, as they are generally experienced today, emerged through an exposure to these (and other) objects over the last century. The intellectual error is to mistake these exteriorized relations as some interiorized identity (either one that is real or a socially constructed, discursive reality), which is now threatened by the arrival of new technologies.

That said, these new technologies clearly offer powerfully disruptive potential to the regular operation of social assemblages. DeLanda explains that assemblages can have tendencies toward both territorialization and deterritorialization (Deleuze & Guattari's terms, 1987). Processes of territorialization "define or sharpen the spatial boundaries of actual territories" and/or "increase the internal homogeneity of an assemblage" (DeLanda, 2006, p. 13). For example, a conference hotel with its various meeting rooms establishes a spatial territory, while registration fees and membership dues serve to increase the internal homogeneity of the conference by other means. However, communication technologies offer a potential for deterritorialization in that they can "blur the spatial boundaries of social entities by eliminating the need for co-presence" (p. 13). In a sense, since the beginning of academic conferences, this has always been the case, inasmuch as one could write a letter to a friend or an article for a journal discussing the events of the conference. More recently one could provide a telephone account, send a fax, or write an email. The difference with social media however is pointed on two accounts. First, the accounts can occur in near real-time with a mobile phone while one is listening to a conference presentation. Second, social media communications can be published to the public Internet where they can be immediately accessed by other conference participants, other experts not attending the conference, and, of course, the general public. In short, the particular character of these information networks allows them to deterritorialize the familiar spaces of a community by carrying its messages to new contexts. Of course, having said that, the next thought would be that there are likely still very few readers for such scholarship. Perhaps, but in the "long tail" communities of the web, even a tiny readership can mean an exponentially greater audience for the scholar. A conference presentation given to 20 audience members might garner several hundred hits when posted to a scholar's blog and be further circulated through Twitter, Google Buzz, Facebook, and similar services. Of course, in the expansion of audience across a network there is also destabilization of the scholarship. When I give the presentation at the conference, I can be reasonably sure that my audience are fellow scholars with whom I share significant contexts. In short, we occupy the same disciplinary territory. In a social media environment, that is obviously not the case.

However, processes of deterritorialization are not simply about making one's work available to a larger audience (for good or bad). More significantly, processes of deterritorialization are ones by which existing territories are remade, sometimes violently. In taking up social assemblage-network theory it becomes possible to investigate how the humanities' exposure to social media results in deterritorializations that are remaking disciplinary territories. It is not a matter of *choosing* to embrace new technologies or to stay with the traditional scholarly practices of the pre-Internet era. Even if one continues to write articles and chapters to be published on paper, as I am doing at this moment, one inescapably does so in the context of social assemblages and networks that include social media. The laptop on which I am writing is connected to flows of information from emails to tweets, as is my mobile phone. I can choose to ignore or turn off my access (as Seesmic alerts me that a friend has updated his/her status), but they are still there. On the other hand, choosing to embrace social media hardly guarantees a positive outcome. Instead, technological changes have the potential to be so disruptive as to be not only exo-disciplinary as Liu suggested but "exohumanist": creating a mediascape where the humanities exposure to new technologies leads well beyond the boundaries of what have defined humanism to date. Social assemblage-network theory offers a means for understanding the rupture social media has created from earlier forms of humanities computing and digital humanities where the disciplinary exposure to new communication technologies leads to a deterritorialization of research methods. As I explore in the next section, even when staying within the familiar humanistic territory of literary studies, the emergence of digital modes of analysis has resulted in some acute challenges to disciplinary identity that can be productively understood by thinking through the assemblages and networks in which that identity emerges.

EXOHUMANIST MEDIASCAPES

One of the primary and early interests of the digital humanities has been the digital preservation of print texts. While this digitization may appear a fairly tame and conservative use of technology, the databases which digitizing produce has led to the study of these texts in new ways. Perhaps the most well-known (and notorious in some circles) digital humanities practice is Franco Moretti's "distant reading." In "Style, Inc.: Reflections on Seven Thousand Titles (British Novels, 1740–1850)" (2009), Moretti undertook a statistical analysis made possible by the growing establishment of a digital collection of literary works. He noted, "in a few years, we will have a digital archive with the full texts of (almost) all novels ever published; but for now, titles are still the best way to go beyond the 1 percent of novels that make up the canon, and catch a glimpse of the literary field as a whole" (p. 134). One can certainly imagine a number of statistical analyses that might be possible in that digital archive. However, there are many critics of the disciplinary future as Moretti's work suggests. In a response to "Style, Inc.," Katie Trumpener (2009) concluded that

> We are, first and foremost, highly trained readers, and some of what we find, in library or bookstore, will show us new ways to think. We can change our parameters and our questions simply by reading more: more widely, more deeply, more eclectically, more comparatively. Browsing in addition to quantification; incessant rather than distant reading: the unsystematic nature of our discipline is actually its salvation. (p. 171)

Trumpener's observation that literary scholars are, "first and foremost, highly trained readers" is fairly astute. For literary scholars the book and the library are the specimen and laboratory. The close reading techniques virtually all literary scholars practice are analogous to the experimental methods that researchers in laboratory sciences learn to master. However, I disagree with Trumpener's assertion that such disciplinary practices are "unsystematic." While traditional literary studies scholarship may be unscientific and qualitative, it is highly systematized, as evidenced by the broad methodological, material, and technological similarities among curricula, dissertations, journal articles, monographs, and so on. Indeed, if there were not a deeply entrenched disciplinary system of scholarly practice, there would not be a basis for Trumpener's description of scholars as "highly trained"; there would neither be anything for Moretti's practice to threaten, nor a discipline that could experience "salvation." I am sure that Trumpener (and almost everyone else) would agree that academic literary studies is disciplinary and as such is systematic. That said, much of the disciplinary conversation and historiography in English Studies has focused on differences within the field, resulting in now familiar arguments that English faculty should "teach the conflicts," as Gerald Graff (1987) suggested in *Professing Literature*. This is one of the qualities that sets the humanities apart from the paradigmatic operation of normal science. As Thomas Kuhn (1962/1996) noted, unlike the situation in science education, the humanities student "is constantly made aware of the immense variety of problems ... [and] a number of competing and in commensurable solutions to these problems" that operate within the discipline (p. 165). The humanities in Kuhn's view do not conform to his theory of paradigms because they do not approach the solution of problems in the systematic way typified by science, and it is likely this difference in systemization to which Trumpener referred. However, in the context of emerging digital media, a deeper level of systemization is clearly visible, one which, among other effects, has enabled the discontinuous, conflicted landscape of humanities research.

Though the humanities have deep histories, going back to Classical Greek philosophy, humanistic scholarship practices have a more recent genesis in the nineteenth century. The literary studies journal article, which remains a standard for scholarly publication in the humanities, begins in the 1880s with the inauguration of *PMLA*, the flagship journal of the Modern Language Association, a few years after the first national MLA conference. Undoubtedly the industrialization of the publishing industry had a dramatic impact on literary studies by not only proliferating the texts scholars would come to study but also by providing an effective technological means for distributing research. The generally solitary nature of reading practices afforded by book technologies certainly suggested that scholars could conduct research independently. Thus even though the need for a library might be viewed as a research expense not unlike that of a laboratory in the sciences, the way humanities scholars employ libraries does not require collective labor. As a result, humanists could undertake their work without direct collaboration with others, requiring only that they could pass the muster of peer review. In short, the assemblage of technologies that established the twentieth-century mediascape of humanistic research naturalized particular research practices. For literary scholars, literary works were, for the most part, produced and consumed independently, on a singular basis, and thus could be studied in much the same way. Even though literary studies has focused on the cultural and ideological contexts of literary practices for the last few decades, those contexts have always been viewed as being mediated through the atomized experience of the single reader with a single text (or perhaps a small collection of texts compared to one another). Until the technologies on which Moretti's work is founded arrived, there was no way to study a significant portion of the literary corpus of any historical period. Similarly, until the arrival of social media, there was no effective means to produce, consume, or study media as a large-scale collaborative phenomenon. If one identifies these particular historical media conditions as integral to the identity of the humanities, then it is understandable that emerging media would be perceived as a threat, as an exohumanist mediascape. Unfortunately, such an identification may also mean condemning the humanities to a particular historical period that is now coming to close.

It is in this context that the arrival of social media has meant, at least for some digital humanists, an opportunity to push the humanities into some radically new form. The "Digital Humanities Manifesto 2.0" was produced collectively by participants in the UCLA Mellon Seminar in Digital

Humanities in collaboration with commenters on the first version of the manifesto. The manifesto (2009) argued "The first wave of digital humanities work was quantitative, mobilizing the search and retrieval powers of the database, automating corpus linguistics, stacking hypercards into critical arrays. The second wave is qualitative, interpretive, experiential, emotive, generative in character." In this characterization, the work undertaken by Moretti, McGann, and others would seem to fall largely in the first wave, with the second wave opening into a new, even less familiar, territory. As the manifesto argued, one of the characteristics of this new territory is the importance of collaboration. Todd Presner (2009), one of the principal architects of the manifesto, noted, "Parts of the document were written by Jeffrey Schnapp, Peter Lunenfeld, and myself, while other parts were written (and critiqued) by commenters on the Commentpress blog and still other parts of the manifesto were written by authors who participated in the seminars. This document has the hand and words of about 100 people in it." This fact in itself demonstrates one of the immediate challenges social media present to humanities scholarship, which relies heavily on attribution and authorship. This collectivity is one of the key features the manifesto identified: "The ant colony and the Ivory Tower, the network and the monastery are both potential places of pleasure, knowledge, and reward within an economy founded on abundance. But we can no longer entrust knowledge creation and knowledge stewardship solely to the latter." In other words, while traditional (i.e., solitary) humanistic scholarship may continue, much of the new work of the digital humanities will require collective action. If there is an exohumanist monstrosity out there, it will likely include radical shifts in our understanding of authorship.

One of the primary challenges in imagining such collaboration lies in the establishment of a new ethos of scholarly interaction. Generally speaking, social media have developed with an anti-hierarchical ethos that encourages "folksonomy" and the "wisdom of the crowds" as opposed to the traditional authoritarian institutional practices that validate the expertise of scholars. As McGann (2008) observed, the intersection of social media and humanities scholars has led to a situation that is often characterized as the meeting of two groups:

a hidebound, arthritic community of pedants obsessed with standards, and a 'bottom-up' world of 'free culture' where work is identified and valued by democratic assent rather than by 'top-down' authority. But here a proverb of William Blake comes to mind: 'These two classes of men are always upon earth and they should be enemies. Whoever seeks to reconcile them seeks to destroy existence' (*The Marriage of Heaven and Hell*). Contradictory imperatives are in play here, each of them representing important values ... there are communities, not least the educational community, that pledge allegiance to both. (p. 86)

McGann's observation that the educational community is designed to mediate in some way between the authoritative world of expert knowledge and broader democratic values recognizes the historical role of public education. The authors of the manifesto struck a different tone, suggesting that the digital humanities "dreams of models of knowledge production and reproduction that leverage the increasingly distributed nature of expertise and knowledge and transform this reality into occasions for scholarly innovation, disciplinary cross-fertilization, and the democratization of knowledge" (Digital Humanities Manifesto 2.0, 2009). In short, one of the dividing lines within the digital humanities, particularly in terms of social media, lies over whether one imagines a continued stratification in the humanities with educators providing a mediating role or a more "distributed nature of expertise." Though McGann is correct in his account that the popular discourse surrounding social media pits "top-down" models against "bottom-up" ones, a more accurate mapping of social assemblage-networks suggests a more multidirectional network where expertise is more granular. This is one of the more visible dividing lines between humanists who would seek to territorialize digital media within the field and those who would use digital media to mutate the humanities. Regardless of where one stands on the issue of where the humanities *should* go, assemblage and network theories offer powerful ways of understanding how the humanities have developed over the last century that do not rely upon abstract notions of ideology or power but rather map the operation of such forces through object relations. This critical methodology is particular important as one moves beyond the relatively familiar, if still contentious, humanistic territory of literary studies into radically new intellectual practices such as microblogging.

MICROBLOGGING MONSTROSITIES

If practices such as distant reading raise questions over appropriate disciplinary practices and social media disrupts the boundaries of pro-fessorial authority, nowhere is the granular, multidirectionality of emerging digital work more visible within the humanities than in the microblogging backchannel at conferences. Much of the informal conversation among academics regarding microblog backchannels has focused on the practice of "tweckling" (heckling speakers via Twitter during their presentations) (e.g., Parry, 2009). As danah Boyd (2009) has discussed regarding her own negative experience with backchannel, "as a speaker, I work hard to try to create a conversation with the audience. When it's not possible or when I do a poor job, it sucks. But it also really sucks to just be the talking head as everyone else is having a conversation literally behind your back. It makes you feel like a marionette." Certainly the affordances of microblogging allow, even encourage, brief, spontaneous responses that may emerge with little consideration, and, as boyd and others have experienced, the social aspect of microblogging can result in a kind of mob behavior where reactionary comments spin out of control. Snarky conference behavior existed before Twitter, but the technology has given rise to a new set of ethical challenges, which one imagines that academics and other professionals ought to be able to rise to meet. However, even when the backchannel is respectful and engaged, it can still be a challenge to a conference presenter. As boyd continues, "Lots of folks have talked about making the stream available to the speaker ... There's no way that a speaker can simultaneously consume a stream and convey a message. Sure, a message every 30 seconds or so, no problem. But a stream? No way."

The recent, mainstream academic attention given to microblogging, conference backchannels has led to several recent studies (McNely, 2009; Reinhardt, Ebnerl, Beham, & Costa, 2009; Ross, Terras, Warwick, & Welsh, 2010). Reinhardt et al. (2009) surveyed participants in several academic conferences and concluded "Communicating and sharing resources seem to be one of the most interesting and relevant ways in which one microblogs. Other microblogging practices in conferences include following parallel sessions that otherwise delegates would not have access to, and/or would not receive such visibility. Content attached to tweets was reported to be mostly limited to plain text and web links" (p. 8). However, Ross et al.'s (2010) examination of actual tweets related to three digital humanities conferences noted that most of the tweets could not be characterized as participating in a dialog. Instead they analysis led them to consider "whether a Twitter enabled backchannel promotes more of an opportunity for users to establish an online presence and enhance their digital identity rather than encouraging a participatory conference culture" (p. 14). This is not necessarily a criticism of microblogging, though perhaps it is a less idealistic outcome than that suggested in Reinhardt et al.'s survey. What it instead indicates is the operation of social media in the establishment of social relations among academics in a field; microblogging becomes a way of reinforcing the loose social ties that exist among faculty who may only meet once a year but have common interests that may one day lead to a closer, more collaborative relationship. Moreover, Ross et al. point out the rhetorical challenges presented by communicating in a new medium: "traditional conversation structures are missing from the Twitter corpus, resulting in a different type of participatory culture; rather than following interactions in an ordered exchange, users are placed in a multidirectional discursive space, where they loosely inhabit a multiplicity of conversational contexts at once" (p. 15). In the twitterstream, tweets go back and forth in dialogue with the @username convention, but these tweets also pass through the public stream and the more limited stream of one's followers. In addition the hashtag convention, which is typically used to organize tweets related to a particular topic, such as a conference, creates an ad hoc community. Rarely would one follow the general stream of tweets. This would be analogous to following a general stream of all blog posts or all YouTube video uploads: it would be effectively impossible. Instead one elects to follow a select number of tweeple (i.e., people with Twitter accounts) and perhaps certain keywords or hashtags (in academic terms those would be tags related to one's field). Even still, following several hundred tweeple is difficult enough, especially in the context of an actively tweeted conference. Hence one encounters the multiplicity Ross et al. describe.

A twitterstream is monstrous in the classical sense of jamming together disparate objects. However, in the context of social assemblage-networks where all objects emerge through relations of exteriority, of exposure, there is no interiorized purity to which monstrosity might be juxtaposed. There is no monster but rather a new potential for being. That said, a "multi-directional discursive space" might present users novel rhetorical challenges. Even as rhetorical practices and styles develop organically within a communication network, the operation of microblogs within the digital humanities remains in flux. As has been discussed, exposure to microblogging mutates the space of the academic conference: presenters can be exposed to immediate feedback (positive and negative); presentations can be disseminated across the web; conversations can be expanded to include people at a distance, both in and out of the discipline in question; and new relations between participants and other interested parties can be formed and maintained beyond the duration of the conference itself. The largely unanswered question that remains is how conferences themselves will respond. Some conferences have created their own Twitter accounts as a way of disseminating official information and engaging in the backchannel conversation. However, it also seems possible that the format of conferences themselves might be reshaped. In the context of blogging and YouTube, it hardly seems necessary any longer for academics to gather simply for the purpose of presenting research and receiving feedback. Instead, it seems entirely sensible that presentation content could be made available online and that the time devoted to panels might instead be spent on discussion, which is the real value of face-to-face meetings. In that context, the productive value of microblogging might be further leveraged to extend those discussions over time and space.

While the current state of academic microblogging, as detailed in these studies, indicates that the technology is mostly used for establishing an identity rather than developing more collaborative research practices. It may be the case that the specific microblogging applications that are currently available will not develop into the site for scholarly production some imagine it might become. The more recent Google Buzz allows for longer posts than Twitter, but it has not taken off in the digital humanities. Setting these particular technologies aside, however, the many-to-many, mobile, digital media, communications network that is emerging suggests a new space and pace for research practices. It is not difficult to imagine a future of near real-time, digital collaboration in either teaching or scholarship. However, even those who choose to remain in more traditional spaces will find their work impacted. Scholars and disciplines who manage to take advantage of these networks should be able to develop means to compose and share research more quickly and more broadly. Compositional practices will certainly develop in response to these changes. Just as Moretti's distant reading offers a new way to study a broad corpus of literary works, social media presents new means to examine quickly a broad range of research that builds a network of human intelligence atop database queries. Where the traditional scholar might conduct research by tracking works cited pages in journal articles, social media allows one potentially to track the real-time reading habits of disciplinary colleagues. Even though scholars may not collaborate in the traditional sense of co-authoring a text, the traditional conversation of cross-citation, which has occurred at the stately pace of journal publication, can now be accelerated as digital humanistic scholarship responds in real-time. What this will mean in terms of a final product is difficult to know, but it would seem likely that a more seamless relationship might emerge between the informal conversations of social media and whatever scholars produce as their research. Indeed it is not difficult to imagine that the notion of final products might be supplanted by a more recursive conversation.

Microblogging and other social media developments present exciting if unsettling possibilities for the digital humanities. These possibilities arise at a time when the role of the humanities in higher education has come under question, when organizations such as MLA have raised their own concerns about their disciplinary futures (Report to the Teagle Foundation, 2008), and the economics of traditional scholarly publication have put university presses at risk. Indeed over the last decade many have echoed Peter Drucker's prediction that the information age would spell the end of higher education. While I would not argue that social media represents a panacea for these problems, emerging technologies do represent a shift in cultural practices that the humanities cannot simply choose to ignore without losing further relevance. If, as has been argued here, one recognizes that traditional humanistic scholarly methods emerged through exposure to a particular set of technological conditions, then the defense of those methods as some pure intellectual practice makes little sense. In that context, rather than perceiving the digital as a threat to some interiorized humanistic identity, these new technological contexts offer an opportunity to build new scholarly practices through sustained intellectual engagement with the possibilities rather than simply accepting practices received from a disciplinary past.

REFERENCES

Boyd, D. (2009). Spectacle at Web2.0 Expo... from my perspective, November 28. Retrieved from http://www.zephoria.org/thoughts/archives/2009/11/24/spectacle_at_we.html

DeLanda, M. (2006). *A new philosophy of society: Assemblage theory and social complexity*. London: Continuum.

Deleuze, G., & Guattari, F. (1983). *Anti-Oedipus: Capitalism and schizophrenia*. In: R. Hurley, M. Seem, & H. R. Lane (Trans.). Minneapolis: University of Minnesota Press.

Deleuze, G., & Guattari, F. (1987). *A thousand plateaus: Capitalism and schizophrenia*. In: B. Massumi (Trans.). Minneapolis: University of Minnesota Press.

Deleuze, G., & Guattari, F. (1994). *What is Philosophy?* In: H. Tomlinson, G. Burchell (Trans.). New York: Columbia University Press.

Graff, G. (1987). *Professing literature: An institutional history*. Chicago: University of Chicago Press.

Hockey, S. (2004). A history of humanities computing. In: S. Schreibman, R. Siemens & J. Unsworth (Eds), *A companion to digital humanities*. Oxford: Blackwell. Retrieved from http://www.digitalhumanities.org/companion/

Kuhn, T. (1962/1996). *The structure of scientific revolutions* (3rd ed). Chicago: University of Chicago Press.

Latour, B. (2005). *Reassembling the social: An introduction to actor-network theory*. London: Continuum.

Liu, A. (2009). Digital humanities and academic change. *English Language Notes*, *47*(1), 17–35.

McGann, J. (2008). The future is digital. *Journal of Victorian Culture*, *13*(1), 80–88.

McNely, B. J. (2009). Backchannel persistence and collaborative meaning-making. SIGDOC'09, October 5–7, 2009, Bloomington, IN.

Moretti, F. (2009). Style, Inc.: Reflections on seven thousand titles (British novels, 1740–1850). *Critical Inquiry*, *36*(1), 134–158.

O'Donnell, J. J. (2009). Engaging the humanities: The digital humanities. *Daedalus*, *138*, 99–104.

Parry, M. (2009). Conference humiliation: They're tweeting behind your back. *The Chronicle of Higher Education*, November 17. Retrieved from http://chronicle.com/article/Conference-Humiliation-/49185/

Presner, T. (2009). Comment. *THAT Camp 09*, June 16. Retrieved from http://thatcamp.org/2009/digital-humanities-manifesto-comments-blitz/

Reinhardt, W., Ebnerl, M., Beham, G., & Costa, C. (2009). How people are using Twitter at conferences. Retrieved from http://www.apo.org.au/research/how-people-are-using-twitter-during-conferences

Report to the Teagle Foundation on the undergraduate major in language and literature. (2008). *Modern language association*. Retrieved from http://www.mla.org/teaglereport_page

Ross, C., Terras, M., Warwick, C., & Welsh, A. (2010). Enabled backchannel: Conference Twitter use by digital humanists. Retrieved from http://www.ucl.ac.uk/dh-blog/?p = 46

The digital humanities manifesto 2.0. (2009). Retrieved from http://manifesto.humanities.ucla.edu/2009/05/29/the-digital-humanities-manifesto-20/

Trumpener, K. (2009). Paratext and genre system: A response to Franco Moretti. *Critical Inquiry*, *36*(1), 159–171.

Zorich, D. M. (2008). A survey of digital humanities centers in the United States. Council on Library and Information Resources, Washington, DC. Retrieved from http://www.clir.org/pubs/abstract/pub143abst.html

SOCIAL MEDIA AS A PROFESSIONAL DEVELOPMENT TOOL: USING BLOGS, MICROBLOGS, AND SOCIAL BOOKMARKS TO CREATE PERSONAL LEARNING NETWORKS

Corinne Weisgerber and Shannan H. Butler

ABSTRACT

While personal learning networks (PLNs) are not new (Warlick, 2009), social media technologies are now enabling us "to fashion new kinds of networks that extend far beyond our immediate location and face-to-face connections, and to grow our networks based not on explicit decisions, but through the ideas of other nodes (people and resources), whose ideas intersect with ours" (Warlick, 2010, para. 5). What is new then, and what is changing the nature of PLNs, is the rapid growth of information and the emergence of new technologies capable of filtering that information and connecting us to others we can interact with and learn from (Siemens, 2008). In this chapter, we discuss the steps involved in building, growing, and maintaining online connections made possible entirely through new

Teaching Arts and Science with the New Social Media
Cutting-edge Technologies in Higher Education, Volume 3, 339–363
ISSN: 2044-9968/doi:10.1108/S2044-9968(2011)0000003020

technologies. We argue that in the context of higher education, PLNs should be viewed as an informal alternative to the more formal professional development programs that are commonplace in K-12 education.

INTRODUCTION

The advent of the social web has created countless new opportunities for collaboration, participation, and innovation. Over the course of the past few years, more and more college professors have realized the pedagogical benefits of social technologies and begun incorporating them into their courses in an attempt to break down classroom walls (Parry, 2010). While we may have encouraged student adoption and in-class use of social media, most academics have not yet fully embraced these technologies for their own professional development and research needs. Years of working in the traditionally cloistered academic environment may have instilled in many a fair amount of skepticism towards ideas of sharing and public pedagogy. Given the ever-accelerating rate of change of technology and the ensuing decreasing shelf life of knowledge, the question that arises is whether this closed model of scholarship still works in the 21st century.

If we are to truly reap the benefits of the social web, why not take advantage of its connective powers for our own professional goals? After all, as education professionals, do we not owe it to ourselves and to our students to keep current with the latest developments in our field of expertise? While the idea of life-long learning may not be new, many academics may be unfamiliar with the concept of using social media technologies to do so in the open by deliberately building personal learning networks (PLNs) designed to connect them to a community of experts. This chapter will demonstrate how to use social technologies to create networks capable of supporting an academic's research as well as professional and curriculum development needs. We also argue that the creation of a culture of collaboration and the active participation in networked academic communities not only advance our own professional learning goals but that they also serve to model the values of openness and connectedness to our students.

What Is a Personal Learning Network?

According to Stephen Downes (2009), a key contributor to the learning theory of connectivism (Siemens, 2005), the term PLN first appeared in a report on career development published by the Treasury Board of Canada Secretariat. In it, PLNs are listed as a best practice example used by the Royal Bank of Canada to support the career development of its employees. PLNs are described as "PC based multimedia learning activities that employees can tailor to a learning map" (Duxbury, Dyke, & Lam, 2000). While educational experts agree that the idea of a PLN is not new (Warlick, 2009), they do, however, contend that "new techniques for organizing digital networked information, have enabled us to fashion new kinds of networks that extend far beyond our immediate location and face-to-face connections, and to grow our networks based not on explicit decisions, but through the ideas of other nodes (people and resources), whose ideas intersect with ours" (Warlick, 2010, para. 5). What is new then, and what is changing the nature of PLNs, is the rapid growth of information and the emergence of new technologies capable of filtering that information and connecting us to others we can interact with and learn from (Siemens, 2008).

In this chapter, we define PLNs as deliberately formed networks of people and resources capable of guiding our independent learning goals and professional development needs. More specifically, in the context of higher education, we view them as an informal alternative to the more formal professional development programs that are commonplace in K-12 education. The importance of informal professional development activities in the realm of teacher education has been noted elsewhere (Barab, Jackson, & Piekarsky, 2006). Critics have long pointed to the problems associated with formal training programs that tend to be "treated as a special add-on event rather than as part of a natural process" (Miles, 1995, p. vii) and performed by "some external change agent who focuses on 'educating' the teachers" (Barab et al., 2006, p. 164). In higher education, those types of programs tend to be administered by a Center for Teaching Excellence, or other campus entity with a similar name. These centers are designed to support faculty in exploring effective teaching and learning methods and typically sponsor workshops and events on instructional design and modern pedagogy. Unlike K-12 education, however, where participation in teacher professional development activities is often mandated by the State, participation by college faculty in the programs offered by such centers usually occurs on a voluntary basis.

This not only means that higher education professional development activities tend to be based on the model of the external change agent – in this case a campus entity educating the faculty – but it also suggests that college faculty may not participate at all in such programs. Some have cited time constraints as a key challenge to overcome in this regard (Futrell, 1994). While teachers in other countries such as Japan and Taiwan "are allowed up to 40% of the school day to interact with colleagues, plan and assess, tutor students, or participate in activities that are professionally enriching" (Futrell, 1994, p. 130), neither K-12 teachers, nor college professors in the United States have the opportunity to devote nearly a fraction of that time to such activities. Considering that busy schedules filled with teaching, research and committee work may prevent faculty from taking full advantage of the professional development opportunities provided by their university, and that other types of formal professional development, such as faculty mentoring programs, remain uncommon across college campuses (Lucas & Murry, 2002), many college professors may be cut off altogether from a formal peer support network. Although their work, by its very nature, tends to be an isolating experience happening behind closed classroom walls (Futrell, 1994; Lucas & Murry, 2002), social media technologies provide a powerful new way to bring down those walls and connect faculty to peers while simultaneously eliminating some of the time constraints associated with professional development activities. As we shall see, new technologies, such as really simple syndication (RSS), now make it possible to let information and resources find a person, rather than the person actively spending time looking for them.

While we recognize the importance of offline connections as an integral part of any PLN, in this chapter, we focus on the steps involved in building, growing, and maintaining online connections made possible entirely through new social media technologies. It is important to note in this regard that PLNs are not defined by any particular social media tool or application. Instead, PLNs can be created using any number of social technologies as long as those technologies have the ability to connect learners and resources and encourage the free flow of information to and from a learner. It is also worth stressing that we view PLNs as more specialized networks than for instance, a person's entire Twitter or Facebook network. While one could argue that everyone in such a network has the potential to provide valuable information to the network owner, our definition of PLNs is much more narrow and strategic. We like to think of them as purposefully built networks designed to support one's self-directed learning needs regarding a particular professional or personal interest. As such, PLNs can be tailored to fit pretty much any learning project.

BUILDING A PERSONAL LEARNING NETWORK

Using Twitter to Create a PLN

Regardless of the technology used, the first step in creating a PLN consists of identifying experts who hold valuable information on a particular topic, or in a particular field, and are willing to share that information. With over 105 million registered users worldwide (Chacksfield, 2010), Twitter, the microblogging platform that allows its users to exchange brief messages of up to 140 characters each, is a good start for any PLN. Many critics consider Twitter a waste of time calling it a "rolling news service of the ego" (Pemberton, 2009, para. 4) and dismissing its users as narcissistic attention mongers. What these critics fail to see, however, are the thousands of registered members who do not utilize the service to comment on their latest lunch or current whereabouts, but instead use it to seek information and advice from a diverse and geographically dispersed group of people. To use Twitter in that second capacity and reap the pedagogical benefits of the service, members must sign up to receive other users' tweets, a process which Twitter calls "following." Choosing whom to follow is an act of identifying people who bring value to your network. As Trinkle (2009) explains, it allows Twitter users to participate in the "collective inquiry into best practices" (DuFour, DuFour, & Eaker, 2008) by choosing "the best practices for their personal interests" (p. 22).

To build your own PLN, you should start by identifying a few people whose work you admire (Fig. 1). These could be colleagues in your field or general area of study, or professionals who work in related fields. Once you have picked people to follow, you need to find out whether they actually tweet. One option to do so is to use Twitter's built-in search engine to locate registered users (click on the "Find People" link on the top right hand side of the page). Alternatively, you could check a person's blog or other social networking profile to look for their Twitter handle – a person's unique user name on Twitter. That handle can then be used to pull up a user's profile page (by entering the following URL: http://twitter.com/username). Immediately under the profile picture, is a follow button, which allows a Twitter user to subscribe to that person's tweets.

The nice thing about Twitter is that unlike other social networking sites, it is not built on a reciprocal relationship model. This means that you can follow any person with a public profile without that person needing to follow you back. Even if a user had decided to protect his or her tweets by

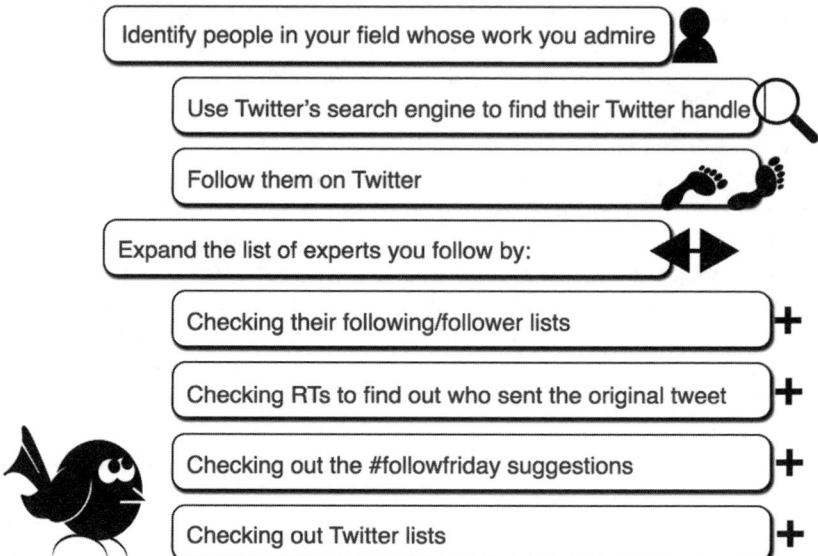

Fig. 1. Setting up Your PLN Using Twitter.

making them private – an option which less than 8% of Twitter users take advantage of (Moore, 2009) – that user's list of people they follow and are followed by, would still be open to public. And those lists, housed in the upper right hand side of a user's profile page, are a gold mine for anyone working on building a PLN.

To move beyond the initial set of experts in your PLN, we suggest going through each person's list of followers, and maybe even more importantly, the list of people they follow. This is one of the easiest ways to identify additional Twitter users who share your interests or who tend to pass along valuable information. A quick look at their latest tweets is usually enough to make that call. What makes this method of discovery even more worthwhile is the fact that those Twitter users have already been vetted by people you trust and that it tends to diversify the emerging network by adding users that go beyond your immediate connections. At this point, your PLN should be filled with at least a handful of professionals who tweet on a regular basis.

When someone in your network re-circulates information they received from another user (called sending out an RT, or retweet), we recommend looking up the person who sent the original tweet. A simple click on their

Twitter handle will pull up their profile page and display their latest tweets. Again, you would want to scan their latest tweets to determine whether you could benefit from the type of information they tend to share and whether you should add them to your PLN.

Other ways to grow your network include checking out suggestions from the people you follow, from Twitter itself, or from third-party recommendation services. Every Friday, thousands of Twitter users all over the world participate in the #followfriday event by recommending people who are worth following to their network (Baldwin, 2009). Twitter also provides two types of recommendations: a suggested user list that can be personalized based on one's interests, and a list of staff picks. The personalized lists are based on algorithms that "identify users across a variety of clusters who tweet actively and are engaged with their audiences" (Elman, 2010, para. 3). Twitter then groups these active users into lists based on their interests and lets "users browse into the areas they are interested in and choose who they want to follow from these lists" (Elman, 2010, para. 3). In addition to those suggestions, Twitter also publishes a set of staff picks, which have been manually selected by Twitter employees and list some of their favorite tweeters. Another service similar to Twitter's personalized suggestion lists is Mr. Tweet (http://mrtweet.com), a tool that identifies influential people in your current network, good followers you are not following back, and communities your network members belong to. All those suggestions are designed to help you identify topical experts Mr. Tweet thinks you might be interested in.

An even easier way to discover new and interesting Twitter users to add to a PLN is to consult a Twitter list, either using Twitter's built-in list feature, or one of the lists published on TweepML (http://tweepml.org/), a service that allows you "to manage and share groups of Twitter users" (TweepML, n.d.). TweepML features a "find a list" search tool, which college professors wishing to build a PLN can use to locate lists of Twitter users in their field. Similarly, they can consult Twitter's list feature by navigating to a user's profile page, clicking the "listed" link below the bio, and browsing through the two types of lists displayed on that page: (a) the lists following the user, and (b) the lists that particular user follows him or herself. Either one could prove to be a treasure trove of connections capable of growing and enhancing one's PLN.

Considering that many organizations have set up, or are in the process of setting up, a presence on the social web, it is also worth considering including organizational Twitter accounts in one's PLN. As Cox (2010) has pointed out, "many educational associations and organizations provide

Twitter updates" (p. 52) and could help academics expand their professional dialogue. In the area of communication for instance, organizations such as the National Communication Association (@NatComm), the Association for Education in Journalism and Mass Communication (@AEJMC), the International Communication Association (@icahdq), and the Public Relations Society of America (@PRSA, @prsanews) are all tweeting regularly about news from the field.

Following the steps outlined above should result in a fairly large Twitter PLN. Keeping up with such a large group of people, however, might prove challenging without proper network tweaking strategies. The first of these strategies is to filter incoming tweets. We recommend getting a Twitter client such as Tweetdeck (http://www.tweetdeck.com), Seesmic (http://seesmic.com), or HootSuite (http://hootsuite.com) to divide the network into smaller groups. For instance, one of the authors is a communication professor who teaches social media and public relations classes, and therefore has a group for PR educators, one for PR professionals, one for social media experts, another one for non-English tweets, etc. By organizing them into categories, the tweets will be neatly displayed in columns, which in turn will make it much easier to scan tweets for relevant information.

Another network tweaking strategy is to "test drive" your Twitter subscriptions for a month or two and to re-evaluate them afterwards. If you find accounts you are subscribing to that do not provide much value to you, simply hit the "unfollow" button. We like to think of our subscriptions as coming with a money back guarantee. If you do not like what you see, simply cancel at no cost to you. Of course, to get the most value out of your Twitter PLN, you will need to repeat the steps described above every now and then to add new voices to your Twitterstream.

Using Blogs and a Feed Reader to Create a PLN

While Twitter provides a great place to start a PLN, it is important to realize that despite its massive growth in 2009 (McGiboney, 2009), Twitter is currently only "used by seven percent of the population, or approximately 17 million Americans" (Webster, 2010, p. 3). Consequently, a PLN built entirely around Twitter would leave out the voices of many an expert who for one reason or another may not be using this microblogging service. To diversify the network, it is therefore necessary to turn to other information streams, such as blogs. According to Technorati (2008), a blog search engine which has monitored the size of the blogosphere since late 2004, there are

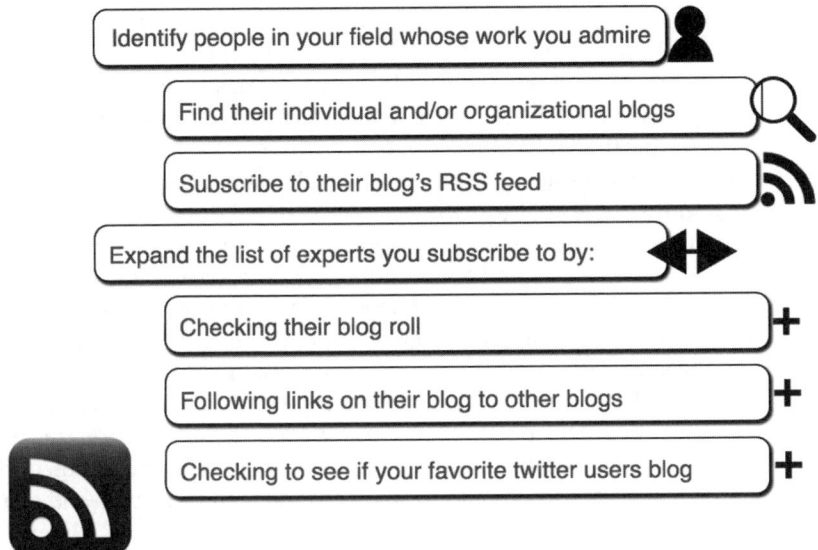

Fig. 2. Setting up Your PLN Using Blogs.

roughly 133 million blogs on the Internet. Of course, depending on a person's interests and learning needs, only a tiny fraction of those blogs may be suitable for inclusion in their PLN. The trick, as before, is to locate the gems.

To do so, most of the steps outlined in the Twitter PLN section still apply. To get started, it is again necessary to begin with a few individuals whose work or research interests pertain to yours (Fig. 2). A simple search, or Google blog search (blogsearch.google.com) should be enough to identify their blogs' URLs, if indeed, they keep a blog. Most bloggers publish what is called a blog roll – a list of blogs they personally recommend. Consulting those lists is a great way to discover new voices to add to a PLN, and as before, includes the additional benefit of having already been vetted by a blogger whose work you value. A related strategy for locating useful blogs is to follow links to other blogs. In other words, when a blogger you like links to another resource on the Net, follow that link to see whether it may lead you to valuable information. Lastly, check to see if the people you follow on Twitter also keep a blog. Although some may argue that this could lead to duplicate information, in most cases, individuals with more than one social media presence use their blogs and Twitter accounts for different purposes.

Locating relevant blogs to incorporate into a PLN may not be that useful though if we cannot devise a simple way to keep track of all the content posted to them. Fortunately, most blogs automatically generate an RSS feed, "which makes it possible for readers to 'subscribe' to the content that is created on a particular Weblog so they no longer have to visit the blog itself to get it" (Richardson, 2006, p. 75). The subscription part of this process is handled by a desktop client, or web-based application, called an aggregator or feed reader. These aggregators check the feed subscriptions at regular intervals, collect any new content they come across, and display all of it nicely packaged on one page. "Then, when you're ready, you open up your aggregator to read the individual stories, file them for later use, click through to the site itself, or delete them if they're not relevant" (Richardson, 2006, p. 76). Most web-based aggregators, such as Google Reader (www.google.com/reader) or Bloglines (www.bloglines.com), are free and allow feeds to be accessed from any computer with an Internet connection. The key advantage of feed readers though, is their ability to turn the chaos of the Internet into manageable chunks of information and to enable Internet users to "read more content from more sources in less time" (Richardson, 2006, p. 76).

The second step in the construction of a blog-based PLN therefore consists of setting up a feed reader and subscribing to the various blogs identified in the previous step. Since the actual subscription process varies by the type of feed reader used, we will not discuss it here, but instead refer the reader to their feed reader's tutorial for detailed instructions. Once all the blog feeds have been added to the aggregator, we recommend making it a habit to check the feeds regularly. Setting your browser's homepage to display your feeds is an excellent way to do so. As before though, the network created this way still needs to be fine tuned periodically to yield maximum benefits. This means revisiting subscriptions every now and then to re-evaluate their relevance to professional development needs, deleting irrelevant feeds, and adding new ones along the way.

Using Social Bookmarks to Create a PLN

Another, probably underused, social technology with regard to PLNs is social bookmarking services (Fig. 3). Social bookmarks allow Internet users to save, categorize, and share online content. Unlike regular bookmarks, which are stored on a computer's browser, these bookmarks are saved in a publicly available online database provided by one of many social bookmarking

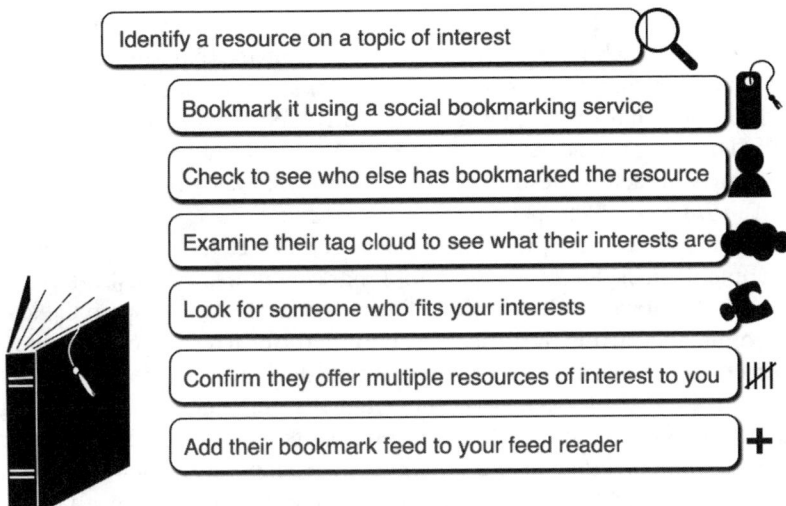

Identify a resource on a topic of interest

Bookmark it using a social bookmarking service

Check to see who else has bookmarked the resource

Examine their tag cloud to see what their interests are

Look for someone who fits your interests

Confirm they offer multiple resources of interest to you

Add their bookmark feed to your feed reader

Fig. 3. Setting up Your PLN Using Social Bookmarks.

services such as Delicious (http://delicious.com), or Diigo (www.diigo.com). Besides enabling its members to save links to online resources, these sites also allow members to describe the content they are saving with user-defined keywords (referred to as tags) and to share that tagged content with others on the Internet. What makes these services social is the fact that they "take all of the entries that are tagged the same way and connect them, and then connect all of the people who posted those links in the first place" (Richardson, 2006, p. 91). In a way, social bookmarking users are net miners who "act independently, gathering what interests them" and tagging the information they come across in a way that makes sense to them. "Collectively, their actions create a Web of networked sites, information, and people" (Fichter, 2004, p. 53). Because of these connective powers, social bookmarking services such as Delicious are invaluable PLN tools (Warlick, 2009) – especially when fed into a feed reader.

Adding a social bookmarking component to a PLN consists of identifying specific tags or users to follow, and subscribing to the tag or user feed with a feed aggregator. In the latter case, when a user whose bookmarks you subscribe to bookmarks a new resource, that resource will automatically be displayed in your feed reader. Similarly, if you subscribe to a specific tag, your feed reader will display any new online content categorized under that tag, regardless of who tagged the content. This means that any time a person adds a new resource to their online bookmarks using the tag you are subscribing to, that resource is automatically shared with you via your feed reader. This is particularly useful from a PLN perspective "because you are connecting yourself to resources that are supported by a recommendation system" (Warlick, 2009, p. 15).

While it is easy to identify tags of relevance to one's professional development needs, locating social bookmarking users to follow takes a little work. What follows is a description of how to complete this step on Delicious. We recommend starting the process by bookmarking a resource on a topic that pertains to your professional development needs or learning goals. Generally, the more specialized the resource, the better the outcome. Next, pull up your Delicious bookmarks and look for the number to the right of the bookmark's title. This number indicates how many other people have bookmarked the same resource on Delicious. Clicking that number will open a page that lists all those users, shows how they tagged the resource, and when they bookmarked it. The underlying idea is that individuals who have bookmarked the same resource you did, may share similar interests and may connect you to other resources of value to you and your learning needs. By clicking on their username, we can see all the bookmarks they have saved publicly. To get a quick sense of a user's bookmarks, click the "tags" link right above their latest bookmark. This will generate a tag cloud – a list of all their tags with the most frequently used tags displayed in a bigger font size. If a user has bookmarked a fair amount of resources, the tag cloud should provide a pretty accurate visual representation of their interests. This visualization should help you determine the user's relevance to your PLN. To verify impressions formed through the study of tag clouds, it is also useful to explore some of the tagged content and determine its quality.

The ideal candidate to add to your PLN is a person who has bookmarked lots of resources on the topic of interest to you. To locate such individuals, you may need to repeat the process described above a few times. Once you have identified a Delicious user to follow, you will want to add that person's bookmark feed to your feed reader. To do so, simply enter the following feed address (http://delicious.com/rss/username) into your feed reader's subscription area by replacing the word "username" with the actual user name of the person whose bookmarks you would like to subscribe to. Similarly, to subscribe to a tag, use http://delicious.com/rss/tag/NameofTag. Of course, the social bookmarking component of a PLN is not limited to Delicious and could be easily adapted to other bookmarking services.

Adding Additional Social Media Technologies to a PLN

While it is possible to build a PLN entirely on a single one of the technologies described in this chapter (i.e., a PLN based solely on Twitter or a social bookmarking service), we believe that integrating various social technologies into a PLN has the potential of making the network more diverse and in turn more powerful. The question of what particular social media tool to integrate into a PLN does not matter as much in this regard as the tool's ability to encourage knowledge sharing among a pool of experts.

By now it should be obvious that regardless of the technology used, the steps involved in creating a PLN always involve identifying relevant resources or experts willing to share valuable information, and figuring out how to subscribe to that information. Although a PLN composed of a Twitter, blogging, and social bookmarking network seems fairly exhaustive and should be more than enough to launch one's online professional development endeavor, there are other social technologies worth a brief mention here. Giga Alert (www.googlealert.com), formerly known as Google Alert and Yahoo Pipes (pipes.yahoo.com) in particular, deserves a quick discussion. Giga Alert is a monitoring service that tracks the entire Internet for search terms or phrases specified by a user. Although the free version is somewhat limited, it nonetheless allows users to set up three automated web or news searches and to subscribe to those searches via RSS. The applications of automated searches to a PLN are obvious. They allow learners to direct resources their way without having to actively scan the Internet for them.

Of course, none of this would be feasible without RSS feeds that have made it possible for information to come find us, rather than the other way round. While RSS feeds have made the information on the Internet more manageable, there are other tools such as Yahoo Pipes, designed to help us manage multiple RSS feeds. Yahoo Pipes is a free tool that lets users aggregate content from around the web, combine multiple feeds into one, and sort and filter the output (Pipes, n.d.) before delivering it to a feed reader. While the details on how to set up a basic Yahoo Pipe would go beyond the scope of this chapter, we briefly discuss its advantages from a conceptual perspective. Rather than limiting an automated web search to Giga Alert alone, Yahoo Pipes allows us to set up an automated search for any keyword or keyword phrase on Twitter (http://search.twitter.com), blogs (http://blogsearch.google.com; http://twingly.com), news (http://news.google.com), Flickr (http://flickr.com/search), etc., and to combine the results from all these searches into a single feed we could subscribe to through a feed reader.

Warlick (2009) refers to such automated searches as the dynamically maintained asynchronous connections of a PLN. While automation of information scanning and retrieval may sound attractive to busy academics, we do agree with Warlick that the "heart of every PLN is its members" and that "the networked learner at the center of the PLN is not merely a destination for information" (p. 15). Instead, we believe that learners need to actively engage their network and give back to it in order for PLNs to live up to their true potential.

MAINTAINING A PLN

The Importance of Sharing

While the first part of this chapter was devoted to a discussion of how to set up a PLN using various social media technologies, it is important to note that the work does not stop there. PLNs, once created, need to be maintained in order for a true symmetrical relationship to develop between a learner and the members of his or her network. Contrary to most academic institutions' fears about openness, we believe that academics must embrace the ideas of sharing research and course material in order to design effective PLNs. Unfortunately, the current competitive nature of most research institutions encourages scholars to keep their projects under wraps until the project is ready for publication. Along those same lines, most publishers will not consider publishing a manuscript that has already been made public. Many researchers may reasonably fear, then, that conducting research in the open may unnecessarily leave them vulnerable to more competition and possibly eliminate them from the highly valued process of publication. We hope that as social media adoption progresses, and as the academy wrestles with issue of publication and tenure, researchers will become more collaborative and less concerned with traditional means of publication. We believe that even in the current system, the benefits of sharing research among scholars far outweigh the potential negative consequences.

The sharing of work is what transforms the Internet from a broadcast medium, based on a linear view of communication into a socially shared interaction, based on a more transactional view. The transactional model of communication makes clear that the construction of meaning occurs not in a one to many, or even sender to receiver process, but rather in a simultaneous process of co-construction of meaning between the sender and the receiver

(Beebe, Beebe, & Ivy, 2007). If we choose not to be a part of the process and merely act as lurkers on the side of the information superhighway, we cannot expect our PLNs to provide much value. To honor the transactive nature of a PLN, sharing content and insights therefore is a must. By providing quality resources, users strengthen and maintain connections with both their followers and those whom they follow.

The benefits of providing content for those who may wish to include you in their PLN are fourfold: visibility, expertise, feedback, and reciprocity. Visibility may, at first glance, seem overly egoistic, and to some extent it may be, but any scholar who attaches his or her name to an article, for some measure at least, participates in the desires of the ego for recognition. Visibility has obvious advantages but also comes with potential drawbacks. Once you have chosen to introduce yourself to an online community by providing content, it will be through that content that your value to that community is assessed. If your queries, comments, and content meet with approval and others find merit in your work and insight, then perceived expertise in your area will likely follow. The more expertise you develop, the more visibility you will get, which in turn increases the likelihood of your content being relayed outside of your current network. This may eventually lead to the acquisition of new followers. We would like to caution the reader though that growing your network of followers, or adding more people to your PLN should never be the main goal. Following a select group of individuals who consistently share quality insights provides far better results than trying to follow everyone. Indeed, research suggests that "it is more influential to have an active audience who retweets or mentions the user" than to have a lot of followers (Cha, Haddadi, Benevenuto, & Gummadi, 2010, p. 1).

The last two benefits are ostensibly aspects of the first two. Making content available inevitably leads to feedback on your work. This feedback is a priceless resource allowing academics to gauge how others react to their ideas as they develop them into peer-reviewed presentations and publications. Reciprocity, the last benefit to be discussed here, is the big driver of social media. It lies at the heart of the gift economy that networked communities thrive on. The reciprocal nature of leaving feedback for those in your PLN acts to maintain your ties with them and can strengthen their perception of your significance within the community. This simple method of sharing tends to drive more and better-targeted knowledge into your learning network and is ultimately what transforms data into information and information into knowledge. "One useful way to distinguish knowledge from data and information is to see them as a continuum in which the material becomes richer, more valuable, more laden with meaning, and more important within a particular context" (Wallace, 2004, p. 139). Without interaction within a PLN, this transformation is unlikely to occur however.

Sharing, as a PLN maintenance strategy, can take on many forms. It can come in the form of short tweets, of more elaborate commentary provided on your own blog, or as a response to someone else's blog post. In terms of content, these comments can range anywhere from applying theories or perspectives from your field onto recent news events, to bring underreported happenings in your discipline to the attention of colleagues, or discuss classroom projects and experiences. Of course, knowing your audience and following current discussion threads and examples are of paramount importance here. Community members may find tangential posts to be off topic for their interests and eventually drop the authors of those posts from their cycle of information generation.

Lagniappes or Creating a Gift Culture

Along the gulf-coast of the United States, the term lagniappe means "a little something extra" and refers to a small gift included by merchants as a "thank you" for their loyal customers. We suggest that periodically providing a lagniappe for your PLN members is a very effective network maintenance strategy. Virtual lagniappes are distilled information applications that allow for instant consumption or dissemination and that provide your network members with an incentive to keep coming back to your blog, Twitter account, etc. These gifts can take on many forms such a presentations, timelines, visualizations, web videos, and podcasts. While there are many services available for gifting, we limit our discussion to some of the sites we have found most beneficial over the years.

SlideShare (http://www.slideshare.net) is a presentation sharing tool that allows users to upload a PowerPoint deck and convert it into a web-friendly format. Whether to their inclination or distaste, many academics are in the business of producing slides for the courses they teach. In many fields that means producing content on a daily basis, as news stories unfold. A timely, well-designed PowerPoint presentation on something of interest in a particular field can be of great value to others who may also be teaching similar courses or studying similar recent events. SlideShare has a nice set of sharing features which allow anyone to embed the presentation in a blog or share it on a number of social media sites such as Twitter or Facebook. Users can also leave feedback on the SlideShare site.

As an alternative to SlideShare, Prezi (http://prezi.com) allows members to create presentations using their proprietary presentation system. Presentations created with Prezi break down the serial nature of PowerPoint shows into a widescreen, panoramic zooming-in-and-out presentation. Prezis can be downloaded to a computer as standalone files so that Internet access is not required for the presentation. Like SlideShare, Prezis are made available on their website and as embeddable files ready for sharing.

Dipity (http://www.dipity.com) is another useful application designed to make historical data more accessible, interesting, and interactive. Dipity allows users to create interactive timelines, which are represented visually along the horizontal axis of the screen. Information, photographs, charts, links, maps, RSS feeds, and even video can be uploaded for any particular date. The finished product allows users to scroll along the timeline and click on any date in order to view a larger version of the data. One of the authors of this chapter has successfully used the site to allow her class to build a collaboratively designed history of the PR profession and share it with students and PR educators in her PLN.

Like Dipity, Many Eyes (http://www.many-eyes.com) focuses on the visual manipulation and representation of data. Many Eyes is hosted by IBM's Visual Communication Lab and provides users with 20 different ways to visualize almost any kind of data from simple graphs to scatterplots, bubble charts, network diagrams, tree maps, tag clouds, and geographic maps. Many Eyes accepts both tabular numerical data and textual data and allows the user to play with the various visualizations styles to find the best way to represent the information. Users can upload a dataset and create visualizations, which may be shared as interactive embeds.

A prime candidate for virtual gifting is a web video exploring a topic in your field of expertise. Creating a well-produced web video is somewhat more time and application intensive than the other gifts we have mentioned, but we believe it to be an excellent PLN maintenance strategy. There are many applications and websites to help produce and host videos, and we briefly discuss a few here. We believe that some of the best applications for creating web videos on the PC platform include Avid Pinnacle Studio and Microsoft Movie Maker; and on the Mac platform, Final Cut and iMovie. For educators who do not have access to desktop video editing software, the following websites offer online editing capabilities: Video Toolbox (http://www.videotoolbox.com) and JayCut (http://www.jaycut.com). For PC users, Avid now offers a free downloadable application similar to its

Pinnacle Studio called VideoSpin (http://www.videospin.com). In terms of hosting the video, the two most popular free video hosting sites are YouTube (http://www.youtube.com) and Vimeo (http://www.vimeo.com). To disseminate the video, we suggest embedding the video on your blog and tweeting the link to the blog post to your PLN.

Apple's iTunes Store now contains an abundance of educational materials in a section called iTunes University, which makes course content from many prestigious universities such as Stanford, Princeton, Oxford, UC Berkeley, and Penn State available for free. Educators can make the podcasts shared on iTunes University a part of their PLN by subscribing to them, or, alternatively, they can add videos or audio content for sharing with other scholars and members of their network. iTunes University even allows users to make content accessible only to a select group of users such as a class or the members of a PLN.

Although the list of virtual lagniappes described here is not exhaustive, the point we have been trying to make is that they can enrich the community of learners who have made you a part of their network. This act of giving back to the community can be seen as an acknowledgement of the responsibility all of us carry as members of other people's PLNs. As Warlick (2009) has argued "working your PLN involves a great deal of responsibility because you are almost certainly part of someone else's network" (p. 16). According to Warlick, "learners become amplifiers as they engage in reflective and knowledge-building activities, connect and reconnect what they learn, add value to existing knowledge and ideas, and then reissue them back into the network to be captured by others through their PLNs" (p. 16). Lagniappes are an important aspect in this process of knowledge amplification.

PLNs AT WORK IN A UNIVERSITY SETTING

Having discussed the steps involved in building, growing, and maintaining an online PLN, we now turn to concrete examples to illustrate how such networks can be used in academe in an effort to support one's research, professional and curriculum development needs. As one of the authors wrote in a blog post:

> If there is one thing that hasn't ceased to amaze me since I started blogging, tweeting, bookmarking and aggregating, it's this thing I've come to refer to as social media serendipity. It can happen any time, anywhere. Sometimes it strikes as I am preparing classes, other times it happens as I am working on research. I may be working on a class

on pitching stories when a new blog post just pops up in my feed reader with a relevant, up-to-the-minute case study to include in my class. Other times, it is a Twitter or Delicious user I follow who will share the perfect example. The reason I call it serendipity is because I didn't ask for it. The information just has a way of finding me. It's as if there were hundreds of research assistants out there scanning the web and bringing the information back to me just when I need it. I honestly cannot remember the last time I spent hours online searching the net for that perfect example to illustrate course material. (Weisgerber, 2009)

To demonstrate that the serendipitous moments mentioned in that post are not isolated events, we will use two examples that happened while working on this very book chapter. The first is an example of an active social search, in which the first author "outsourced" a question to her Twitter PLN. Unable to find reliable statistics on the percentage of Twitter users who keep their tweets protected (information she needed to complete a section of this chapter), she turned to her Twitter network for advice (Fig. 4).

Within hours, she received the information she needed and had another member of her PLN retweet the question, thereby enabling it to travel beyond the author's own network (Fig. 5).

The second example was the result of a passive social search. In this case, the author did not actively seek out any information from her PLN, but simply received pertinent information from her network without actually asking for it. While taking a short break from this chapter, she pulled up her Twitter client and noticed that one of the members of her "educators" Twitter group had posted a link to wmtools.net, a tool for assessing the credibility of websites (Fig. 6). Since she just finished grading a social media audit project which, among other things, asked students to verify the

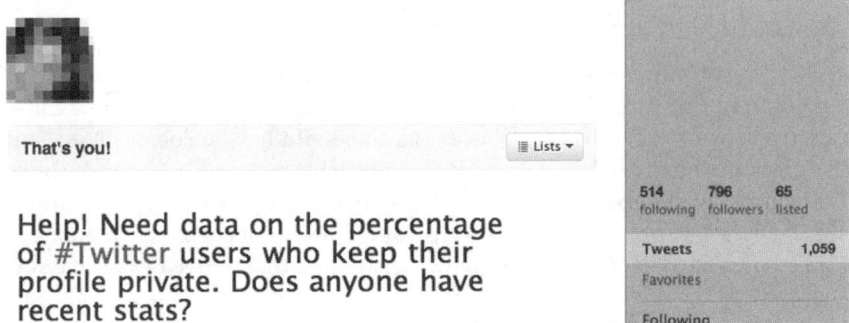

Fig. 4. Active Social Search Using a Twitter PLN.

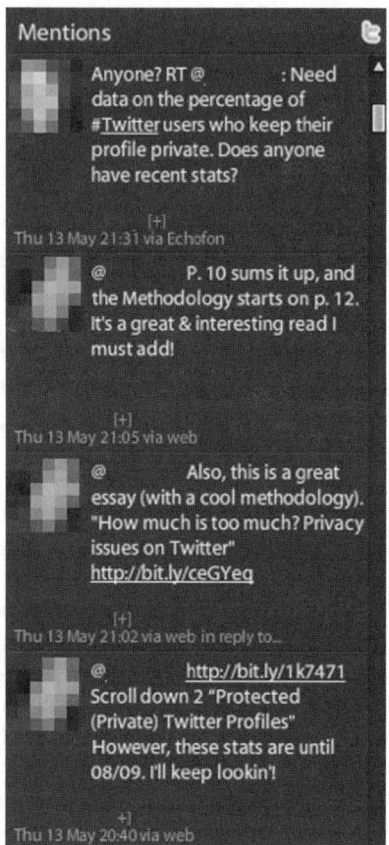

Fig. 5. Responses to an Active Social Search on Twitter.

credibility of various websites, she checked out the site, found it relevant, bookmarked it on Delicious, and added the URL to the assignment description for use in future classes. Coincidentally, the social media audit assignment thus revised was an assignment she received a few months ago from another member of her PLN! These two examples nicely illustrate how the Twitter component of a PLN can contribute to one's research endeavors and curriculum development activities through either active or passive social searches.

Another, more deliberate, attempt to incorporate PLNs into curriculum development efforts includes the first author's call to crowdsource the

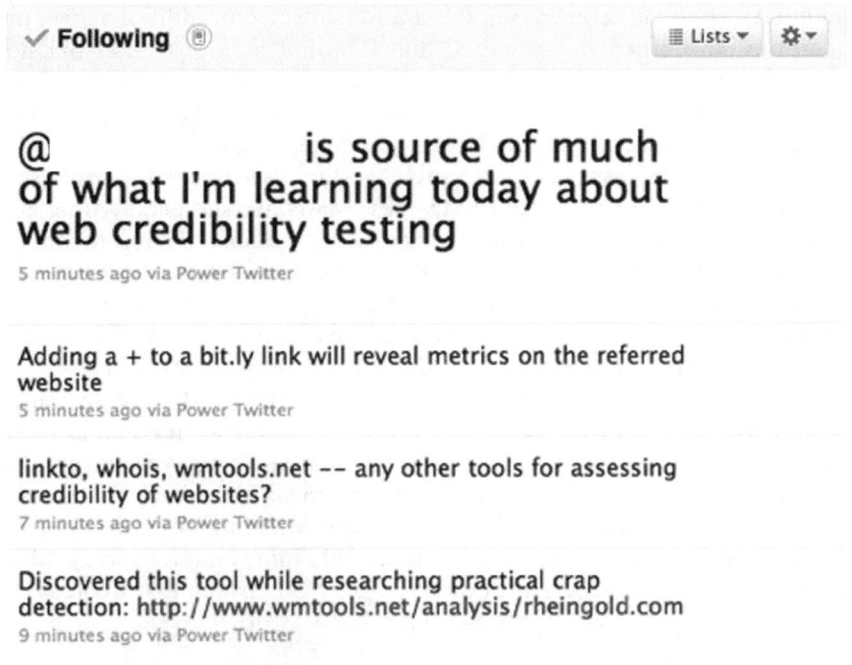

Fig. 6. Passive Social Search Using a Twitter PLN.

syllabus for her social media for PR class. Knowing that there were several colleagues both in and outside the United States working on developing similar classes, the author blogged about her idea to collaborate on a syllabus, launched a wiki to facilitate the effort, shared the link to the wiki on her Twitter network, and asked for input from both educators and professionals in the field. The result was a collaboration between several professors who shared ideas, assignments, and reading material, and professionals who offered hands-on real-life projects for students to work on. The success of this endeavor illustrates the need to actively engage your network and give back to it through the act of sharing.

Once you have fine-tuned your PLN, you will notice that the serendipitous moments described in these anecdotes actually occur with great frequency. We use one last example from the first author's personal experience to make that point. About a year ago, the author came across an interesting article on Twitter. Since the article seemed relevant to research she was working on at the time, she bookmarked it and set it aside to read at a later point. As she read it later on, she annotated it with additional research questions. In particular, she noted that it would be nice to find out how many of the videos uploaded to YouTube each month actually receive a significant number of views. The next day, one of the Delicious users whose bookmarks she subscribes to, bookmarked an article published in *Slate Magazine,* which answered that exact question. Then, after finishing the section of the paper that dealt with these statistics, she checked her feed reader only to stumble upon a paper from HP's Social Computing lab on the success dynamics of 10 million YouTube videos-a perfect fit for her research and another example of a passive social search. Of course, this doesn't replace traditional forms of library research, but it has the capacity of greatly supplementing them.

It is important to note in this regard that the quality of those searches – both active and passive – depends to a large extent on the quality of the PLN. While we recommend screening for shared interests when deciding who to include in your PLN, that does not mean their backgrounds and values need to match yours. In fact, to avoid casting your network too narrowly and possibly missing out on valuable information sources, we do suggest including members with diverse backgrounds and experiences.

CONCLUSION

Most academics would agree that professional development in pretty much any area of education is a lifelong process (Barab et al., 2006; Futrell, 1994). This awareness among educators for the need to become adaptive experts who "change their core competencies and continually expand the breadth and depth of their expertise" (Bransford, Derry, Berliner, & Hammerness, 2005, p. 49) does not always translate into practice. As Hammerness, Darling-Hammond, and Branford (2005) have pointed out, in order to successfully prepare effective teachers, "the concept of lifelong learning must become something more than a cliché" (p. 359). In this chapter, we have discussed PLNs as an informal, 24/7 learning tool capable of moving professional development into the 21st century and away from that cliché.

Academics, as we have seen, often lack the time and/or incentive to participate in formal professional development activities, yet, simultaneously need to keep up with changes in their respective fields to stay relevant in both their research and teaching and prepare students for career paths that do not yet exist (Ketelhut, McCloskey, Dede, Breit, & Whitehouse, 2006).

These demands position the type of PLN we have described here as the perfect solution for an academic's professional development needs by casting the professor "as a member of a professional community and as a lifelong learner, focusing upon collegial, career-long development" (Hammerness et al., 2005, p. 383).

REFERENCES

Baldwin, M. (2009, March 6). #FollowFriday: The anatomy of a Twitter trend. *Mashable.* Available at http://mashable.com/2009/03/06/twitter-followfriday. Retrieved on May 8, 2010.

Beebe, S. A., Beebe, S. I., & Ivy, D. K. (2007). *Communication principles for a lifetime* (3rd ed.). Boston: Allyn & Bacon.

Barab, S. A., Jackson, C., & Piekarsky, E. (2006). Embedded professional development: Learning through enacting innovation. In: C. Dede (Ed.), *Online professional development for teachers: Emerging models and methods* (pp. 155–174). Cambridge, MA: Harvard Education Press.

Bransford, J., Derry, S., Berliner, D., & Hammerness, K. (2005). Theories of learning and their roles in teaching. In: L. Darling-Hammond & J. Bransford (Eds), *Preparing teachers for a changing world: What teachers should learn and be able to do* (pp. 40–87). San Francisco, CA: Jossey-Bass.

Cha, M., Haddadi, H., Benevenuto, F., & Gummadi, K. P. (2010). Measuring user influence in Twitter: The million follower fallacy. Paper presented at the 2010 International AAAI Conference on Weblogs and Social Media (ICWSM). Available at http://an.kaist.ac.kr/~mycha/docs/icwsm2010_cha.pdf. Retrieved on May 12, 2010.

Chacksfield, M. (2010, April 15). Twitter boasts of 105 million registered users adding 300,000 users a day. *TechRadar UK.* Available at http://www.techradar.com/news/internet/twitter-boasts-of-105-million-registered-users-683663?src=rss&attr=all. Retrieved on May 10, 2010.

Cox, E. J. (2010). Twitter for all: Expanding professional dialogue with social media. *Library Media Connection, 28*(5), 52–53.

Downes, S. (2009, October 5). *Origins of the term 'personal learning network.'* Available at http://halfanhour.blogspot.com/2009/10/origins-of-term-personal-learning.html. Retrieved on May 10, 2010.

DuFour, R., DuFour, R., & Eaker, R. (2008). *Revisiting professional learning communities at work: New insights for improving schools.* Bloomington, IN: Solution Tree.

Duxbury, L., Dyke, L., & Lam, N. (2000, April 8). *Career development in the federal public service-building a world-class workforce. Treasury Board of Canada Secretariat.* Available at http://www.tbs-sct.gc.ca/pubs_pol/partners/workreport-PR-eng.asp?printable=True

Elman, J. (2010, January 21). The power of suggestions. *Twitter Blog.* Available at http://blog.twitter.com/2010/01/power-of-suggestions.html. Retrieved on May 12, 2010.

Fichter, D. (2004). Tools for finding things again. *Online, 28*(5), 52–56.

Futrell, M. H. (1994). Empowering teachers as learners and leaders. In: D. R. Walling (Ed.), *Teachers as leaders: Perspectives on the professional development of teachers* (pp. 119–135). Bloomington, IN: Phi Delta Kappa Educational Foundation.

Hammerness, K., Darling-Hammond, L., & Branford, J. (2005). How teachers learn and develop. In: L. Darling-Hammond & J. Bransford (Eds), *Preparing teachers for a changing world: What teachers should learn and be able to do* (pp. 358–389). San Francisco, CA: Jossey-Bass.

Ketelhut, D. J., McCloskey, E. M., Dede, C., Breit, L. A., & Whitehouse, P. L. (2006). Core tensions in the evolution of online teacher professional development. In: C. Dede (Ed.), *Online professional development for teachers: Emerging models and methods* (pp. 237–263). Cambridge, MA: Harvard Education Press.

Lucas, C. J., & Murry, J. W. (2002). *New faculty: A practical guide for academic beginners.* New York: Palgrave.

McGiboney, M. (2009, March 18). *Twitter's tweet smell of success.* Available at http://blog.nielsen.com/nielsenwire/online_mobile/twitters-tweet-smell-of-success. Retrieved on July 18, 2009.

Miles, M. B. (1995). Introduction. In: T. Guskey & M. Huberman (Eds), *Professional development in education: New paradigms and practices* (p. vii). New York: Teachers College.

Moore, R. J. (2009, October 5). *Twitter data analysis: An investor's perspective.* Available at http://www.techcrunch.com/2009/10/05/twitter-data-analysis-an-investors-perspective. Retrieved on May 14, 2010.

Parry, M. (2010, May 4). Most professors use social media. *The Chronicle of Higher Education.* Available at http://chronicle.com/blogPost/Most-Professors-Use-Social/23716. Retrieved on May 8, 2010.

Pemberton, A. (2009, February 22). A load of Twitter: Feel the need to tell everyone everything you're doing all of the time? Then tweeting is for you. *The Sunday Times.* Available at http://women.timesonline.co.uk/tol/life_and_style/women/the_way_we_live/article 5747308.ece. Retrieved on May 12, 2010.

Pipes: Rewire the web. (n.d.). Available at http://pipes.yahoo.com/pipes. Retrieved on May 5, 2010.

Richardson, W. (2006). *Blogs, wikis, podcasts, and other powerful web tools for classrooms.* Thousand Oaks, CA: Corwin Press.

Siemens, G. (2005). Connectivism: A learning theory for the digital age. *International Journal of Instructional Technology and Distance Learning, 2*(1), 3–10. Available at http://www.itdl.org/Journal/Jan_05/Jan_05.pdf.

Siemens, G. (2008). Learning and knowing in networks: Changing roles for educators and designers. Paper presented at the University of Georgia IT Forum. Available at http://it.coe.uga.edu/itforum/Paper105/Siemens.pdf

Technorati. (2008). *State of the Blogosphere 2008.* Available at http://technorati.com/blogging/state-of-the-blogosphere. Retrieved on July 18, 2009.

Trinkle, C. (2009). Twitter as a professional learning community. *School Library Monthly, 26*(4), 22–23.

TweetML: Share groups of Twitter users. (n.d.). Available at http://tweepml.org. Retrieved on May 5, 2010.

Wallace, P. (2004). *The Internet in the workplace: How new technology is transforming work.* Cambridge, U.K.: Cambridge University Press.

Warlick, D. (2009). Grow your personal learning network: New technologies can keep you connected and help you manage information overload. *Learning and Leading with Technology, 36*(6), 12–16.

Warlick, D. (2010). *The art & technique of personal learning networks or "a gardener's approach to learning."* Available at http://davidwarlick.com/wiki/pmwiki.php/Main/TheArtAmpTechniqueOfCultivatingYourPersonalLearningNetwork

Webster, T. (2010). *Twitter usage in America: 2010. The Edison research/Arbitron Internet and multimedia study.* Available at http://www.edisonresearch.com/twitter_usage_2010.php. Retrieved on May 11, 2010.

Weisgerber, C. (2009, July 25). *Social search in academic research.* Available at http://socialmediaprclass.blogspot.com/2009/07/social-search-in-academic-research.html. Retrieved on May 10, 2010.

MICRO-BLOGGING AND THE HIGHER EDUCATION CLASSROOM: APPROACHES AND CONSIDERATIONS

Gavan P. L. Watson

ABSTRACT

This chapter offers reflections on the successes and failures of integrating the micro-blogging platform Twitter into a first-year university class. Twitter, intended as a way to answer the question "What are you doing?" is now used in originally unexpected ways. Broadly speaking, Twitter's popularity can be traced to three factors: conversation between users; a decentralized ecosystem of third-party applications; and as a result, the distributed nature of the users. Adopted by educators in higher education, Twitter has been used as: an object for study, a tool to communicating classroom announcements, as a way to enable student to reflect on their learning, a chance to get instant feedback from students, and as the specific tool used to facilitate in-class conversations. The ongoing use of micro-blogging also appears to have an ability to change the social dynamics of a classroom, expanding the social of the classroom beyond the physical. While identifying Twitter's limitations, the chapter outlines the most significant outcome from the author's integration of Twitter: an

Teaching Arts and Science with the New Social Media
Cutting-edge Technologies in Higher Education, Volume 3, 365–383
Copyright © 2011 by Emerald Group Publishing Limited
ISSN: 2044-9968/doi:10.1108/S2044-9968(2011)0000003021

evolution of blended learning, proposed as a plesiochronous learning model, where learning occurs outside the classroom, with learner and instructor in different places but occurring at (virtually) the same time.

CHAPTER HIGHLIGHTS

- Micro-blogging, where users can distribute short messages using various platforms, including instant messaging software, SMS messages on mobile phones, websites, browser plug-ins or stand-alone programs (Java, Song, Finin, & Tseng, 2007), has emerged in 2006 as a communication form.
- Twitter, both a website and a platform, is the most popular micro-blogging service with over 180 million unique visitors a month.
- Twitter's uses have emerged in unintended ways.
- Twitter's popularity can be traced to three factors: the organic emergence of conventions that facilitate conversation; the decentralized ecosystem of third party applications allowing for the proliferation of uses and the updating of tweets away from a desktop computer; and as a result, the distributed nature of the users.
- Twitter, as a tool to use in the higher education classroom, gained prominence in 2008.
- Twitter has been used in higher education classrooms as an object for study, a tool to communicate classroom announcements, as a way to enable student to reflect on their learning, a chance to get instant feedback from students and as the specific tool used to facilitate in-class conversations.
- The ongoing use of micro-blogging also appears to have an ability to change the social dynamics of a classroom, expanding the social of the classroom beyond the physical.
- I then outline how I integrated Twitter into a first-year class with an enrollment of approximately 100 students and reflect on the successes and failures.
- I chose Twitter for its low technological threshold: students could benefit from the technology without having to buy extra hardware.
- Twitter was initially framed as a way for students uncomfortable with asking a question in a large classroom.
- Unintended uses of Twitter appeared, including: communicating directly with students for instant-feedback; using Twitter as a kind of "super-clicker"; a way to provide background or further information on course topics outside of class time (forming the beginning of a personal learning

environment); a mobile form to communicate while on field trips; and most significantly, an example of the evolution of blended learning that I propose is an example of a plesiochronous (versus asynchronous learning) learning model.

- Integrating Twitter into the course did mean taking risks as far as losing control in the classroom, but only minor issues emerged and were dealt with using Twitter.
- There are larger considerations for using Twitter in the classroom. As it is a company in addition to being a social media platform, increased efforts to monetize the stream of tweets may negatively impact its use in the higher education classroom. This becomes a larger philosophical question of the continued corporatization of the higher education classroom.
- This may force the expansion of open-source micro-blogging platforms that can be used by higher education institutions to support student learning, in the model of Moodle.
- The success of integrating micro-blogging in the classroom may have less to do with what it is used for (as its flexibility is well-documented) and more to do with being open to the organic uses that emerge.

THE LARGER CONTEXT OF MICRO-BLOGGING

Given the prefix, it is easy to see how micro-blogging might be described as a truncated effort to blog. Blogs, the short-form term for weblogs, are now a familiar feature of our experience on the Internet and have been claimed by educators as a pedagogical tool (Ferdig & Trammell, 2004). These websites, which now range in content, scope, and traffic, emerged as a coherent genre of publication, if you will, in the late 1990s and early 2000s. While varying in content, blogs can best be identified by their style of content, often primarily textual, organized by dated entries in reverse chronological order.

Both blogs and microblogs fall within a broad category of websites that are often called Web 2.0, a tongue-in-cheek reference to the convention of sequentially numbering new versions of software. There was never a new "version" of the Internet, rather, these sites are defined as "a group of internet-based tools and technologies with a strong social component" (Kompen, Edirisingha, & Monguet, 2009, p. 33) that "operate on networks or the Web versus software that users install and use on an individual hard drive" (Bianco, 2009, p. 303). They are, in other words, a suite of networked applications that encourage social interaction, significantly different in nature from the websites that had existed to date.

Micro-blogging is a form of communication where users can distribute short messages using various platforms, including instant messaging software, SMS messages on mobile phones, websites, browser plug-ins, or standalone programs (Java et al., 2007). What initially differentiated micro-blogging from blogging proper is the forced brevity: rather than entries of hundreds of words, micro-blogging limits an entry total to hundreds of characters. While this limits the depth of content, the shorter entries act to lower the barrier "of time and thought investment for content generation" (Java et al., 2007, p. 2). As a consequence, while someone who blogs may add an entry over the scale of days, this scale is compressed for microbloggers, and content is often added many times in a day. Arguably, this is the initial appeal for those who start micro-blogging: rather than just being some abridged form of sharing, the lower barrier to entry opens up possibilities to share a different kind of content over a different scale of time than would appear on a blog.

While Twitter may be the best-known micro-blogging platform, it is not the only one. Tumblr, Plurk, Squeelr, Beeing, Jaiku (add a .com to these services' names to visit their website), Emote.in and Identi.ca all offer different interpretations and implementations of micro-blogging. Yet common between them is the low barrier to share short bursts of information.

In this chapter, however, I focus on Twitter as it has emerged as the micro-blogging tool that has captured the imagination of the higher-education crowd for use in classrooms (Young, 2008, 2009) and the platform that I used as a course director. While Twitter's popularity has increased since its introduction in 2006, and its use as a tool in the higher education classroom has been reported, its adoption by course directors is still in its naissance. A 2009 survey of 847 lecturers at the University of Portsmouth, UK, for example, showed that Twitter had been adopted for use by only 7% of the faculty population (King, Duke-Williams, & Mottershead, 2009). It is worthwhile, for those interested in investigating micro-blogging and considering how to integrate it into the classroom, to outline the factors that led to Twitter's popularity, ensuing power and subsequent suitability as a higher education technological teaching tool.

Twitter, a Website and a Platform for Micro-Blogging

The most popular of the reported micro-blogging platforms, in early 2010, Twitter reported that it has 105 million registered users with traffic reaching a 180 million unique visitors a month (Kincaid, 2010b). While this popularity means that Twitter has reached a certain level of infamy for navel-gazing, as a corpus Twitter's collection of microblog entries, each limited to 140 characters and called tweets, has been deemed important enough to be preserved by the Library of Congress (Goddard, 2010). Twitter's original stated purpose was to answer the question "What are you doing?" in 140 characters or less. Rather than just using it as a service to answer that specific question, a 2007 study showed that user interactions fell into four broad categories: daily chatter, conversations, sharing information, and reporting news (Java et al., 2007). Reflecting users' various uses of the platform, in 2010 Twitter changed this prompting question to "What's happening?" This question, however, does not even fully capture just what Twitter is about. Users can choose to follow other users and, in turn, can be followed by others. Rather than just broadcasting information, a majority use Twitter to interact with others and despite the apparent messiness of the tweets "short, dyadic exchanges occur relatively often, along with some longer conversations with multiple participants that are surprisingly coherent" (Courtenay & Herring, 2009, p. 1).

This interest in conversation, then, coupled with the platform's flexibility – or lack of rules combined with users' creativity – has allowed various user-generated conventions to emerge which helps facilitate these exchanges. This vernacular includes the ampersand symbol to address a reply to another user (e.g., a reply to me, as my username is gavatron, would look like this: @gavatron); adding the prefix "RT" to another user's tweet to attribute the source (e.g., re-tweeting my tweet that "I love Twitter" would look like "RT @gavatron I love Twitter"); and the use of hash-tags, the addition of # to any word, to create a *de facto* index (e.g., #AERA2010 is added to a tweet to denote the content as being related to the American Educational Research Association 2010 conference). While first appearing organically with loose rules for use, some of these innovations have been so widespread in their adoption that they have been officially incorporated into the platform. For example, re-tweeting is now an act that can be accomplished on the Twitter website and attribution recognized without the explicit need for an "RT."

Beyond unintended conventions, unintended uses have emerged as well. Twitter opened its application programming interface (API) to third-party developers. Similar to how learning a new language opening new opportunities to communicate, access to APIs means that the data that drives Twitter is accessible to applications not written by the company behind Twitter. With third-party access to this information, Twitter no longer is *just* a website. Rather, it is a wide suite of tools that allow for the retrieval and updating of the 140-character messages. Again, in early 2010,

Twitter reported receiving 3 billion requests daily through APIs and stated that 75% of their traffic comes from these third-party applications (Kincaid, 2010b). This proliferation of third-party applications allows users to leverage the technology in unanticipated ways. For example, objects now tweet their status – asteroids that are near-misses with the Earth (@LowFlyingRocks), a house plant needing watering (@pothos) and even London's Tower Bridge (@TowerBridge) announce relevant updates. In practice, this ecosystem of applications built to send and receive tweets means that users are not just limited to a web browser to update or read entries. Tweets can be composed and received, for example, in standalone applications on a computer, as text messages in cell phones and as instant messages. Without this explicit connection between the user and desktop computer to tweet, coupled to the platform's popularity, some of the more powerful uses of Twitter emerge.

Twitter's Power of Distributed Users

This power appears to come from the distributed nature of Twitter users, most often contributing their own 140 character views on an event. In this sense, Twitter has become known as tool to transmit real-time information where users' tweets are recognized as: a form of on-the-spot citizen journalism (Bianco, 2009); a tool for political organization ("EDITORIAL: Iran's Twitter revolution," 2009); or even as a means to detect earthquakes (Lubeck, 2009). Three factors have combined to allow this on-the-spot updating to occur: the portability and penetration of mobile phones, coupled with the ease with which tweets can be sent as text messages using this platform. So important was the perceived role of Twitter as a tool for organizing during the 2009 Iranian presidential election, that the company put off a critical network upgrade to a time of day that was expected to cause the least amount of interruption (Stone, 2009) in Iran. Twitter, in these examples, has become more than the simple act of sharing what you had for lunch. While that kind of triviality still exists, when tweets are aggregated, its networked users can act as a powerful force that flattens and redefines relationships of power.

These uses are not without fault. While the speed with which information can flow using Twitter is not in doubt, the effectiveness of relying on tweets as a source of correct information might be. Given its nebulous organization of users, the information spread throughout the network is not infallible and can lead to mis-information being amplified, such as the alleged death of Canadian singer Gordon Lightfoot (Saxberg, 2010), reported and re-tweeted in early 2010 (the incorrect information spread so widely, so quickly that Lightfoot's management was required to issue a press release to the contrary).

Yet, for a platform limited to 140 character updates, Twitter has proven to be lightweight and flexible enough to allow for these originally-unintended uses to emerge. As the larger Web 2.0 movement has been described as offering "'ground-up' capacities for social action, productivity, collaboration, and interaction" (Bianco, 2009, p. 304), micro-blogging, as seen in the case of Twitter, broadly offers these same possibilities. Twitter works in this way because of the characteristics identified: the organic emergence of conventions that facilitate conversation; the decentralized ecosystem of third party applications allowing for the proliferation of uses and the updating of tweets away from a desktop computer; and as a result, the distributed nature of the users. As I now turn to the use of Twitter in the higher education classroom, the significance of these factors play an important role in the adoption of Twitter as a technology to support and enhance student learning.

MY EXPERIENCE IN INTEGRATING TWITTER IN THE HIGHER EDUCATION CLASSROOM

Inspired by Rankin's (2009) use of Twitter in higher education I decided to try to integrate Twitter while planning for the Fall 2009 iteration of a course that I had developed and instructed while a PhD student in the Faculty of Environmental Studies at York University, in Toronto, Canada. First, a little about the course: The Natural History of the Greater Toronto Area is a first year course open to any major. As a result, the enrollment is about 100 students from various academic disciplines and backgrounds. The teaching team consisted of me, the course director, and two other graduate students acting as teaching assistants. The content focuses on exploring the natural history of the Toronto and is designed, in part, as an exercise in experiential education: as a consequence half the classes are field trips. The balance of classes are lectures held in a typical raked, fixed-seating lecture hall.

The Plan

began to plan, I spent time reflecting how I could best use Twitter in class. I saw 140 characters as a limitation but also as liberating: since was integrating the use of micro-blogging, I felt that the character limit constrained the scope of the technology's inclusion and as a consequence, I did not worry about biting off more than I could chew. I first turned to other course directors' experiences of using Twitter. Searching on-line, I found that micro-blogging generally, and Twitter specifically, has been used as a tool to support learning in courses across a range of disciplines, including: communication (Parry, 2007; Sweetser, 2008), public relations (Russell, 2008), English (Young, 2008), aerospace engineering (Zax, 2009), education (Wheeler, 2009), nursing (Skiba, 2008), language learning (Ullrich et al., 2008), consumer science (Young, 2009) anthropology (Wesch, 2008), and history (Rankin, 2009).

In these higher education classrooms, Twitter was used in various ways. Parry (2007), Sweetser (2008), and Russell (2008) used Twitter to facilitate an assignment: students were expected to sign up for an account, use it for a set period of time and then reflected on their use. Eckford (2010) set up a class-specific Twitter account and used it to make course-specific announcements, with the benefit of hearing back from students. Twitter has also been used as a way to facilitate reflection on learning, where students were expected to tweet after class what they learned or found difficult in a particular lecture (Young, 2008). Other course directors have used it as a way to facilitate class discussions (Rankin, 2009) or complement in-class experiential learning (Wesch, 2008). Micro-blogging has also been used in on-line learning situations to create the opportunity for free-flowing, just-in-time interactions (Dunlap & Lowenthal, 2009b), often seen as lacking in this setting; short and informal, these interactions occur in-between the log-ins common in learning management systems (Dunlap & Lowenthal, 2009a).

Socially, micro-blogging allows for a kind of out-of-classroom interaction that was previously unavailable in learning management systems, strengthening interpersonal relationships between faculty and students (Dunlap & Lowenthal, 2009a). Echoing this finding for off-line courses, Parry (2008) found that its use can change the classroom dynamics through what's been called the "sixth-sense" (Thompson, 2007) of Twitter, where individual updates have a cumulative effect and create a peripheral awareness; this, in turn, creates a different kind of classroom community and helps facilitate better (more respectful, more participation) classroom discussions.

To date, then, Twitter has been used in higher education classrooms as an object for study, a tool to communicating classroom announcements, as a way to enable student to reflect on their learning, a chance to get instant feedback from students and as the specific tool used to facilitate in-class conversations. The ongoing use of micro-blogging also appears to have an ability to change the social dynamics of the classroom.

peripheral awareness that comes from following classmates' updates changes the existing impression, in the case of online learning, that a social space does not exist or, in the case of the classroom, that the social space begins and ends within the physical classroom.

Reading others' experiences helped narrow down what I did and did not want to do. I decided that, this first time through, I would use Twitter as a compliment to regular classroom interactions rather than replacing them. So, I finally settled on initially introducing it as a kind of awareness tool. The example that I was going to give students was that they could use it as another way to ask me questions while lecturing. Since I was lecturing in a hall with wireless networking and two projectors, I was going to project the Twitter website on one screen and my lecture slides on the other. I imagined that student could post questions, most likely from their laptops, to lecture content and I would get a chance to clarify concepts on the spot.

But key in my plan was that I was open to see how it evolved from my initial expectations – and I told the students so. I helped frame it as a trial by calling it "The Great Twitter Experiment of 2009." So, with that intention in mind, I went about including information on Twitter in the syllabus and planned to introduce it in the first class by describing what Twitter was, what I wanted them to do with it, and some of the conventions of conversation that I have described earlier.

I do think it is fair to analyze the objects required by students to use technology in the classroom. While an ever-increasing percentage of students bring laptops into the classroom, I knew that not all students would have a laptop. Asking, for example, students to use a technology in the lecture hall that *requires* a laptop is still a significant barrier, as not all students at York have mobile computers. Other instructors who have used Twitter in the lecture hall have had students work in groups of three to five (Rankin, 2009) generating and answering discussion questions. In this situation, the requirement for one laptop per student is obviously diminished. As I intended it as a tool for a broader kind of discussion, Twitter, as a micro-blogging platform, was appealing because students could send and receive tweets on their mobile phones – an object that a majority already had in their pockets. Even so, sending individual text messages can cost a few cents, so for those students who do not have text messages packaged with their mobile service, *requiring* micro-blogging in class – such as the case with a class discussion – can cost additional money. An issue that others have identified (Young, 2008) when discussion through Twitter is integral, this is a barrier that Rankin (2009) solved by allowing students without a laptop or phone to write out their question or comment and having a TA with a laptop enter for them. Finally, where laptops and cell phones are sometimes seen as distracting objects to be excluded from the classroom, I wanted to necessitate, in a sense, their use for legitimate classroom activities. I believed that through their inclusion in classroom activities, I could change the adversarial relationship that a course director sometimes has with extra-curricular texting or browsing.

The First Day

I opened a new Twitter account specifically for the course and during the first lecture, asked every student to sign up for a Twitter account themselves and follow my course-specific account. I also asked them to add their username to a specific wiki page I had setup on the course's Moodle site. Students' privacy was an issue that I addressed and I told students that they could, if they were concerned, protect their updates (tweets are free for all to view if an account is not protected). In other examples, such as Parry (2008), instructors have required students to follow themselves and each other. Parry had less students – 30 compared to my 100 – and so following all students was probably seen as a reasonable expectation. I, however, just encouraged them to follow the other students in the course. I let them know that they could not only tweet from their computer, but they could tweet from and get tweets from me sent to their cell phones through SMS. I tweeted myself. I asked them to send me a tweet as homework. And I left it at that.

Blank Faces

Interestingly, as I introduced Twitter that first day and how I hoped to use it, there were a surprising number of blank faces looking back at me and some vocal criticism of the plan.

"We can't ask questions in the classroom anymore?" one student asked.

"No," I assured them, "This is for people who might be uncomfortable with asking questions in a large classroom. If you want to put your hand up, I have no problem with that."

"Will we be marked on this?" asked another.

I had not really thought about this. "Part of the evaluation is participation. If you participate through the use of Twitter, I will take that into consideration when calculating your mark."

"How can you know who we are when we could use any username?" a quick-thinking student replied.

Taking a moment I realized this is where the wiki came into play. "You'll add your real name and username to the wiki, so I'll keep track of you that way."

I was particularly interested in the lack of initial enthusiasm on the part of students. Not surprisingly, perhaps, students were interested in the extrinsic benefits – Does this count for marks? – for them making the effort to sign up for a Twitter account. On the first day of class, only a handful of students had a Twitter account. This echoes Kompen et al. (2009) who found a majority of students in their study were not familiar with and did not use the majority of Web 2.0 services, like Twitter. Those who had an account and were active users were quick to come to my assistance in class and on-line, suggesting that students at least give it a try. Lots has been made of the generational difference between those born, according to Palfrey & Gasser (2008), before and after 1980. So-called digital natives, as a mythical group are supposed to hold the skills to successfully use the digital networked technologies that a tool like Twitter represents. I, frankly, expected more enthusiasm from the class at the idea of using micro-blogging as a different way to engage with me and the course content. The possibility that Twitter could be used as a tool for learning did not initially appear to students and they seemed to need enticing to give the "experiment" a try. In the future, I would take some more time to introduce Twitter in the classroom. Specifically, since I introduced the whole experiment during the first day in the lecture hall, I would consider using some of this classroom time to book a computer lab. In that space, I would get students to sign up and get them using the platform to learn the vernacular of @replies and #hash-tags by doing it, right away.

At the end of the semester, the specific account I set up had 72 followers. This means that, in hard numbers, with enrollment dropping to 92, I had about a 78% sign-up rate. Consequently, because I did not feel like there was a barn-burning adoption of Twitter and I never explicitly mandated it, when I eventually calculated participation I did not use it as an evaluation criterion. More significantly, though, was my first-hand experience that echoes some of the critique (Bennett, Matonand, & Kervin, 2008) of this idea of a monolithic generation of digital natives.

The Classroom Setup

As I mentioned before, I was fortunate to have a two-projector setup in the lecture hall. This enabled me to project the Twitter website in a browser window on one screen and my lecture slides on the other. I asked students to add the #envs hash-tag to any course-related post so we could all follow the conversation. I would, then, specifically project only the tweets with an #envs hash-tag in class. I also asked the TAs to follow the #envs hash-tag and draw my attention to relevant comments or questions when necessary.

I planned, if uninterrupted by the TAs, to take a break while lecturing and see what, if anything with the #envs hash-tag had popped up while I was lecturing. I also own a smart phone with a Twitter client installed that notifies me when someone @messages me. I did not introduce this explicitly in the first class, but it emerged as being important later in the course.

What Emerged: The Classroom

As initially planned students used Twitter to ask questions. They, not surprisingly, also began to see it as a way to communicate directly with me (e.g. "hi! what's the link for the course website?"). As such they sent me messages seeking clarification and giving feedback. With regards to class-related questions, I found it faster than email, with the added benefit of all followers hearing the answer to the same question. Rather than waiting a few weeks or even to the end of the course to see how students were doing, I especially liked the instant-feedback I received. This meant that I changed my approach to reflecting on my teaching practice: rather than waiting to see feedback at the end of the course and deploy corrections for the next iteration, I took the feedback and felt like I could be more nimble in deploying quick course corrections as the course was ongoing.

I also found I began to use Twitter as a kind of "super"-clicker. Clickers have been offered, at York, as a technological support to course directors. Purchased by students, they are used to test comprehension and seek feedback by course directors. Students select, through the clicker, an appropriate answer to a question and up to four answers projected on a slide. The course director, then, could evaluate how the class answered a question, and if follow-up was necessary. Clickers have been promoted as a way of increasing engagement and interactivity in large classrooms (Jones, Henderson, & Sealover, 2009).

In my mind, however, they are limited in two important ways: students are forced to choose one of four pre-determined answers and new clickers cost approximately $50. Using Twitter as a way to seek feedback from students, I could ask questions that required more than selecting one of four multiple choice answers. Now, students often chose to put their hands up to answer these questions, but students answered questions I posed to the class through Twitter. And significantly, I was not as concerned about polling the entire class for an answer. The concern about cost might appear to be misplaced given that using Twitter in the classroom could costs as much as the price of a laptop or the price of a mobile phone, plus monthly service. In my experience, students already are using a laptop or have a mobile phone; they would have to send 500 text messages to incur a $50 charge. Thus, it appears as though in practice that the cumulative cost of text messages in class come nowhere close to the cost of a clicker.

Unexpectedly, Twitter was also a great platform to provide further information based on classroom experiences (e.g., "While not the same species of cicadas we heard today, here's a neat video on 17-year cicadas: http://bit.ly/1fZxvY"). Outside the classroom, as I found newspaper articles or blog postings related to course content, I would tweet about them. This use, in a sense, is an expansion of Twitter's use outside the classroom: sharing news and information. Personally, I use Twitter to follow the individuals and organizations with interests I am curious about. As a result, I have found that the other users I follow bring relevant content to my attention that I would not have come across otherwise. In this sense, my personal use of Twitter supports my informal learning in these interests.

Twitter, through users' ability to follow and be followed by others, has been identified as creating a kind of personal learning environment or PLE. PLEs are "a user-built, personalized set of tools – not necessarily digital ones – that are used to manage content and interactions, and support the learning experience" (Kompen, et al., 2009, p. 33). With my students following me, and tweeting related content out of class, I was facilitating the start of a PLE related to the course content. Obviously, this is an area that I could expand in the future, for example, by providing a list of naturalists to follow on Twitter.

The unanticipated emerged as well. King et al. (2009) reported that resistance, on the part of course directors, to use Twitter and other Web 2.0 platforms comes from the obscured place these platforms play in their personal or professional life and the fear of losing control in the classroom. Projecting students' tweets does mean putting some of the control of the classroom in their hands. This did not come without consequences. I was lecturing about the difference between plants and trees and used the term "woody structures." What would have likely been a snicker between friends became this tweet, broadcast to the classroom on the projector: "Size does matter when talking about woody structures." Mildly amusing, the students reacted to it as you might expect and snickering spread across the lecture hall. Rather than verbally chastising the class for the inappropriate comment, I decided to use the technology to provide the correction. During a break, I @replied to the student: "About woody structures ... humorous, yes, but inappropriate." So in reaction to the projected tweets, there were some examples of the "look at me, look what I can do" mentality from students, similar to the aping seen when people know they are on live television. Nothing too subversive emerged, however, and any loss of control in the classroom due to Twitter, I addressed using Twitter itself.

What Emerged: Field Trips

Again, unexpectedly, having a smart phone with a Twitter client that announced when I received @replies became a significant tool for assisting with the logistics of the field trips. Students began tweeting me about field trips to seek clarification about what time or where we were to meet (e.g., "which woodlot are we meeting at? Having a hard time finding it"). With the phone's ability to geo-locate itself, I was even able to provide the class' exact location for a late-comer. In one sense, this was good as students who were truly lost could get a quick update from me.

The downside to this is an extension of a larger criticism that I hear from other course directors: students do not read instructions. These tweets were asking the kind of questions that were already provided on the course website. So is it good to have a way for students to get in touch? Regardless of whether or not they read the instructions they were lost and having them participate was important enough that helping them get to the right place was something I was willing to do. At the same time, I am sure there are course directors out there who would find this especially infuriating. It was disruptive, at times, to have a phone buzzing in my pocket. An easy solution would have been to give a TA the phone and let them answer any incoming messages.

What Emerged: Outside the Classroom

Significant for my practice of teaching are the "outside the classroom" experiences that appeared through students' use of Twitter. Late in October, for example, during a time the class was not meeting, I received this tweet from a student: "Are the pine trees by the Ross bldg Tamarack, because the needles are changing colour and are falling off?" The short answer was yes, they were. More significant was this spontaneous identification of the tree as it was the illustration of this student achieving two of the course's larger learning objectives: (1) differentiating between and identify common plants and animals found in Southern Ontario and (2) fostering personal habits of observation. As far as learning goes, to have a student demonstrate the skills introduced in the classroom, outside the classroom was a dream come true. Twitter facilitated this conversation outside of class and I was able to support this learning by replying back to the student the successful identification. This example, facilitated by Twitter, appears to mark a continuing evolution of blended learning in undergraduate education, where asynchronous learning occurs at the same time but student and instructor are in different places.

Now facilitated through the use of networked computers, the separation of teacher and learner was identified by Keegan (1996) as one of five key characteristics to distance education. While place is implicated in these asynchronous learning networks (hence the "distance" in "distance education") learning across time is a hallmark of distance education (asynchronous meaning, literally from the Greek, "not with time") and the separation that has emerged as described by Keegan. While there are examples in asynchronous learning networks, such as real-time chats, where all users are coordinated in time, place is often just a consequence of users' asynchronicity. Not so with this example. It is a model of a plesiochronous learning model, where the identification and confirmation occurred, effectively, in real-time, separated by place. It was the combination of Twitter's low barrier to communicate, mobility and third-party ecosystem, characteristics described as key in the service's wide adoption rate, which combined to facilitate this learning moment.

LARGER CONSIDERATIONS FOR TWITTER'S USE

Twitter specifically, and micro-blogging more generally, have been shown to offer significant opportunities for communication and learning in the higher education classroom. These opportunities are not without some larger considerations. I have outlined some of these considerations, such as the technological barriers for students, earlier in the chapter. While Twitter appears to be the flavor of the moment and central to the majority of first-hand experiences outlined here, including my own, it is important to remember that it is a company *as well as a* web service.

As I summarized earlier, the sense of community that the web service appears to enable has been celebrated for, among other things, its perceived ability to decentralize power; a body of tweets important enough to be saved in the Library of Congress. While your tweets never disappear, they do have an expiry date, of sorts. Owing to the expansion in popularity of the service, tweets are currently searchable for approximately 1.5 weeks (McBride, 2010). This raises questions about the service's data portability and, in reaction to this expiry date, third-party services do exist that allow users to archive their tweets (Perez, 2009). Course directors will have to weigh this limitation when considering Twitter as their micro-blogging platform of choice.

Additionally, in early 2010, Twitter, the company, continues to look for ways to monetize its stream of 140 character messages (Jutras, 2010). For Twitter, this monetization is proposed in the form of contextual paid-for tweets, called "promoted tweets," in search results and on the Twitter website (Kincaid, 2010a). As the physical campus of a university continues to show evidence of its ever-increasing sponsorship and research agendas show a flavor of corporatization (Sidhu, 2009), educators will have to decide if they are comfortable with this creep into their teaching practice.

True, there are micro-blogging alternatives. Identi.ca, for example, is similar to Twitter but built on an open-source (echoing the "free as in speech, not free as in beer" ethos of the free software movement) platform available under a GNU license. Yet, it lacks the larger ecosystem of third-party applications. As such, many of the uses outlined in this chapter become more difficult, or impossible to implement using the service as it exists today. Given, however, its open-source nature, Universities (or any other interested party) interested using micro-blogging to support student learning could install the code and create a version on their own networks with an expanded list of features to match Twitter's. This, while perhaps appearing idealistic, is not without precedent: Moodle, a learning management system with over 50,000 registered sites as of April, 2010, ("Moodle Statistics," 2010) is a GNU licensed open-source learning management system, built in reaction to a closed-source proprietary learning management system ("*Background*," 2009).

As micro-blogging in higher education expands and becomes more widely adopted and if Twitter continues to be the micro-blogging platform of choice, critical thought will need to be given to the status and limitations of the service as a private company. In my experience these limitations are not significant enough *not* to use Twitter as a micro-blogging platform. Yet, as the service expands and changes, the act of micro-blogging in higher education may be better served by a larger institutional investment of time an effort in an open-source alternative.

MY EXPERIENCE, IN SUMMARY

In retrospect, I discovered that there was not one right way to use Twitter in the classroom. I like this. For me the best uses of Twitter emerged over the course of the semester *through its use*. This speaks to course directors being flexible and open to the unexpected if choosing to integrate micro-blogging in the classroom. This echoes associate professor of educational psychology and technology at Michigan State University, Punya Mishra's claims that there are no educational technologies, per se, rather there are technologies that exist that need to be repurposed for each instructor's needs (Zax, 2009). In my case, the initial idea of using micro-blogging as a safer way of asking questions in class morphed into a bunch of different things:

- a vehicle for immediate feedback, from student to instructor and instructor to students
- a chance for (mild) student subversiveness
- a plesiochronous learning model based on *almost* synchronized communication that allowed students to:
  ○ get last-minute (even geo-tagged) updates about the course
  ○ continue the learning process outside the classroom and still get feedback from the course director
- and a way for students uncomfortable with speaking in a class of 100 to have voice

So while I did not create the kind of community of 72 students such as those I see between some of the Twitter contacts that I have, I deem this experiment as anything but a failure. As I was writing these reflections many weeks after the course had officially ended, I quickly returned to my course Twitter account and sent out this message: "Hey #ENVS 1010c students! Writing about using Twitter last semester. Do you have any thoughts? Did you like it/hate it/not care?" In return, I got the following tweets: "i thought that it was useful at times, but that overall, not everyone would benefit from using it"; "i thought it was really interesting and added a small but powerful element in engaging students in the class" and "I for one loved it … i was tweeting with a purpose." While a certain ambiguity exists from each individual's experience, the fact that I was quickly and easily able to communicate with students after the official learning experience had ended, continues to show me that Twitter has its unexpected uses.

Micro-blogging then, as a tool for communication, learning and collaboration has a great deal to offer the higher educator: it can be many "things": a plesiochronous learning model; a kind of "super-clicker"; a way to facilitate discussion; a instantaneous student-feedback tool; or the focus of a student's investigation. Clearly as we, a collective body of educators, continue to experiment with the use of micro-blogging in the classroom, its flexibility in use and, in the specific case of Twitter, its larger ecosystem of third-party applications, unexpected uses will continue emerge. We will be challenged then to ensure that these uses, and the platforms that support them, fit within our philosophic approach to teaching and learning.

REFERENCES

Background. (2009, June 23). Available at http://docs.moodle.org/en/Background

Bennett, S., Matonand, K., & Kervin, L. (2008). The "digital natives" debate: A critical review of the evidence. *British Journal of Educational Technology, 39*(5), 775–786.

Bianco, J. S. (2009). Social networking and cloud computing: Precarious affordances for the "prosumer". *Women's Studies Quarterly, 37*(1&2), 303–312.

Courtenay, H., & Herring, S. C. (2009). Beyond microblogging: Conversation and collaboration via Twitter. Paper presented at the 42nd Hawaii International Conference on System Sciences, Waikoloa, Big Island, Hawaii.

Dunlap, J. C., & Lowenthal, P. R. (2009a). Instructional uses of Twitter. In: P. R. Lowenthal, D. Thomas, A. Thai & B. Yuhnke (Eds), *The CU online handbook* (pp. 45–50). Denver: University of Colorado Denver.

Dunlap, J. C., & Lowenthal, P. R. (2009b). Tweeting the night away: Using Twitter to enhance social presence. *Journal of Information Systems Education, 20*(2), 129–135.

Eckford, A. (2010, February 4). Social media and the modern day classroom. Available at http://andreweckford.blogspot.com/2010/02/social-media-and-modern-day-classroom.html

Ferdig, R., & Trammell, K. (2004). Content delivery in the 'blogosphere'. *THE Journal (Technological Horizons in Education), 31*(7), 12–16.

Goddard, J. (2010, April 17). Twitter archive to be stored by Library of Congress. *The Daily Telegraph.* Available at http://www.telegraph.co.uk/news/worldnews/northamerica/usa/7601281/Twitter-archive-to-be-stored-by-Library-of-Congress.html

Java, A., Song, X., Finin, T., & Tseng, B. (2007). Why we twitter: understanding microblogging usage and communities. Paper presented at the Proceedings of the 9th WebKDD and 1st SNA-KDD 2007 workshop on Web mining and social network analysis, San Jose, CA.

Jones, S., Henderson, D., & Sealover, P. (2009). "Clickers" in the classroom. *Teaching and Learning in Nursing, 4*(1), 2–5.

Jutras, L. (2010, April 18). How will the Twitterati deal with the ad men?, *The Globe and Mail.* Available at http://www.theglobeandmail.com/news/technology/personal-tech/lisan-jutras/how-will-the-twitterati-deal-with-the-ad-men/article1538798/

Keegan, D. (1996). *Foundations of distance education* (3rd ed.). London: Routledge.

Kincaid, J. (2010a). Twitter execs address the big question: Monetization. Available at http://techcrunch.com/2010/04/14/twitter-execs-address-the-big-question-monetization/

Kincaid, J. (2010b). Twitter has 105,779,710 registered users, adding 300K a day. Available at http://techcrunch.com/2010/04/14/twitter-has-105779710-registered-users-adding-300k-a-day/

King, T., Duke-Williams, E., & Mottershead, G. (2009). Learning and Knowledge Building with Web 2.0 Technologies: Implications for Teacher Education. Paper presented at the 2009 Knowledge Building Summer Institute, Palma de Mallorca, Spain.

Kompen, R., Edirisingha, P., & Monguet, J. (2009). Using Web 2.0 Applications as supporting tools for personal learning environments. In: M. D. Lytras, P. Ordonez de Pablos, E. Damiani, D. Avison, A. Naeve & D. G. Horner (Eds), *Best Practices for the Knowledge Society. Knowledge, Learning, Development and Technology for All* (pp. 33–40). Berlin: Springer Berlin Heidelberg.

Lubeck, M. (Producer). (2009, December 14). Shaking and Tweeting: The USGS Twitter Earthquake Detection Program. *USGS CoreCast.* [Podcast] Retrieved from http://www.usgs.gov/corecast/details.asp?ep = 113

McBride, M. (2010, March 21). Things every developer should know Retrieved April 21, 2010, from http://apiwiki.twitter.com/Things-Every-Developer-Should-Know#6Thereare-paginationlimits

Moodle Statistics. (2010, April 21). Available at http://moodle.org/stats/

Palfrey, J., & Gasser, U. (2008). *Born digital: Understanding the first generation of digital natives.* New York: Basic Books.

Parry, D. (2007). Twitter away your weekend. Available at http://outsidethetext.com/trace/38/

Parry, D. (2008). Twitter for academia. Available at http://academhack.outsidethetext.com/home/2008/twitter-for-academia/

Perez, S. (2009). 10 Ways to archive your tweets. Available at http://www.readwriteweb.com/archives/10_ways_to_archive_your_tweets.php

Rankin, M. (2009). Some general comments on the "Twitter Experiment". Available at http://www.utdallas.edu/~mar046000/usweb/twitterconclusions.htm

Russell, K. M. (2008). "48 Hours of Twitter" class assignment. Available at http://teachingpr.blogspot.com/2008/01/48-hours-of-twitter-class-assignment.html

Saxberg, L. (2010, April 10). Lightfoot's demise greatly exaggerated. *The Ottawa Citizen.* Available at http://www.ottawacitizen.com/entertainment/Lightfoot + demise + greatly + exaggerated/2785850/story.html

Sidhu, R. (2009). The 'brand name' research university goes global. *Higher Education, 57*(2), 125–140.

Skiba, D. (2008). Emerging technologies center: Nursing Education 2.0: Twitter & tweets. Can you post a nugget of knowledge in 140 characters or less? *Nursing Education Perspectives, 29*(2), 110–112.

Stone, B. (2009). Down time rescheduled. Available at http://blog.twitter.com/2009/06/down-time-rescheduled.html

Sweetser, K. D. (2008). Teaching tweets. Available at http://www.kayesweetser.com/archives/80

The Washington Times. (2009, June 16). EDITORIAL: Iran's Twitter revolution. *The Washington Times.* Available at http://www.washingtontimes.com/news/2009/jun/16/irans-twitter-revolution/

Thompson, C. (2007). Clive Thompson on how Twitter creates a social sixth sense. *Wired Magazine, 15*(7). Available at http://www.wired.com/techbiz/media/magazine/15-07/st_thompson

Ullrich, C., Borau, K., Luo, H., Tan, X., Shen, L., & Shen, R. (2008). Why web 2.0 is good for learning and for research: principles and prototypes. Paper presented at the Proceeding of the 17th International Conference on World Wide Web, Beijing, China.

Wesch, M. (2008). Teaching with Twitter. Available at http://mediatedcultures.net/ksudigg/?p = 170

Wheeler, S. (2009, January 2). Teaching with Twitter. Available at http://steve-wheeler.blogspot.com/2009/01/teaching-with-twitter.html

Young, J. R. (2008). Forget E-Mail: New messaging service has students and professors atwitter. *Chronicle of Higher Education, 54*, A15.

Young, J. R. (2009). Teaching with Twitter: Not for the faint of heart. *Chronicle of Higher Education, 56*, A1.

Zax, D. (2009). Learning in 140-character bites. *Prism* (8). Available at http://www.prism-magazine.org/oct09/tt_01.cfm

ABOUT THE AUTHORS

Robert Bodle received his Ph.D. in Critical Studies from the University of Southern California's School of Cinematic Arts. His research focuses on the social, political, and ethical implications of networked media (social media and networks, social reporting, alternative media, mobile and convergence culture, internet governance, information ethics, and new media literacies). As assistant professor of Communication Studies at the College of Mount St. Joseph, Bodle designs and teaches a digital media curriculum that includes Social Media and Social Change, New Media and Society, Human Rights in the Digital Age, New Media Ethics, and Visual Communication. His research appears in the *Journal of International Communication, Information, Communication & Society*, and the book collection *The Ethics of Emerging Media: Information, Social Norms, and New Media Technology*.

Boris H. J. M. Brummans is an associate professor in the Department of Communication at the University of Montreal. He earned his Ph.D. from Texas A&M University. His research looks at the organizational communication practices of members of humanitarian organizations, ranging from the Buddhist Compassion Relief Tzu Chi Foundation in the Republic of China (Taiwan) to Doctors without Borders in Africa. His work has been published in edited books and journals, such as *Communication Monographs, Human Relations, Journal of Applied Communication Research*, and *Qualitative Inquiry*.

Ryan Busch is an elearning innovator and online educator. In 2009, Ryan founded the digital marketing, social media, and e-learning consulting firm Higher Ed Gadfly. As the founder of Higher Ed Gadfly, Ryan developed marketing and e-learning innovations for various industries. Ryan has been using social media ideologies in the development of a new approach to workforce development for a large Boston-based health care network (a 50,000 employee organization) under a substantial grant from a large philanthropic group. In 2007, prior to establishing Higher Ed Gadfly, Ryan was the founding director and chief architect of StraighterLine (noted by Clayton Christensen, author of Disrupting Class: How Disruptive Innovation

Will Change the Way the World Learns, as a disruptive innovation). While at StraighterLine, Ryan oversaw successful course reviews and approvals by the Distance Education and Training Council (DETC) and earned nods from the Software Information Industry Association (SIIA) as both a 2009 CODiE finalist for Best Virtual School Solution for Students and a participant in the SIIA Innovation Incubator program. Before developing StraighterLine Ryan led the marketing development efforts for a number of educational technology innovations developed at the University of Phoenix as part of the research and development group Apollo Publishing & Learning Technologies and later as a part of the academic affairs department as Learning Content Projects Director. He is currently a member of the online faculty of the College of Humanities and College of Natural Sciences at University of Phoenix and an occasional adjunct for other universities as well. In November 2010, and after the writing of his chapter in this volume, Ryan was offered and accepted the presidency of eduFire.

Shannan H. Butler is assistant professor of Communication at St. Edward's University in Austin, Texas. Shannan conducts research in the area of visual communication, new media, pedagogy, and rhetorical criticism of visual media. He was invited to present his work on visualization at the 2010 *Horizon Report* general meeting. Together with his colleague, Dr. Corinne Weisgerber, he has just finished editing a special issue on communication education in the age of social media for the *Electronic Journal of Communication*.

Aimee deNoyelles is an adjunct professor in the liberal arts program at Ivy Technical Community College in Indiana. Her research interests include computer-mediated communication, immersive technology, and virtual social identity. Ms. deNoyelles received her M.A. in academic psychology from East Carolina University in North Carolina and is currently pursuing her doctoral degree in instructional design and technology at the University of Cincinnati.

Helen Farley is an interdisciplinary research academic with significant experience in diverse fields including architecture, veterinary science, journalism, philosophy, and studies in religion. She has received her B.V.Sc., M.A., M.Ed., and her Ph.D. In 2009, she joined the newly formed the Centre for Educational Innovation and Technology (CEIT) at the University of Queensland as a Research Fellow. She is currently a lecturer in virtual worlds at the Australian Digital Futures Institute at the University of Southern Queensland. She was CEIT's virtual worlds expert and has worked in many Multi-User Virtual Environments (MUVEs) including Second Life, Twinity,

OpenSim, and Project Wonderland. Her own project, UQ Religion Bazaar in Second Life, has been nominated for many awards and has attracted significant national and international media attention. She is currently researching the use of the Nintendo Wii to facilitate authentic 3D movement in 3D virtual environments.

Tricia M. Farwell received her Ph.D. from Arizona State University and is an assistant professor in the School of Journalism in the College of Mass Communication at Middle Tennessee State University. She is the critical and cultural division research co-chair for the Association for Education in Journalism and Mass communication and the entertainment and sports section secretary for the Public Relations Society of America. Her research interests include product placement, media depictions of practitioners in advertising and public relations and social media in advertising and public relations. She is the author of *Love and Death in Edith Wharton's Fiction.*

Monica Flippin-Wynn is assistant professor at Jackson State University. She received her Ph.D. from the University of Oklahoma in Communication Studies with an emphasis in race and ethnicity. Her research interests include social media, student learning, interpersonal communication, race and ethnicity, and intersections.

Scott Grant has been involved in Second Life since mid-2007 and has been teaching in world since the beginning of 2008. He has established an island in Second Life dedicated to the teaching and learning of Chinese language and culture. An economics graduate of Monash University, Scott also has a Master's degree in translation studies. In his 20s he studied in China for four years and was the recipient of two Australian government scholarships. After 10 years in industry, 6 years of which were spent negotiating and implementing aid projects in China, Scott moved to Monash University where he has been teaching Chinese language and culture for more than 10 years. He is also Deputy Director (Shanghai) of the Monash University Chinese Incountry Program, which sends an average of 200 students to Shanghai and Beijing each year to undertake intensive studies in Chinese language and culture.

Howard M. Gregory, II, is currently pursuing a graduate degree in the Information Architecture Knowledge Management (IAKM) program at Kent State University (Kent, Ohio) with a focus on Knowledge Management. He holds an undergraduate degree in Computer Information Systems and is currently the Simulation Educator for the William G Wasson M.D. Center for Clinical Skills Training, Assessment and Scholarship, part of the

Northeastern Ohio Universities Colleges of Medicine & Pharmacy (NEOUCOM) located in Rootstown, Ohio.

Lora Helvie-Mason is currently employed at Southern University at New Orleans. She is an assistant professor of Communication Studies whose research involves the understanding of education as a culture and communication as an avenue to matriculating successfully through that culture. She explores technology and instruction through a communicative lens. Her work has emphasized the marginalized voices within education, including female faculty members, minorities, and inmates. As she teaches communication courses in public speaking, advanced public speaking and debate and argumentation, Lora utilizes a reflective pedagogy that gives the students opportunities to shape and direct their learning experiences. With a bachelor's degree from Purdue University in Agricultural Communication and minors in Communication Studies and International Studies, Lora's journey in education has truly been an interdisciplinary approach. She completed her Master's degree in Communication Studies from Ball State University in 2003 and graduated with her Ed.D in Adult, Higher, and Community Education with a Cognate in Communication Studies in 2007.

Jennie M. Hwang is an assistant professor in the Communication Studies Department at California Polytechnic State University. She earned her Ph.D. from the University at Buffalo, SUNY. Her research looks at the social implications of communication technologies, Internet use and youth well-being, and online collaboration. Her work has been published in edited books and journals such as *New Media & Society*, *Journal of Applied Communication Research*, and *Academic Exchange Quarterly*.

Annie Jeffery has extensive experience in Educational Technology and Instructional Design and has undertaken research projects into areas such as learning activities, learning object repositories and metadata, web 2.0 tools, learning styles and intellectual property rights. She is an experienced virtual worlds researcher and educator and currently teaches two classes at Boise State University in Second Life. Her virtual worlds research interests are in sensory pedagogies, collaborative pedagogies and learning styles. She has a degree in Archaeology and two graduate degrees; one in Theoretical Archaeology and the second in Interactive Multimedia Production. She currently has a virtual worlds paper in print as Hardaker, G., Jeffery, A., and Sabki, A., "Learning Styles and Personal Pedagogy in the Virtual Worlds of Learning," which will be included in the work tentatively entitled "Style Differences in Cognition, Learning, and Management: Theory, Research, and Practice."

Andy Jones has taught for the University Writing Program, Technocultural Studies, and/or the Department of English at UC Davis since 1990. A Coordinator of Teaching Programs for The Center for Excellence in Teaching and Learning, Andy coordinated the yearly Summer Institute on Teaching and Technology and the weekly meetings of the Faculty Mentoring Faculty Program. The host of "Dr. Andy's Poetry and Technology Hour" on KDVS-FM, Andy has given talks at many academic and writers conferences on using media in the classroom, interactive teaching, and thinking and writing creatively. In 2006 Andy was presented the UC Davis "Educator of the Year" Award by the Associated Students of UC Davis.

Nick Pearce was a research associate on the Digital Scholarship project at the Open University. This project audited and promoted the use of digital tools for research, teaching and public engagement. His research interests lie in trying to understand the relationships between academic practice and technology use, and in particular how this relationship differs across disciplines. He is now a Teaching Fellow at Durham University, in the UK. He writes a blog at http://digitalscholar.wordpress.com

Alexander Reid is an associate professor and Director of Composition and Teaching Fellows in the English department at the University at Buffalo where his research focuses on social media, rhetoric, and pedagogy. His book, *The Two Virtuals: New Media and Composition* won honorable mention for the W. Ross Winterowd Award for best book in composition theory in 2008. His blog, *Digital Digs* (alex-reid.net), won the John Lovas Academic Weblog award for scholarly contributions in the field of rhetoric and composition in 2008. His articles are published in journals such as *Kairos: A Journal of Rhetoric, Technology, and Pedagogy*; *Computers and Composition*; *Culture Machine*; and *Theory & Event* as well as in several book collections, most recently *The Handbook of Public Pedagogy*.

Geoffrey Roth is assistant professor of Journalism at Hofstra University. He has a career in media spanning 30 years, most of which has been spent in news writing, production, and management in local television news departments. He has developed two websites for local television stations and oversaw the digital transition of several TV newsrooms.

James Schirmer is assistant professor of English at the University of Michigan-Flint. His research and teaching interests include first-year and advanced composition, technical communication, social media analysis and videogame studies. His most recent publications are "The Personal as

Public: Identity Construction/Fragmentation Online" in The Computer Culture Reader, edited by Judd Ethan Ruggill, Ken S. McAllister and Joseph R. Chaney, and "We All Stray From Our Paths Sometimes: Morality and Survival in Fallout 3" in the forthcoming Network Apocalypse: Visions of the End in an Age of Internet Media, edited by Robert Glenn Howard. He received his Ph.D. in Rhetoric & Writing from Bowling Green State University in 2008.

Sheila Scutter is the Director of Foundation Studies in the medical program at James Cook University, in Queensland Australia. This focus of her research is on innovative and effective teaching approaches to improve learning outcomes, with a particular emphasis on student transition to higher education. She has also held positions as Dean: Teaching and Learning in the Division of Health Sciences and Dean of Health Sciences, Higher Colleges of Technology, United Arab Emirates.

Kay Kyeongju Seo is assistant professor of Instructional Design and Technology at the University of Cincinnati. Seo earned her Ph.D. in Instructional Technology at Utah State University. Her research interests revolve around socio-cognitive development in online immersive virtual environments, multimedia-assisted problem-driven instruction, constructivist approaches to educational simulations and microworlds, and student interaction in computer-mediated communication.

Natalie T. J. Tindall is assistant professor at Georgia State University. She earned her Ph.D. from University of Maryland. Her research interests include: diversity in public relations and strategic communication, intersectionality, feminist thought, philanthropy, and fundraising.

Charles Wankel is associate professor of Management at St. John's University, New York. He holds a doctorate from New York University. He serves at Erasmus University, Rotterdam School of Management, on the Dissertation Committee and as Honorary Vice Rector at the Poznań University of Business. He has received numerous awards from the Academy of Management, including the 2010 Service Award of its Organizations and the Natural Environment Division. His recent books include *Educating Educators with Social Media (2011)*, *Higher Education Administration with Social Media: Including Applications in Student Affairs, Enrollment Management, Alumni Relations, and Career Centers* (2011), *Cutting-Edge Social Media Approaches to Business Education* (2010), *Being and Becoming a Management Education Scholar* (2010), *Emerging Ethical Issues of Life in Virtual Worlds* (2010), and *Global Sustainability as a*

Business Imperative (2010). He is the leading founder and director of scholarly virtual communities for management professors, currently directing eight with thousands of participants in more than seventy nations. Charles has taught in Lithuania at the Kaunas University of Technology (Fulbright Fellowship) and the University of Vilnius (United Nations Development Program and Soros Foundation funding). Invited lectures include Distinguished Speaker at the Education without Border Conference, Abu Dhabi and Keynote speaker at the Nippon Academy of Management Conference. Corporate clients include McDonald's Corporation's Hamburger University and IBM Learning Services. Pro bono consulting assignments include total quality management programs for the Lithuanian National Postal Service.

Richard D. Waters received his Ph.D. from University of Florida and is now an assistant professor of public relations in the Department of Communications in the College of Humanities and Social Sciences at North Carolina State University. His research interests include relationship management between organizations and stakeholders, the development of fundraising theory as it pertains to charitable nonprofits, and the use of Web 2.0 technologies by nonprofit organizations. He is a former fundraising practitioner and consultant to healthcare organizations in Northern California. His research has been published in nonprofit management and public relations journals.

Gavan P. L. Watson received his Ph.D. from York University, Toronto, Canada and is now an Educational Developer in Teaching Support Services at the University of Guelph. An experienced naturalist and environmental educator, he developed and taught a first-year experiential natural history course while a Ph.D. student. His doctoral research was located at the intersection of environmental education, environmental philosophy and animal studies and focused, in part, on how the technologies used to observe and collect data about the natural world come to shape perspectives of it. In his role as Educational Developer, Gavan continues to research and support the thoughtful use of technology in the higher education classroom to support student learning.

Corinne Weisgerber is assistant professor of Communication at St. Edward's University in Austin, Texas where she teaches social media, interpersonal communication and public relations classes. Corinne has been studying computer-mediated communication and how people develop and maintain online relationships since the beginning of her doctoral studies at the

Pennsylvania State University. Her research interests include new media, pedagogy, and interpersonal communication. She developed one of the first social media for public relations classes – a course which explores emerging social media technologies and studies their application in contemporary PR practice and which she has been teaching since 2007.

Linda Wilks is a Research Associate in Digital Humanities in the Faculty of Arts at The Open University. She is currently involved in assessing the roles of digital elements in current and recent academic research in the Faculty of Arts, as well as in encouraging experimentation by faculty staff in the use of new technologies. Linda also promotes awareness of and debate about the impact of digital technology on the nature and scope of research questions and research project design by organizing a programme of research seminars featuring highly-regarded guest speakers.

SUBJECT INDEX